*Once more search undismay'd
the dark profound*

Where Nature walks in secret...

Mark Arkenside
Pleasures of Imagination
1744

Photo by Gavin Newman

Published by the

Cave Diving Group
c/o Wells & Mendip Museum
8 Cathedral Green, Wells,
Somerset, BA5 2UE
www.cavedivinggroup.org.uk

Supported by the

British Cave Research Association
www.bcra.org.uk

First published in the UK 2010

Copyright ©: Cave Diving Group, 2010

ISBN 978-0-901031-07-5

All rights reserved. No part of this publication may be reproduced or transmitted in any form or by any means, electronic or mechanical, including photocopying, recording, or any information or retrieval system, without prior permission in writing from the Cave Diving Group. All rights in photographs and other illustrations remain with the photographers and illustrators, as individually acknowledged, and are used here with permission

Design, artwork and retouching by

Mark *Gonzo* Lumley

The Creative Edge
7 Langleys Lane,
Clapton, Radstock,
Somerset, BA3 4DX
T. 01761 419246
mark@creativeedge.me.uk

Printed by

The Complete Product Company Ltd.,
The Dairy, Pinkney Park, Malmesbury,
Wiltshire, SN16 0NX
T. 01666 841148   F. 01666 841014
sales@tcpc.co.uk
www.tcpc.co.uk

Wookey Hole

75 Years of Cave Diving & Exploration

Jim Hanwell
Duncan Price
Richard Witcombe

In memory of
James Gordon Ingram-Marriott
and
Keith Martin Redmond Potter
who lost their lives whilst diving in the Great Cave.

Contents

- iv **PREFACE**
  by Duncan Price and Richard Witcombe

1935-1985

- vi **INTRODUCTION**
  by Jim Hanwell

- 1 **Chapter 1**
  EXPLORING WOOKEY HOLE AND MENDIP'S STREAM CAVES
  by Jim Hanwell

- 29 **Chapter 2**
  FOREWORD, DISCOVERY AND DIVERS' OBSERVATIONS
  by Herbert Balch, Graham Balcombe and Penelope Powell

- 47 **Chapter 3**
  'MOSSY' POWELL AND THE BASE CAMP
  by Frank 'Mac' Brown

- 57 **Chapter 4**
  'D-DAY', THEN 'BUNG' TO 'INNOMINATE'
  by Jim Hanwell

- 91 **Chapter 5**
  SACRIFICES TO THE WITCH?
  by Ted Mason

- 103 **Chapter 6**
  PHOTOGRAPHING THE NINTH CHAMBER
  by Bob Davies

- 109 **Chapter 7**
  IN THE WATERS UNDER THE EARTH
  -THE THIRTEENTH CHAMBER OF WOOKEY HOLE
  by Bob Davies

- 119 **Chapter 8**
  MIXTURE BREATHING AT WOOKEY HOLE
  by Oliver Wells

- 133 **Chapter 9**
  TWENTY-FIVE YEARS AND THE TURNING POINT
  based on Mike Thompson

- 149 **Chapter 10**
  CHANGING GEAR
  based on Fred Davies

- 163 **Chapter 11**
  AN INDEPENDENT AIR
  based on Oliver Lloyd

- 181 **Chapter 12**
  UP TO THE EIGHTEENTH
  by Dave Savage

- 187 **Chapter 13**
  SUCCESS IN THE SEVENTIES
  by Brian Woodward

- 211 **Chapter 14**
  DRY TO WOOKEY NINE
  by Brian Prewer

219	**Chapter 15** **THE GREAT CAVE SURVEYED** *by Willie Stanton*	
227	**Chapter 16** **PHOTOGRAPHY AND OTHER FINDS** *by Peter Glanvill*	
239	**Chapter 17** **FARTHEST BY FARR** *by Martyn Farr*	
247	**Chapter 18** **UP TO THE WAY ON?** *by Rob Harper*	
253	**Chapter 19** **ANOTHER YEAR OLDER AND DEEPER IN DEPTH** *based on Rob Parker and Bill Stone*	
265	**Chapter 20** **BACK TO BASICS IN ST. CUTHBERT'S** *by Alan Butcher and Stuart McManus*	
277	**Chapter 21** **THE SHAPE OF THINGS TO COME** *by Oliver Wells*	

1986-2010

289	**Chapter 22** **LIGHTER AND FASTER** *by John Cordingley*	
293	**Chapter 23** **CLIMBING HIGHER** *by Alex Gee*	
301	**Chapter 24** **DIGGING DEEPER** *based on Mike Barnes, Pete Bolt, Tim Chapman and Clive Stell*	
309	**Chapter 25** **WET TO TWENTY-FOUR** *by Duncan Price*	
313	**Chapter 26** **THE EASTWATER OPTION** *by Richard Witcombe*	
321	**Chapter 27** **MAKING 'WOOKEY'** *by Gavin Newman*	
329	**Chapter 28** **FULL CIRCLE** *by Duncan Price*	
341	**ACKNOWLEDGEMENTS**	
342	**BIBLIOGRAPHY**	
345	**INDEX**	

Photo by Martyn Farr

Preface

This is the story of the first seventy-five years of cave diving and exploration at Somerset's famous Wookey Hole Cave, or Caves, as successive show cave managements have preferred to call their attraction. Since the pioneer 'hard hat' dives of 1935, generations of cave divers have explored the River Axe upstream from the Third Chamber in the show cave discovering nearly 4 km of both submerged and above water cave passages to reach the present limit which lies 90 m underwater (26 m below sea level) and 1 km north-east of the cave entrance.

This book was conceived in 1985 to mark the Golden Jubilee of that underwater exploration. Between 1985 and 1987 Jim Hanwell collected the reminiscences of those involved with cave diving in Wookey Hole Cave together with a vast collection of relevant illustrations. The stories comprised either first-hand accounts by the divers themselves or Jim's own re-telling of each participant's contribution based upon personal knowledge and interviews. The material was carefully edited and arranged to form the first twenty-one chapters of this book.

Sadly, however, no one volunteered to take on the final task of publishing the book as originally envisaged, and the project faltered. Only thirty-three un-illustrated copies of the text were kindly produced by Dave Turner, perfectly bound and distributed to the original contributors. Then, as has often been the case over the years with exploration under Mendip, success in one cave inspires breakthroughs in another. So it was that publication by the Wessex Cave Club of *Swildon's Hole: 100 years of exploration* in 2007 provided the ideal spur: Wookey Hole Cave is, after all, the 'other end' of Swildon's in many respects, not least regarding the history of cave diving.

The principal editors of the Swildon's book, reinforced by a new generation of cave divers, have thus taken on the challenge of up-dating the accounts of upstream exploration of the subterranean River Axe in time for the seventy-fifth anniversary of the first Wookey Hole dives in 2010.

A deliberate choice has been made to retain the contents of Jim's work in its original form, with only light editing and correction where necessary. Regrettably, many of the characters mentioned are no longer with us, through infirmity, old age or mishap, and footnotes have been inserted in the text at appropriate points to record their passing. Further chapters have been added to recount the events and achievements of the subsequent twenty-five years in Wookey Hole Cave and other relevant sites. The latter

Duncan Price in Chamber Nine
Photo by Mark 'Gonzo' Lumley

include some of the swallet caves on the plateau above where cave diggers and divers have been working just as hard to close the gap between these feeders and the great resurgence at Wookey.

The new book thus divides into two parts (known colloquially to the editors as the 'Old' and 'New Testament' respectively), comprising the original twenty-one chapters covering the period 1935 to 1985 compiled by Jim, and the recently written chapters from twenty-two onwards which deal with the period from the mid-1980s to the present day.

As Jim explains more fully in his introduction, the coverage of the various phases of exploration is by no means uniform, as the individuals concerned have been allowed to tell their stories in their own, sometimes idiosyncratic, way. By the same token, the terminology and measurements used – Imperial or metric – reflect the author's and not the editors' preferences. This may annoy some, but it more faithfully represents the atmosphere and thinking of the time, and adds to the colour and richness of the overall story.

A huge number of people have helped in the compilation of this book, not only the direct contributors, including photographers, but those who have helped make its publication possible. Martin Grass has acted throughout as the book's 'Commercial Manager' and the British Cave Research Association has put forward money, appropriately drawn from Graham Balcombe's substantial bequest to them, to fund its printing. The striking design and layout of the book is a testament to the artistic flair of graphic designer and caver, Mark 'Gonzo' Lumley. A full acknowledgement of all who have given their time and energy to this project appears at the end of the book. The editors owe them a great debt of gratitude for without them the book would never have seen the light of day.

*Jim Hanwell on the hillside above Wookey Hole.
Photo by Roger Cookman, 2002*

Introduction

On 23rd August 1935, the *Wells Journal* and *Somerset and West of England Advertiser* featured two notable firsts, as follows:

THRILLING ADVENTURE IN WOOKEY HOLE CAVES
Divers Brave The Depths of Hidden Waters
NEW CAVERNS DISCOVERED
SUCCESSFUL BROADCAST BY B.B.C.

During the past fifty years much exploration work has been undertaken in Wookey Hole Caves, but no attempt to explore the secret depths of these famous caverns has aroused such interest as the daring exploit undertaken on Saturday evening last, when two divers, one a woman, walked the hidden bed of the underground River Axe probing the mysteries of the Mendip underworld.

The divers were Mr. Graham Balcombe, and Mrs. Penelope Powell, and their daring adventure in exploring the unknown caves of Wookey Hole was sponsored by the Mendip Nature Research Committee of the Wells Natural History and Archaeological Society...

At 10.30 the B.B.C. announcer, Mr. Francis Worsley, from his commentator's box started to speak to the thousands who were listening to what must have been the most thrilling outside broadcast ever arranged...

He said: 'Here we are, 600 feet underground in the famous Wookey Hole Caves. The sounds you can hear going on mean that the exploration party is ready to try out this daring feat of exploration. We are standing now in the Third Chamber... This is as far as the public can go, but the caves and river go on for a long way beyond... When the water has been low people have been through on a raft to a Fourth Chamber and then on through another arch to a Fifth. Beyond that no one has ever been and only divers can get there. That is the object of this exploration... I hope to get Mr. Balcombe to... speak direct to us from under the water when he reaches territory where no one has ever been before. He has a special microphone in his helmet... Balcombe has gone 160 feet and is going further. Mrs. Powell is being belayed up and is passing cable to him. There is about 15 to 18 feet of water where they are... Balcombe has got to the entrance to the Sixth Chamber and hopes to find it a real chamber, that is one that has air space above the water, but we shall not know anything about it until we hear from him. We are going to try to get through to him now and get him to tell us from the actual site what he has found...'

The 'Escape', or Resurgence, of the River Axe at Wookey Hole (showing the weir constructed about 1852 that raised the water level throughout the Great Cave).
Photo by Antoinette Bennett

The telephonist asked him to tell listeners what he had seen: 'Hello, Balcombe,' he calls... 'Yes,' came back the voice of the leading diver... 'I have now dropped another five or six feet, which I can tell by my ears. You can check the gauge on the shore. I am apparently in the Seventh Chamber now. We have passed through the Sixth, which has a large water space but only a small air surface. Ahead of me I can see a further air surface which looks promising...'

So, in a fanfare of publicity, began the Stories of the Wookey Hole Cave Divers for the next fifty years to the present. These pages celebrate the Golden Jubilee of the first ever cave dives, and the book itself is presented to Wells Natural History and Archæological Society for their Centenary in December 1988. For those who pioneered Cave Diving, Wells Museum was a special place at which to meet, make friends and plan to do what most thought to be impossible.

Such recruits to this fascinating work first traverse the footsteps trod by the explorers of 20, 30 or 40 years ago, thrilled by the novelty and difficulty of the game, and then yearn for fresh fields to conquer.

Thus it was that the deeply flooded chambers of Wookey Hole beyond the visitors' limitations appear to offer possibilities surpassing anything yet attempted.

Herbert Ernest Balch,

Wells Museum, 1936.

So wrote the man who led the older generation of cave explorers under Mendip, fascinated by stories of new underwater discoveries told by 'young enthusiasts eager to push back the confines of the unknown.'

The waters under the earth are a final frontier to map in our continuing exodus to promised lands. To find fresh fields to search and learn about is reason enough. To have fun and friendships in doing so makes it all worthwhile. The stories in this book tell how exploring submerged cave passages began here at Wookey Hole in Somerset on the southern slopes of the Mendip Hills. Britain's oldest known river cave has seen the youngest and perhaps most daring means of exploration develop. Over the past fifty years or so, Wookey Hole Caves have been the cradle and nursery of cave diving. It is such a fleeting period of time that the few involved know each other and have shared the action. They give their own accounts in the pages that follow.

Our blend of both story-telling and history is deliberate. So, those wishing to get on with the action of the former should now skip to the first chapter about the work of the pioneers of cave diving. Those, on the other hand, preferring to prepare more for what lies ahead may find the rest of this Introduction a help. As in cave exploring and cave diving especially, the adventurers and academics support each other in their own ways. Many thrive on both as time passes.

One hundred years ago, explorers were enticed by the world's most remote and highest places. There were still plenty of unknown and unvisited parts of our planet's surface to discover. More distant travel and climbing skills were the major challenges. To survive and return with new knowledge completed an expedition and wetted the appetite for more. Yet, in being beckoned afar and attracted aloft, most adventurous spirits missed the underground world hidden beneath them back home. Despite centuries of mining and tunnelling under the earth, few chose to venture far into the natural caverns that were known. These mysterious netherworlds were forbidding and only the largest were visited, mainly as curiosities. Cave exploration is a comparatively new challenge hardly on the threshold of its centenary in Britain, and in other countries where limestone caverns abound even more, though often in remote places. Organised caving began in the 1890s here and Edouard Alfred Martel's epic descent of Gaping Gill deep beneath the Yorkshire fells in 1895 brought a new blend of sport and science from France called spelæology.

It was long after ways to fly in the sky had become commonplace that we acquired the nerve and means to breathe without restraining air lines under the sea or in rivers. Even today, more is widely known about outer space and the stars than about the greater part of our planet's surface that forms the sea bed. Going upwards has appealed more than going downwards. Underwater exploration was first undertaken to do specific jobs, and penetrating into flooded tunnels from which there is no escape upwards if things go wrong is akin to cave diving in many ways. The dives made by Henry Fleuss and Alexander Lambert in total darkness into the flooded Severn Tunnel to close a door in 1880 and William Walker's replacement of decaying beech timbers with bags of cement beneath Winchester Cathedral from 1906 to 1911 are examples of the tasks that commercial divers in many countries have carried out in confined spaces underwater. Cave divers, then, not only at Wookey Hole but also elsewhere, can feel a certain satisfaction in being part of such a tradition while exploring caves. But, to combine the sports of both caving and diving is quite new and within the living memory of those who pioneered it in their youth.

Not until that other doyen of France, Jacques-Yves Cousteau, invented his famous aqualung in the 1940s did the sport and science of undersea exploration open up for subaqua enthusiasts. Had Cousteau's invention come a decade earlier and had no war followed, cave diving would undoubtedly have been differently conceived and developed. But history is not served by 'ifs' and the pioneers of cave diving some fifty years ago were on their own. Every generation since has followed the same tradition of making the most of what was available in their day. Technology has tended to lead techniques and impose limits on what could be found.

So, the stories told here do not begin in earnest until the mid-1930s. Whilst the purpose in exploring the waters under Mendip has been simple enough, the problems to be overcome by cave divers have been complex. There is still much to do. Those faced with today's limits will read in the following pages how their predecessors took on similar unknowns before each forward push. There is a familiar ring about every gain of new ground to reach the next challenge. Whilst the caves found and mapped through diving are the most priceless legacy of all, this book hopes to add to their value by recording how they were discovered by those actually involved at the time. I will now set the stage and introduce the main contributors.

In our case, recent history has very ancient roots within the rocks of Mendip. It has taken many millenia for subterranean waters to enlarge weaknesses in the hard beds of limestone into underground river channels. Rocks and rainfall have shaped much of the surface above too. At times, streams unable to sink into the limestone have run off the hillslopes to erode spectacular features such as Cheddar Gorge, one of Britain's most notable and visited landforms. Rugged cliffs tower 400 feet above the road that winds up the Gorge from the village onto the Mendip plateau. Here is a distinctive open landscape with old mine workings, scattered farms and a few small hill villages such as Priddy. Once isolated and bleak, Priddy is now easily reached and within an hour's drive south of the great cities of Bath and Bristol. Beneath the parish lie the deepest and longest caverns in the south of England. But their exploration did not begin until their narrow and blocked entrance passages were dug open by the first cavers after the turn of the century. Nowadays, cavers can crawl and climb down over 400 feet beneath the village, following the streamway that heads for Wookey Hole over two miles away until constricted and submerged passages known as sumps bar the way on.

Tourists who venture into the paved and electrically lit caves at Cheddar and Wookey Hole discover some of the attractions that lure cavers underground. In Wookey Hole Caves, barely a couple of miles from the ancient cathedral City of Wells, they see the River Axe escaping from under Mendip. They can follow the river upstream through great vaulted chambers in the red conglomerate rocks that surround the hills, then walk through low passages and tunnels until the roof plunges below the water's surface. All the streams flowing through the caves at Priddy have joined somewhere beyond to feed the River Axe. Only Wookey Hole cave divers can tell the stories that begin here. And being surrounded by water and rock clearly doubles the skills that such explorers have to master. For the pioneers, of course, this presented completely unknown dangers to be overcome.

The river caves of Mendip, and Wookey Hole in particular, play a very special part in the history of cave diving world wide. It all started here in 1934-35 and this book has been compiled to celebrate the Fiftieth Anniversary of diving in Wookey Hole Caves. Any editor and compiler of history able to talk to and know as friends those who have actually made the stories to be told is especially privileged. By good fortune, this editor has also lived throughout the same period of time in the area concerned, and in the village of Wookey Hole in particular. For most of us, caving and cave diving have been purposeful pastimes in search of pleasure rather than pecuniary gain; yet no less professional in outlook for all that, it is hoped. Those who have worked hard to find new caverns will understand the real rewards. Likewise, I know that, for all the time and information given freely by those who have contributed, this book itself is sufficient recompense for everyone's labours. To tell a wider audience about something few have experienced has to be worthwhile.

As the Contents show, some chapters have been written by those involved at the time. Other contributors have preferred to reminisce and refer to personal notes, letters, diaries and contemporary logs. They have allowed me to quote freely from such authentic sources in compiling chapters based on their work. So much will be evident from the differently printed passages in the text. Even the narrative that links original extracts stems mostly from countless yarns with those concerned on Mendip and elsewhere over the years. All that differs is the style of telling from one generation to another. So, we move forward with each phase of exploration as it was and try to capture the spirit of the times in a pursuit that depends so much upon its personalities.

The book is profusely illustrated with photographs and surveys. Many of the former are treasures from private collections and few have been published before. Full credits and acknowledgements are given in the captions for each figure and plate within the text. Books cited, other helpful literature and sources of information are also included within the story where appropriate. For those who wish to follow them up, the concluding bibliography explains where they may be found. Because caves are hidden places and their exploration is only seen by those who participate, others must glean what they can from the maps made and reports produced. To cavers in particular, pictures and written records are invaluable; the next best thing to being actually there oneself.

During a span of fifty years, the usage of terms and sometimes even names are bound to change. New and developing pastimes cannot easily follow one convention, and this is particularly the case with caving at Priddy and Wookey Hole. Contrasting approaches have been used in both conversation and records. In river caves such as Swildon's Hole at Priddy, cavers usually refer to stretches of the stream flowing through open passages by numbers written as words (e.g. Swildon's Four), whilst intervening sumps that have to be dived are simply designated by either Roman or Arabic numerals (e.g. Sump IV or Sump 4). This distinction has not been followed in Wookey Hole Caves for good historical and practical reasons; not least because it has often taken several years to progress underwater through one of its much longer and deeper sumps to a new air space or chamber. In this book, I have chosen to have numbers written as words rather than risk confusing readers with both methods. Some passages would appear very cluttered with numbers otherwise. Only when the book focuses solely upon dives in Wookey Hole Caves after 1970 have I allowed the divers' preferred shorthand into print. By then, cave diving phrases and terms will be familiar enough.

Different names pose fewer problems but should be cleared up now. Swildon's Hole takes an apostrophe since its name supposedly derives from St. Swithin; and is often used as the contraction 'Swildon's' which rolls more easily off the tongue and is much more common (with or without the apostrophe). The similar contraction of 'Wookey' used widely by cavers is less happy and can be a great source of confusion to visitors. Wookey is a village down the Axe valley, and anyone going there to find a cave will be disappointed. Wookey Hole is quite a separate village upstream but is not the cave either! The underground passages formed by the River Axe should properly be called Wookey Hole Cave even if this seems rather a tautology. Added to this conundrum has been the various owners' preference to use 'Wookey Hole Caves'. They do so to distinguish the caverns shown to tourists from those only accessible to cave divers; hence, the Old Show Cave to the Third Chamber, the New Show Cave through to the Ninth Chamber and the Divers' Cave beyond. There are clear attractions in the plural for publicity purposes and it appears often here too. It is only in the later chapters that we lapse into the more convenient and familiar name Wookey.

Another significant change during the later stages of our story has been the move from Imperial standards of measurements to metric units of the *Systéme International* (S.I.). Most chapters adopt the former particularly since such quantities were those specified for the equipment used at the time. Purists who raise an eyebrow at the occasional use of both systems should realise that the divers concerned had to contend with such a mix in their day. Only in the later chapters dealing with the modern record breaking dives have metric measurements been used. In these, writers use new conventions such as going down to, say, 'minus 45 metres' or '-45 m'.

My own chapters tell how Graham Balcombe and Jack Sheppard were introduced to cave exploring and then pioneered cave diving on Mendip. Firsthand glimpses of those exciting days before the Second World War are given from the log compiled about the helmet, full dress and hand-pumped air supplies used on the first dives that took place at Wookey Hole Caves in the summer of 1935. Another delightful scene-setting chapter about these dives and caving in the mid-thirties is written by Frank 'Mac' Brown, then Manager of the Great Cave. The War understandably influenced most aspects of cave diving for years afterwards; particularly the availability of breathing apparatus and the military-like approaches to diving operations. That much will be evident from the fourth chapter and the coded 'ops' described. Ted Mason, responsible for the very first underwater 'excavations' within a cave and many books on such subjects, recalls the archæological discoveries made by divers. Bob Davies follows with two more Wookey Hole firsts; the photographing of new finds, and the most remarkable escape story of all in cave diving. Bob is now Benjamin Franklin Professor of Molecular Biology at the University of Pennsylvania, Philadelphia, in the United States[1]. Oliver Wells was to follow him across the Atlantic, but not before he had pushed the limits of cave diving even further and deeper beneath Mendip. Oliver is better known to the scientific world for his contribution to the important treatise entitled *Scanning Electron Microscopy* (1974).

Inevitable changes occurred within cave diving during the early 1960s. These are covered in three chapters compiled from original sources and the reminiscences of Mike Thompson, Fred Davies and Oliver Lloyd. The Cave Diving Group responded too, and more enthusiasts with new equipment and ideas took up the challenge. Mike Boon and Steve Wynne-Roberts played a large part in developing equipment to push constricted sumps whilst Mike Wooding, Dave Drew and Dave Savage from Bristol University developed their own way of doing things. A kitted-up cave diver in the mid-sixties looked very different from his predecessors. But, of course, he had the same purpose as Dave Savage relates in his chapter about finding the way on and upwards beyond the deeply submerged Fifteenth Chamber in Wookey Hole Caves. Dave, who became Deputy Headmaster of a large school in Salford, Manchester, was the first to reach what remains even to the present the limit of exploration along the Swildon's Hole streamway under Priddy.

As Bob Davies knows only too well, thirteen is a lucky number at Wookey Hole. Here, the thirteenth chapter by Brian Woodward follows suit with 'Success in the Seventies'. Brian lectures and researches in pharmacology at the University of Bath. Maps of the new discoveries made by divers allowed the owners of Wookey Hole Caves, under Graham Jackson's imaginative management, to drive tunnels into the majestic Ninth Chamber for tourists to see more of the system too. Brian Prewer and Willie Stanton record these times and the big boost this gave to Wookey Hole as a major tourist attraction in the West Country. Brian is well known locally for his professional expertise in electronics and commitment to cave rescue work. Willie is acknowledged as a major authority on the hydrogeology and geomorphology of the area having published more about the caves of Mendip than most[2]. Peter Glanvill, who hails from Chard in south Somerset and now runs the family's medical practice, brings the further reaches of the Great Cave to life through his pictures and the story of how he got them.

Martyn Farr, already known internationally for his own books on cave diving and participation in expeditions from his native South Wales to remote corners of the world, contributes a chapter on his three record breaking dives up to October 1982. Even he has likened Wookey Hole to 'a caver's Everest'. His assaults, filmed and networked on television have pointed the way ahead with methods borrowed from those who scale Himalayan peaks. Rob Harper is one who has taken

[1] Bob Davies died in 1993, aged 73. [2] Willie Stanton died on 30 January 2010, aged 79.

up the challenge to climb underground and add a new dimension to cave diving. Ascending sheer rock faces in the dark after long, deep dives underground has to be met with good humour and companions who know the score. Jibes urging Rob to carry his veterinary's humane killer on such trips have prompted suitable ripostes!

Summer 1985 saw the fiftieth birthday of cave diving at Wookey Hole and the last deep push up the subterranean River Axe to date. Rob Parker from Bristol and Bill Stone from America teamed-up for this technically complex dive which involved camping underground beyond the reach of most cave rescuers. At the time of writing, both are diving together again on a return fixture to Florida's very deep Wakulla Springs: a submerged 'Grand Canyon' in pure white limestone and a far cry from Mendip's murky and numbing waters. Back home at Priddy, Alan Butcher and Stuart McManus have good cause to know more than most about the power of such water in the saga of sump pushes at the end of St. Cuthbert's Swallet.

Finally, it is back to the future. By one of those coincidences that enliven history, Oliver Wells' illustrious grandfather 'H.G.' once taught at the village school in Wookey as a young man. Without Oliver this book would not have matured and his credentials to write his own account of 'The Shape of Things to Come' in cave diving are clear enough. In our twenty-first and final chapter, he foresees cave diving coming of age in ways undreamed of by those who started it all fifty years ago.

Times change, and so do we. But our stories have a familiar recurring challenge that has been met equally by every generation of cave divers in the waters under Mendip. More than 240 Wookey Hole Cave Divers over the first 50 years and many others mentioned throughout this book are thanked for their interest and support. I have never met one who was not convinced that a way on underwater could be found, and their accounts are presented here with a similar degree of optimism. The two pioneers of cave diving, Graham Balcombe and Jack Sheppard, have kindly given me priceless records and personal memories as well as great consideration in preparing my own chapters. Particular thanks are due to Richard West for typing the original text. Dave Turner, another enthusiastic Mendip caver, converted Richard's efforts into a publishable format at the time. All the oldest photographs, mostly irreplaceable originals, were expertly copied by Eric Lewis. The editor is also appreciative of the encouragement and friendship of the owners and successive managers at Wookey Hole Caves over the years, and the cave guides who have their own stories to tell. Olive Hodgkinson, Graham Jackson, Frank McBratney and, currently, Peter Haylings have been especially helpful and Alf Stapleton is remembered as 'the voice' of the Great Cave.

As the son of a Mendip farmer myself, I feel it particularly appropriate to record the thanks of all cave divers to the landowners at Priddy who have allowed us to explore their caves. Generations of the Main family at Manor Farm are gratefully acknowledged for countless kindnesses concerning trips down Swildon's Hole. Fred Davies and Ray Mansfield corrected and improved the final draft whilst Willie Stanton helped throughout, and put the finishing touches with a keen eye and much appreciated consideration. Chris Hawkes, Linda Wilson, Brian Prewer, Ken Pearson and my wife Judy, proof read and improved the manuscript.

The debt owed to my willing helpers and the main contributors of chapters is, perhaps, best met if this book generates more enthusiasm for pushing the present limits of cave diving and making new discoveries in the waters under Mendip.

Jim Hanwell
Wookey Hole
August 1987

H.E. Balch
Pioneer of Mendip Caving.
Illustration by Ethel. M. Balch

1935-1985

The conglomerate cliffs at the head of the Ravine behind which the Great Cave lies. Photo from Wookey Hole Caves, 1930s

Chapter One

Exploring Wookey Hole and Mendip's Stream Caves

by Jim Hanwell

Springs have a particular fascination for they are more than sources of water from under the earth. The word also evokes the sentiments penned by Mark Arkenside in *Pleasures of Imagination* (1744):

Once more search undismay'd the dark profound

Where Nature walks in secret...

Such is our quest here. Where better, then, to start a collection of stories that focus upon unravelling the secrets of undergound rivers than the spring of waters escaping from the oldest documented cave in Britain – Wookey Hole.

Fig. 1.1
A 'Picturesque View' of Wookey Hole Cave, by 'Botanista Theophilus', 1757 (see Chapter 15)

A B is part of the Hill ; C the Cave's Mouth ; C D the firſt *Antrum* or Vault ; D E the ſecond E F the third and laſt ; where F is the Hole thro' which the Stream comes and runs thro' the Rock from G to A ; from whence it deſcends to the Paper Mill at H ; at I is the City of *Wells* ; and K L the Top of *Mendippe-Hills*.

Today we know that the subterranean River Axe at Wookey Hole gathers its waters from several stream caves under Mendip. The longest and deepest of these are: Eastwater Cavern, Swildon's Hole, and the more recently explored St. Cuthbert's system. But this knowledge has been hard won and we must not scoop our challenge here so soon, save to note that the exit of the Axe from Wookey Hole is our entrance to cave diving; a means of exploration that is found around the world today. And yet, to follow the story fully, we must also go with the flow of the stream from its head and include the early work in Swildon's. The link between Wookey Hole and Swildon's is my source, as it is for other chapters, and we must go to and fro between both caves as the divers have done.

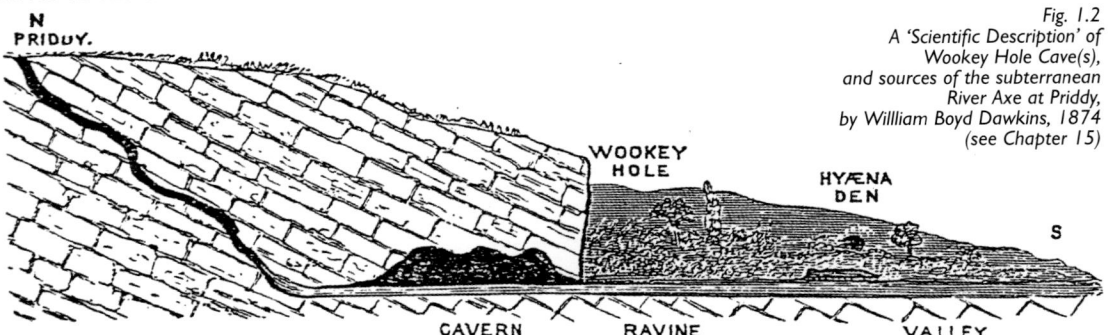

Fig. 1.2
A 'Scientific Description' of Wookey Hole Cave(s), and sources of the subterranean River Axe at Priddy, by Willliam Boyd Dawkins, 1874 (see Chapter 15)

I also chose this introductory title to compare the pioneering roles of Ernest Baker and Herbert Balch in exploring caves with that of Graham Balcombe and Jack Sheppard in the field of cave diving. My purpose is to set the stage through its personalites; what they planned to do and did. It is as much to do with people as the places. However, I am wary of story telling that relies only upon memories and histories that interpret the past from today's realities. Those who set about the task of developing cave diving had no such constraints at the time, and fewer qualms, it seems. Such exploration was simply the way ahead. As Graham Balcombe put it recently:

When we first reached the pool in Swildon's Hole where the roof became submerged I immediately thought that by diving through we could explore what lay beyond.

That is how those who discover things set out when most stop and turn back unrewarded. And, in caving especially, like minds usually accomplish more by sharing and, maybe, daring each other a little. So it was that Jack Sheppard, equally fascinated by the Swildon's Sump, was eventually the first to make the historic dive through to the unexplored passages beyond. As Jack's own story about this follows later as a climax to the chapter, let us go back to the original challenges inherited fom Baker and Balch.

In borrowing the above title from the classic *Netherworld of Mendip* written by E.A. Baker and H.E. Balch and published in 1907, I also thought how appropriate it was that it followed a chapter simply called 'Cave Exploring as a Sport'. The widely travelled Ernest Baker was an unashamed champion of the sporting aspects of caving whilst the local man, Herbert Balch, saw that the natural science of spelæology had more to offer. Both, in fact, seem happy enough with either purpose, for discovering new caves became a shared challenge.

But a science that has discovery as its principal object, and hardships and adventure as its natural concomitants, is bound to attract as many sportsmen as scientists. The geographical might be called the sporting sciences.

This novel view might surprise some modern geographers, but is an acceptable statement of the subject's adventurous traditions. And few would disagree with the following reasons for caving put so succinctly by Baker:

Many of these ancient water courses are now dry, but others are still traversed by streams and present the explorer with most formidable obstacles. The complete exploration of any cave system would involve the tracing out of all its passages from the point where the streams enter the earth to the point of exit.

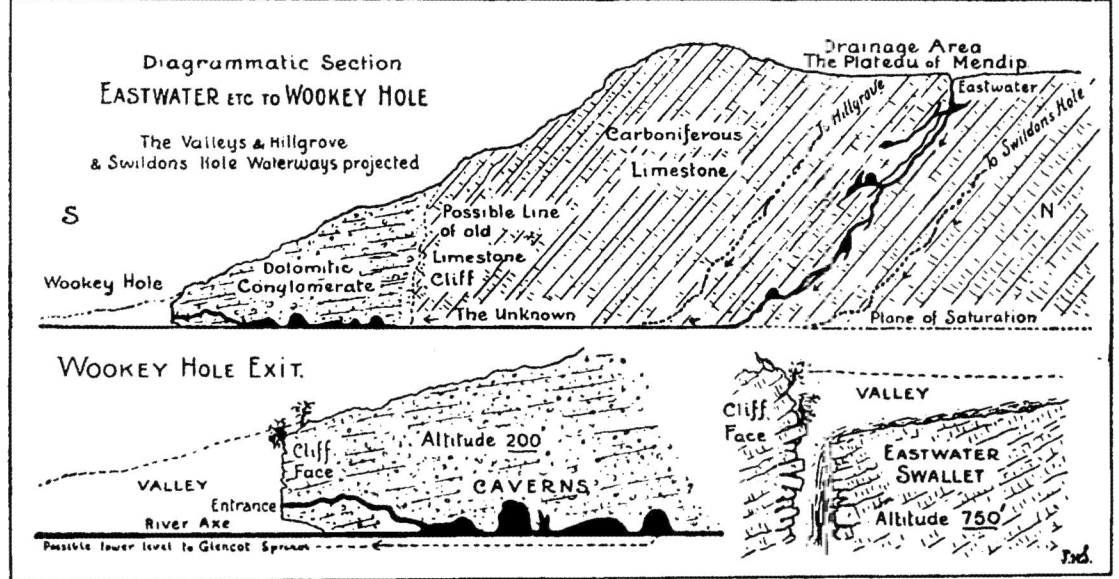

Fig. 1.3 A 'Typical Limestone Water System' on Mendip, by J. Harry Savory, 1914 (see Chapter 15).

To 'follow the stream' underground from either end summarises the appeal. Baker, again, highlights the following examples that have since preoccupied cavers in this country:

> Take two of the most important cave problems still awaiting solution, one in Yorkshire, the other in Somerset. A large beck is precipitated into the abyss of Gaping Ghyll, 360 feet deep, and emerges from an opening in the hillside a mile away, close to the mouth of Ingleborough Cave, which was itself an earlier exit... The other problem is that of Wookey Hole, the cave in Britain which has the longest history, and which is still yielding interesting discoveries. A number of streams disappear into the earth on the Mendip plateau, 2 miles away and 700 feet above, and find their issue in the source of the Axe at Wookey Hole.

Now the Gaping Gill to Ingleborough Cave through-trip was finally accomplished in 1984 after years of painstaking explorations; a fine achievement already well documented by Howard Beck in *Gaping Gill: 150 Years of Exploration* (1984). The task on Mendip would seem to lag way behind. Despite being the cradle of cave diving, much the same challenge here remains for some fortunate explorers well into the future to accomplish. The geology of Mendip is chiefly to blame for this, and we will see why as the unfinished story unfolds. Martyn Farr concludes his review of *Wookey Hole: the Caves Beyond* (1985) by claiming them to be probably 'the ultimate challenge to cave explorers'; even as 'a caver's Everest' so far as organisation and planning are concerned. We shall see these aspects, too.

News of Norbert Casteret's daring dive of a sump in the caves of Montespan in the Pyrenees during August 1922 by holding his breath took time to reach British cavers. His celebrated book *Dix Ans Sous Terre* was not published in France until 1933, and the market in Britain for an English translation was not met until *Ten Years Under the Earth* appeared in 1939. In those days, the international grapevine among cavers was not what it is today, even if those willing and able had wanted to emulate the intrepid Casteret. Rather, even fellow countrymen were astonished by the feat, including the father figure of speleology and doyen of cave exploration from France, Edouard Alfred Martel. Read his introduction to Casteret's classic. The realisation that further cave exploration on Mendip would require people to dive submerged passages dawned with the 1930s. It also became possible because the right personalities and practical support came together around the same time. The challenge favoured a few young Mendip cavers prepared and able to recognise their chance.

Fig. 1.4
A 'Trap' or Sump, portrayed by J. Harry Savory, 1914

Understandably, the prospect of cave diving had not even occurred to the earlier pioneers who were at pains to avoid water as much as possible. Although borrowing the term 'siphon' from French spelæologists, they preferred their own more forbidding word 'trap'. Who knew what pitfalls awaited those venturing into them? The graphic example drawn by J. Harry Savory that illustrates Herbert Balch's own great monograph, *Wookey Hole: its Caves and Cave Dwellers*, published in 1914, is based upon their observations of what happened when constricted passages in Eastwater Cavern and Swildon's Hole flooded. In exploring these swallets at Priddy, particularly in what is now the Upper Series of Swildon's, they had found routes over the top of these obstacles. Naturally enough, high-level bypasses became the only alternative; if one could not be found, the trap or sump became 'terminal'.

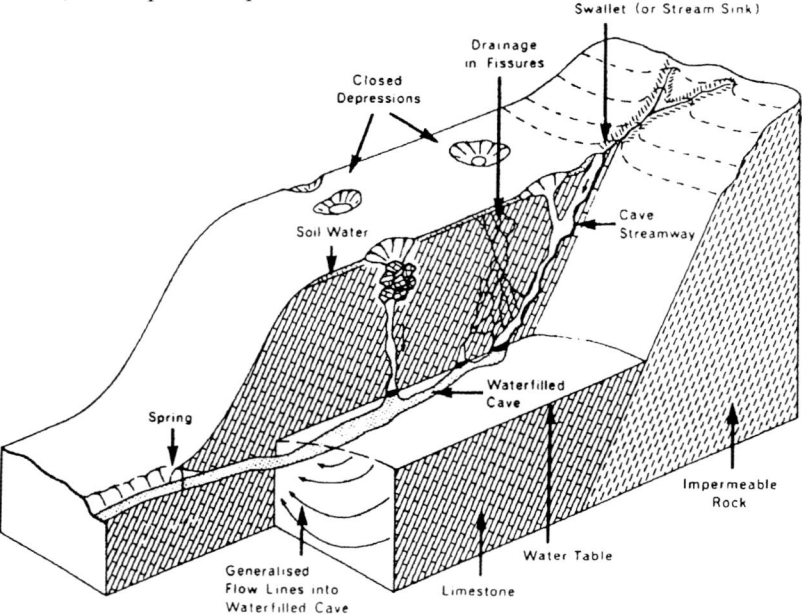

Fig. 1.5
A 'Waterfilled Cave' on Mendip, and related surface features, by Tim Atkinson, 1973

Even by 1930, Balch did not foresee that cavers would shortly take up diving. In an article for the *Somerset Year-Book of 1930*, entitled 'The Transformation of Wookey Hole', he writes:

> *It has now been possible to explore the inner and hidden chambers, beyond the third, and rarely visible. Passing through the archway, with the river drained as low as possible, we floated up the stream, and passing the steep bank which marks the first portion of the hidden cave, lying down in the boat we came through a still lower archway, and floated out on the deeper waters of the chamber... It is obvious that here the water wells up vertically and no further progress is possible unless a way can be found through the rocks overhead, to pass this water barrier.*

Fig. 1.6 (opposite page)
Plan and Section of Wookey Hole Cave, showing the subterranean River Axe entering from the right via the Fourth and Fifth chambers, 1904.
Drawn by J. Harry Savory from sketches made by Herbert Balch and Reginald Troup
(Balch entered these chambers by boat in 1903, and deduced that the leak in the river bed shown in the Fifth Chamber fed the Glencot Spring further down the Axe valley)

For once the usually prophetic Balch was soon to be proved wrong; even though his dry upper passage existed and was wide open, it was not discovered until 1973 as described later. The growing number of tourists to visit the freshly paved and electrically lit caves that he had helped to create by Easter 1927 included a new generation of sporting cavers. Their interests and enthusiasms reflected the times, for these young people enjoyed unprecedented opportunities to become proficient swimmers

EXPLORING WOOKEY HOLE AND MENDIP'S STREAM CAVES

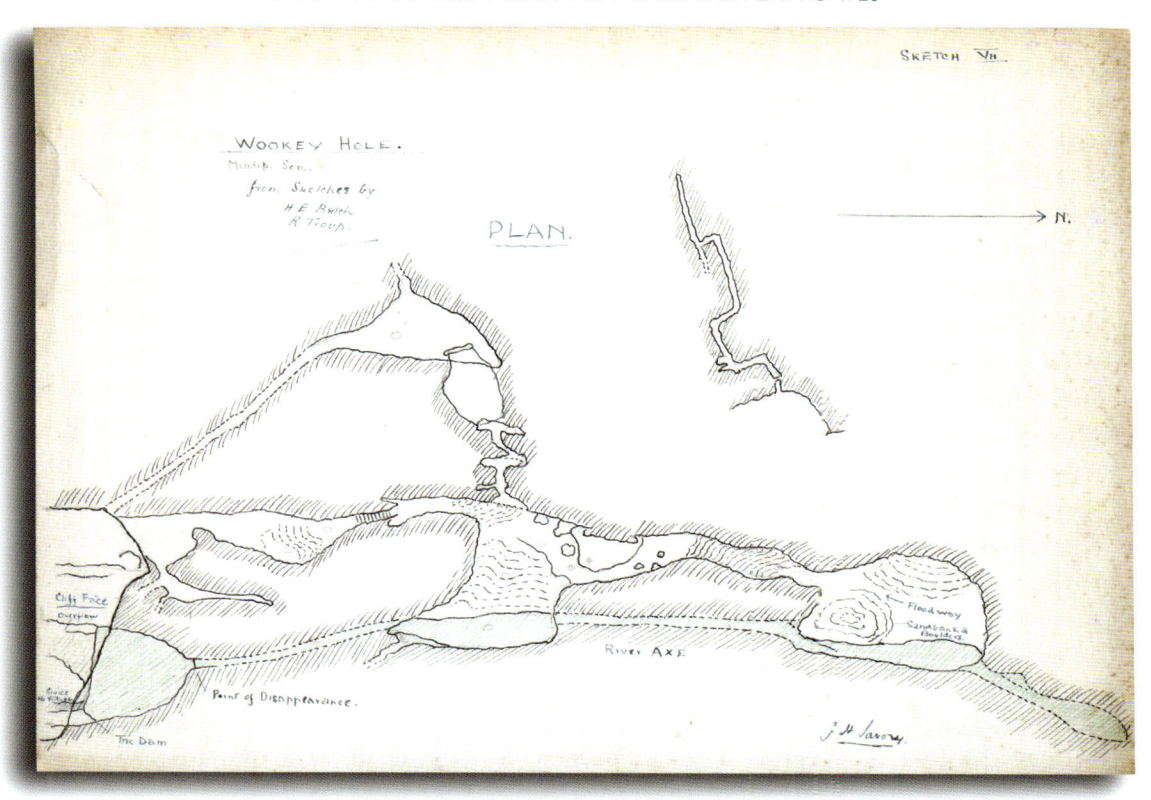

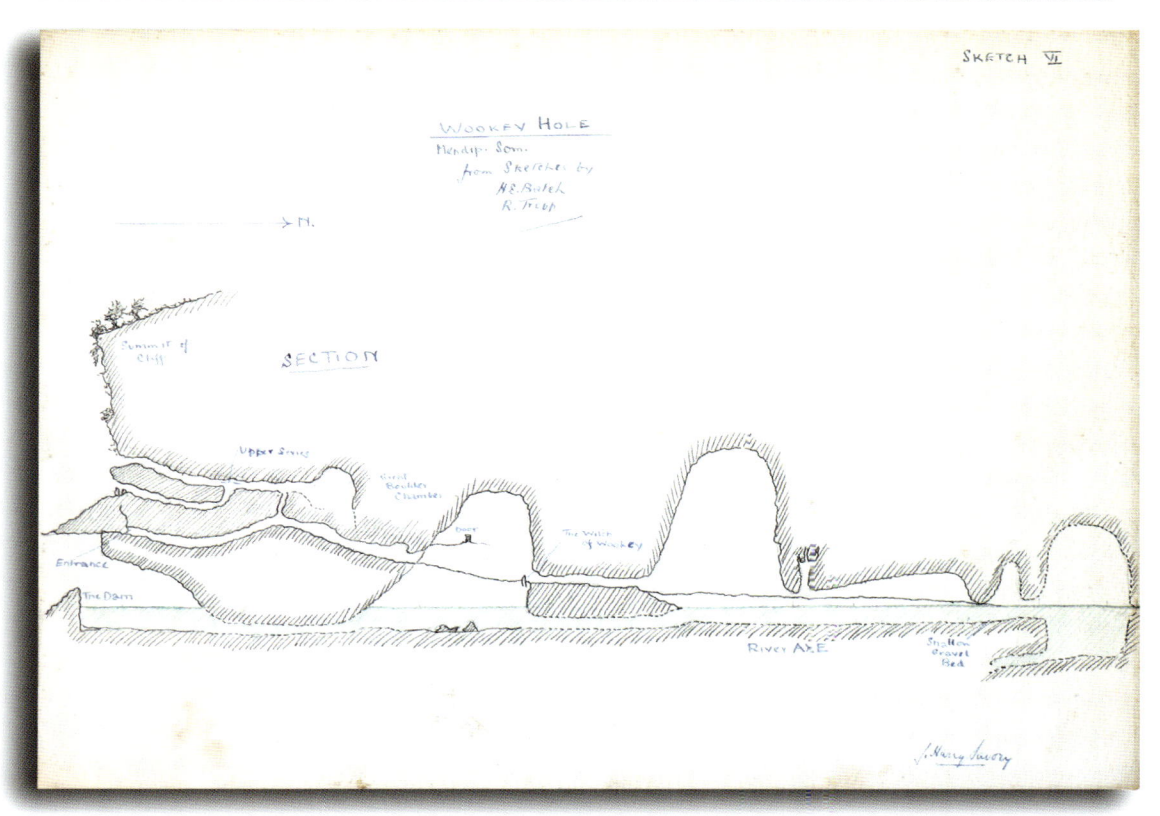

and tour the countryside in search of adventure. Wookey Hole's own swimming pool opened on a chilly Easter in 1934 and crowds came to bathe in the so-called 'Witching Waters'. Advertisements claimed it to be the only heated open-air pool in Somerset, but it quickly acquired a reputation of being fed by cold water direct from the caves! The publicity ploy had backfired to the advantage of the local lads in the know, and those hardy enough not to care and even enjoy a chilly dip at Wookey Hole. They were outdoor types keen on cycling, walking and climbing. The more affluent sported motor cycles, or maybe even a car! Nor is it just coincidence that fifty years of cave diving at Wookey Hole is shared with the jubilee of the Caravan Club. The era of exploring the countryside had arrived.

Fig. 1.7
Herbert Balch revisits the Fourth and Fifth chambers by boat in 1926.
Photo from Wells & Mendip Museum

In a little known article for the Wookey Hole Old Scholars' Association *Magazine* covering 1924 to 1932, Balch describes how the cave was turned into a tourist attraction and notes that over 200,000 visitors had been shown round since the opening in 1927. No doubt he got authentic figures from the Association's President, no less than the owner of the Great Cave, Gerard William Hodgkinson. Much later, in a booklet entitled *Wookey Hole: the Cave of Mystery and History*, published by West Advertising of Bath, Hodgkinson himself recalls:

> When I was 8 to 10 years old, I used, with two friends, to spend hours exploring the upper chambers, inaccessible to visitors, and the main part of the cave now seen by the public and I was so thrilled and enthralled by the grandeur of the chambers, the beauty of the silent, mysterious, subterranean river, and its glorious stalactite formations, that I told my father that one day I would show it to the world. His reaction was definitely negative, so I never broached the subject again. Some time after he died, however, I decided to illuminate it and show it to visitors. I started by installing electric flood lighting, an idea I got from seeing quarries working at night, and Wookey Hole was the first cave in the world to be floodlit.

Fig. 1.8
Guidebook and poster publicising the Great Cave following its opening as a tourist attraction in 1927, the latter painted by Lillian Chapman, 1929

The claim about floodlighting is somewhat wide of the mark, of course. But, no matter here for the darkness beyond remains the main challenge; one which attracted a few at the beginning. They had to organise themselves.

Herbert Balch was one of the youngest members to join Wells Natural History and Archæological Society when it formed in December 1888. He was soon to become the acknowledged expert on the few caves known on Mendip after being taken down Lamb Leer Cavern by Thomas Willcox. As the manager of the Priddy lead works, Willcox also introduced Balch to the geology of Mendip, and the success of this grounding is still evident today in that the extensive Thomas Willcox Collection of geological specimens is a major exhibit in the museum that Balch set up in Wells under the auspices of WNHAS. When the local lads formed a caving group in January 1906, called the Mendip Nature Research Club, Balch played a mediating role and helped it to merge with the senior body as the Mendip Nature Research Committee of the WNHAS in 1908. It was a period of great activity in early cave exploration, too, for, in addition to his archæological excavations in Wookey Hole Caves, Balch was also closely involved with the discoveries of Eastwater Cavern and Swildon's Hole. He was caught up with Ernest Baker's enthusiasm for the sport and very much taken by Edouard Martel's promotion of the science of spelæology when the great man visited Mendip from France in the summer of 1904.

In 1912, university students in Bristol founded a small 'Spelæological Research Society' which grew into the University of Bristol Spelæological Society after the First World War. The 'spelæos' caved mainly in the Burrington Combe area though reinforced the Wells men for significant pushes, especially down the Swildon's Hole streamway. These academic links were strengthened when Balch was conferred an honorary Master of Arts degree at Bristol University in 1927 for his major contributions to cave archæology. Once again, the science of caving had the edge and research interests were expected from members of the MNRC and UBSS. It was this climate that helped to create the Wessex Cave Club in November 1934 and the Bristol Exploration Club shortly afterwards in June 1935. Both were unashamedly 'sporting clubs', and many leading cavers of the day became members of these new groups as well as the older ones. Balch himself became the President of the Wessex Cave Club.

When Balch was an energetic 60-year old preparing the move to his beloved Museum in Wells, the young local cavers of the Mendip Nature Research Committee were led by Charles Wyndham Harris, then in his early twenties. Although Harris and his kindred spirits always remained loyal to Balch and the MNRC, they were frustrated by a growing preoccupation in the group to find out more and more about the known caves. The excitement and adventure of exploring new discoveries are bound to appeal more to young cavers. Nothing had changed and the revival of the search for unexplored caves had begun. On Mendip, this invariably means either digging your way in or, perhaps, diving into the unknown. Much interest was aroused by sites around the area's highest hills, such as the Cow Hole dig near Blackdown and the Waldegrave Swallet dig between North Hill and Stock Hill. 'Digger' Harris was the leading light. The nickname stuck after his digging team had re-opened the collapsed entrance of Lamb Leer Cavern in 1934; a cave beneath Harptree Hill that had been first found by miners in 1674, then closed and lost until its rediscovery in 1880.

Fig. 1.9
'Digger' Harris, a founder member of the Wessex Cave Club in 1934, on the roof of Harris and Harris offices at 14 Market Place, Wells. Photo from Molly Allwright

But 'Digger' was also especially tempted by the so-called terminal sump at the bottom of Swildon's Hole some 400 feet beneath Priddy. It had been first reached when the ever popular streamway was extended on the August Bank Holiday push in 1921. Ten years had elapsed since this epic discovery during the drought summer chosen. Here there was more hope and certainly more of a challenge.

Apart from the sump at the bottom of Swildon's the only other possible way on was a high-level one beyond the beautiful 'upper grottoes'. But this would require damaging some of the best formations yet found on Mendip, so Balch forbade pushing beyond them; a stricture that forced attention to be focused on the streamway itself, and the sump. Wet trips to Lower Swildon's became frequent and had widespread sporting appeal. With something of a reputation for succeeding with lost causes, Harris dug away at the sediment in the sump with an ice axe in 1931. Undeterred by a lack of success, he returned the following year to make the first serious attempt to get through. It was now 1932 and Balch was sufficiently in awe of this exploit to mention that Wyndham Harris had been the first to try passing the sump in his final book, *Mendip: Its Swallet Caves and Rock Shelters*, originally published in 1937.

With a score still to settle with the terminal sump in Swildon's, Harris became an important local contact and correspondent among Mendip cavers of the day. Recently married and articled to his father's practice of solicitors in Wells, parties would sometimes meet at his house and, no doubt, trips to dig and dive unpromising places had much to do with his own enthusiasm for pushing new ground. Contemporaries from Wells remember Wyndham Harris as an infectious personality and full of new ideas. How fortunate it was that he teamed up with Graham Balcombe and his close friend Jack Sheppard at such a fruitful time, for Balcombe was already a notable cragsman in the Lake District and had been potholing since 1933 with the Northern Cavern and Fell Club in the Yorkshire Dales.

Fig. 1.10
Graham Balcombe takes a rare 'back seat' behind Martin Trower at The Grange, the first HQ of the Wessex Cave Club in 1934.
Photo by Hywell Murrell from 'Digger' Harris in Wells & Mendip Museum

From the moment that Graham Balcombe first arrived in Somerset in 1933, this story, for the next twenty or so years, progressively focuses upon his achievements. After the first successful upstream explorations at Wookey Hole in 1935 and Jack Sheppard's historic downstream dive through the Swildon's Sump the following year, Balcombe was destined to take over the show and this book would not exist now but for his foresight. Nor could almost half of it have been written had he not kept detailed records of the fruitful years to follow through the forties. His distinctive hand dominates early Cave Diving Group publications, as evident in subsequent chapters. Jack Sheppard himself still thinks of Graham Balcombe as the founding father of cave diving:

> *Because he had the enthusiasm and determination to wish to form a cave diving group and the organising ability to do so.*

Today at over 80 years old, Graham Balcombe is living proof that those who make history also have the time to write it[3]. At this great age, he has even mastered a word processor and prepared accounts of his cave diving on disk for posterity. One day this unique personal record, thoughtfully and pointedly entitled *A Glimmering in Darkness*, will be published to give a fuller appreciation of early cave exploration and pioneer work in sumps[4]. In our commemorative book, however, we can avoid downplaying the significance of Balcombe's initiative, for he himself is apt to take a more matter-of-fact view of it all. As he recollected recently to me:

Fig. 1.11
Jack Sheppard, at the time of his first attempt to dive Sump One down Swildon's Hole in 1934.
Photo from Jack Sheppard

> *Cave diving began when we could breathe underwater to explore submerged passages. I simply set the ball rolling down a slope that was already there.*

Some of Graham Balcombe's background will set the stage to the first ever such dives in Swildon's Hole leading to those at Wookey Hole in 1935.

Francis Graham Balcombe was born in Manchester in 1907 and, as a young Lancashire lad, learnt to be independent whilst exploring disused buildings around his home. Another sign of independence was his adoption of the name 'Graham'. Later, when his father was posted

---

[3] Graham Balcombe died on 19 March 2000, aged 93.

[4] Published posthumously in 2007.

to Ireland on telegraphy duties the rest of the family moved to Mumbles near Swansea, South Wales, to stay with his grandmother. Here Balcombe had greater scope to travel farther afield, often spending whole days away discovering the delights of the Gower. The sea caves and cliffs there took his fancy and he and his friends would scramble on the stacks off Mumbles Head at low tide. When the tide came in, they would be cut off from the 'mainland'. Alone on their 'island', they could explore their discovery and, of course, dream of future conquests. Maybe it was here that Graham Balcombe acquired his daring for going beyond submerged passages underground. These early images of rock and water certainly became his 'yardstick' whilst exploring caves.

As a young Boy Scout back in Lancashire at Southport, he was taken fell walking in the nearby Lakes during the 1920s. Later, 'when released from the troop discipline' he graduated to scrambling the crags. On an early trip, he once climbed Napes Needle alone, only to be expelled from the group concerned for breaking their rule against solo climbing!

Climbing became a passion and he was already very experienced upon going to work for the Post Office during the depression years. Indeed, Balcombe continued his love affair with the Lakeland crags during an early posting to Reading, Berkshire, by joining a mountaineering group which later became the famous Tricouni Club. It was then that Graham Balcombe met Jack Sheppard, a London office colleague, and struck up a lasting friendship which especially flourished as a climbing partnership. During lunch breaks at work, both would ascend a radio mast to eat their sandwiches, chat and plan. Not every trip met with Jack's approval, however:

Graham's passion for climbing led him to climb any likely building in sight, which included the Police Station at Baldock, Herts, a climb that I did not follow!

Traversing along corridors was another test of his climbing skill, and the rage of the Officer in Charge at Portishead Radio Station, near Bristol, on finding a line of footmarks near to the ceiling was understandable for Graham had gained access through the coal yard.

One Christmas whilst staying at Seathwaite in the Lakes they chanced across members of the Northern Cavern and Fell Club camping nearby and were persuaded to try potholing. The bug bit instantly. Post Office duties brought Graham Balcombe to Highbridge, Somerset, in 1933, not far away from his parents' new home at Buckland St. Mary. His growing interest in potholing led him to read H.E. Balch's books about Mendip and, like so many others, he visited Wells Museum to meet the great man. Balch introduced him to 'Digger' Harris and another enduring friendship through caving began. Rather like Ernest Baker, whose own book *Caving* had just been published (1932), Graham Balcombe brought a similar keen edge to the Mendip scene. Jack Sheppard's role and determination were equally significant. In his chapter about 'The Cave Village of Wookey', however, even the sportsman Baker was resigned to the view that beyond the existing sumps:

...the waterlogged region in between will probably remain for ever impenetrable.

Whilst the prospect of cave diving was unimaginable to Ernest Baker and Herbert Balch, the latter's close links with the Hodgkinsons at Wookey Hole and the owners of Swildon's Hole, Priddy were vital. Moreover Balch, who had worked through the ranks of the Post Office in Wells, eagerly championed Graham Balcombe and Jack Sheppard in their ambitious quest. They had much in common professionally, of course, and it is pertinent that many of their helpers at the time were young friends and colleagues employed as Post Office engineers in Somerset.

First loves are not left easily, on the other hand, and climbing was still important to Balcombe. As Edward Pyatt notes in *A Climber in the West Country* (1968), before long Graham Balcombe had ascended new routes on Churchill Rocks, which are steep limestone slabs beside the main A38 road crossing Mendip that someone with such energy could not resist. Later, in 1936, he was to advance British climbing by establishing the now famous Piton Route in the Avon Gorge, Bristol; so named because it involved 'the first concentrated use of pitons in this country'. He had just returned from a climbing holiday in the Austrian Alps during the summer and had been won over to using artificial aids, like the piton and karabiner, in the homeland of 'ironmongery'. Jack Sheppard recalls that this trip was the second in a series of international climbing meets, the first of which had taken place the previous year in Britain when he and Graham had been selected as the two climbers to represent the UK in the Lake District.

A glowing tribute to Graham Balcombe's climbing exploits in the 1930s appears in the popular magazine *Climber and Rambler* (June 1977) under the apt title: 'A Meteor of the Thirties'. Its author, Ken Smith, introduces his subject thus:

> The requirements for pioneering the hard routes of the day were courage determination and sheer ability. One such pioneer of this period was Graham Balcombe.

The same could be said of his impact upon caving and the explorations at Wookey Hole in particular.

The catalogue of Graham Balcombe's achievements during the three years following his arrival in Somerset remains impressive by any standards. Nearly all were shared with Jack Sheppard and, underground, it was often Jack who set the pace and made the vital 'breakthrough'. Jack was born in London in March 1909 and subsequently graduated from University there as an engineer. He wryly recalls 'being thrown in at the deep end' by Balcombe on severe Lakeland crags:

> The process started with a camp, in February, in the winter snow and ice beside Goatswater for climbing on Dow Crag.

Graham, for his part, was experienced enough to recognise someone with talent and a head for heights, particularly when taking lunch at the opposite ends of the cross-bar at the top of a tall mast! Radio mast to rock face was a natural transition. Indeed, Jack Sheppard not only climbed and caved, but also had a lifelong ambition to fly. On retiring, aged 62, from the Post Office he realised this ambition in style by becoming a qualified pilot from no less an airfield than the famous Biggin Hill. Not many can have done as much after retiring, and even fewer can have reached a high enough standard to undertake 'spin' training, no longer compulsory for a pilot's licence. Yet Sheppard elected to learn this procedure; one which entails climbing to a suitable altitude where the instructor stalls the aircraft. As it then falls in a rapid spiralling drop, the pupil must pull the plane out of the spin before hitting the ground!

> This was most enjoyable, and the sight of Brands Hatch racecourse spinning rapidly and getting nearer far too quickly was a real thrill.

Fig. 1.12
Jack Sheppard leads a Hard Severe route on Great Gable in 1933.
Photo from Jack Sheppard

Still an energetic and amiable man in his late-seventies, Jack and his wife are renovating an old house in Snowdonia, North Wales. He still walks the hills and holds special memories of the Swildon's Hole streamway, particularly Sump 1. It was here that Jack made caving history on 4th October 1936. But such a prize had first to be won, the hard way.

Both Balcombe and Sheppard were taken to probe at the terminal sump in Swildon's Hole by 'Digger' Harris in January 1934; but the sump was not to yield easily. Although no way on down the streamway was found, 1934 proved to be a formative period, notable for the first use of underwater breathing apparatus specially designed to be taken down a cave. A new approach to cave exploration had arrived, but not before every effort to open up a dry route had failed. Some measure of how difficult this task promised to be at the time is evident in that Balcombe was prepared to spend one epic 37-hour trip trying to blast away the submerged roof of the sump later in January. Until Sheppard arrived with the explosives on the Tuesday afternoon, Graham Balcombe had been alone there since entering the cave at 3.30 p.m. the day before, hand-drilling a shot hole and trying to get some sleep on a gravel bank upstream. This solo effort was unprecedented in British caving and, even with today's protective clothing, has rarely been equalled.

Fig. 1.13
Frank Frost, cave photographer, concentrates on his other speciality preparing explosives for a cave dig in 1934.
Photo by Ivy Frost from the Wessex Cave Club

On 17th February 1934 the first diving party went down Swildon's with a home-made respirator. Graham Balcombe himself takes up the story of this historic trip and captures the spirit of the times in the *Log Book* of the Northern Cavern and Fell Club for 1934:

Fig. 1.14
Graham Balcombe (wearing his 'Bicycle Respirator') and 'Digger' Harris at Sump 1 in Swildon's Hole on 17 February 1934. Photo by Frank Frost from 'Mac' Brown – see Fig. 3.9 Chapter Three

Further notes on Swildon's Hole Somerset 1934

After the disappointment of the earlier attack with jumpers and gelignite, hope was never really given up. A sneaking idea that something could be done still lurked in our minds, eventually to form itself into a concrete idea.

There must be a way on big enough to crawl through or almost 'must' so if by any means we could crawl through it perhaps an obstruction would be found easier of removal than the barrier massif. Hence after toying a long time with the idea, the risks of diving seemed to grow less and less until quite justifiable, and a Heath Robinson respiration outfit with 40 ft. of garden hose was finally constructed and tested out in the domestic bath. For the benefit of those who may consider rubber hose a means of air supply for human consumption, it may well be mentioned that half an hour breathing through this foul smelling medium is enough to turn the strongest rather green.

The respirator itself was of very simple construction. The seat tube of a lady's cycle forming the principal member; this was cut down to suitable dimensions, 'raspberry' valves fitted to either end, and the curved member hacked off to take the mouthtube. Connection to the hose pipe, a face-strap to hold the mouthpiece in situ, a nose-clip, swim-goggles, a headlight, and a rope round one ankle completed the equipment save in one detail, just about as much 'guts' as the average man can summon to his assistance. The ghastly noises emanating from the devilish gear have to be heard down in the bowels of the earth, in misty dim-lit surroundings, to be appreciated to the full extent. But to the history of the job: three attempts were made to locate the exit and when found, the respirator failed to respond at the

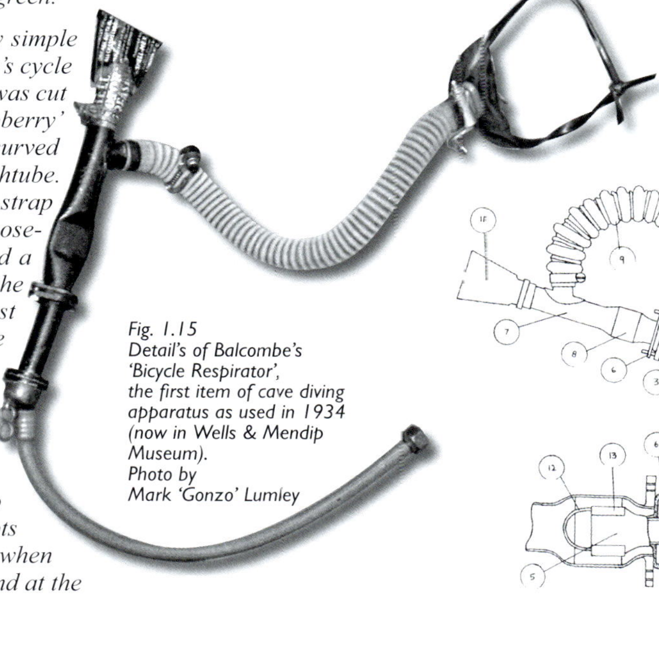

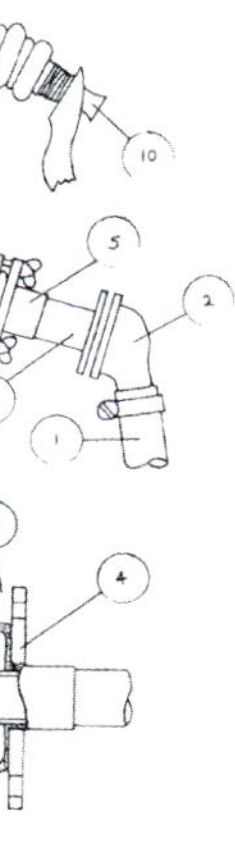

Fig. 1.15
Detail's of Balcombe's 'Bicycle Respirator', the first item of cave diving apparatus as used in 1934 (now in Wells & Mendip Museum).
Photo by Mark 'Gonzo' Lumley

depth necessary, and it was impossible to pass through without inviting serious consequences. An attempt was made by Jack Sheppard, but, alas, the hose had been badly fixed on re-assembly and came adrift at the furthest point reached – about 20 ft. under the rockshelf – it is thanks to his exceptional underwater experience that I am not writing these notes in the 'In Memoriam' column.

Now two things had been learned from this escapade; first, that the respirator must be pressure fed and, second, that waterproof clothing is needed as the low temperature of the water, coupled with the blood circulation impaired by the inevitable nervous apprehension, is more than the ordinary mortal can stand.

So, with a record dash for the surface of the earth to try and restore our dangerously chilled bodies to normal warmth, the second phase in the attack closed. But no, there was an aftermath. The excited tongues of visitor members wagged too rapidly and too loudly, the wily Pressmen pricked up their ears and foul calumnies appeared in the Western Press, over which we had better draw a veil.

Third phase opened with assault and battery. A charge of 10 lb. gelignite was laid against the roof of the newly-found arch and fired on time delay at 1 a.m. A dull rumble as of distant thunder disturbed the countryside and the slumbering village shook and trembled. A party went down next day with ill concealed excitement to view the wreckage, but there was none, or only a flake looked a bit loose and the mud of the tidal wave was plainly evident. Jack Sheppard, the most intrepid of the advance trio, attacked with a crowbar and suddenly woof! splosh! The lights went out and time stood still or nearly so as something like the whole roof fell down before us, almost scraping our knees and then drenching us with the splash. A deathly silence followed, no-one dared speak, until, the spell broken at last, we assured each other that we were untouched, and then lit up. About twenty tons of rock had peeled off the roof and now lay half buried in the mud of the pool. Thus was our objective brought a little nearer.

Another trip was arranged, and loaded with 30 lb. of 'jelly' we wormed our way down to the pool and planted a shot in the mud at the far end in the hopes that it might dislodge the supposed obstruction. Only a tidal wave resulted. Another and larger shot was then fixed under the archway and shot off. It was evident from previous experience that it was quite safe to stay below during the fireworks, and really it seemed that more disturbance was caused at the surface than below. We even managed to keep one of the many candles alight when a shot went off, though the air surged violently up and down the passage in which we were ensconced. It appeared later that this shot went off during evensong in the village church above our heads. Rumour hath it that the hassocks jumped six inches off the floor. The congregation thought perhaps that the Judgement Day had indeed come and afterwards according to our information the Vicar was heard to exceed his allotted vocabulary of 'Dear me, tut ! tut !'

But we are straying. When the fumes had subsided a little, the damage was inspected. The object of our attack was untouched, solid and immovable but the adjacent rib of rock had shed an enormous pile of blocks and had utterly changed the configuration of the final chamber.

Alas! thus doomed to this another disappointment we retreated once more to think it over. The project was announced at the time as 'officially abandoned', but 'Hope will spring eternal' as the poets have said and we hope to have another look at it some time later in the year.

Better that we leave it a while and let the spirit of peace once more settle on Mendip. Let the Press reports of earthquakes in the West be forgotten, and let the inhabitants replace their broken crockery before we venture forth again to the next attack.

'My former hopes are fled

My terror now begins,

I feel, alas! that I am dead

In trespasses and sins.

Ah! Whither shall I fly,

I hear the thunder roar

The law proclaims destruction nigh,

And vengeance at the door.

I see or think I see
A glimmer far away
I'll gaze upon it as I run
And watch the rising sun.

(After Cowper).'

So, they called it a day following the last-ditch blasting at the sump on 18th March 1934. After much more wear and tear fifty years later, it is somewhat ironic that during the drought summer of 1984 water levels fell enough to give an unprecedented open air link through Sump 1 in Swildon's. But the mid-thirties were wetter years, and the pebble bank beyond the sump had not been worn down. The very next diving attempt on 17th November 1935 had to be postponed because torrential rain gave rise to floods. Even the cave entrance was under water.

Better breathing apparatus was required and its effective use underground would need determined teamwork with larger support parties in deep caves such as Swildon's. During an outstanding fortnight's holiday in June 1934, the Balcombe-Sheppard partnership was strengthened in the Lakes. Ken Smith says that they 'burst upon the slumbering Lakeland scene', for Graham Balcombe led no less than seven new Severe to Hard Very Severe routes on Great Gable and Scafell Pike; enough 'to justify his place in the folklore of British climbing'. Such successes on the crags put the team in good heart and would lead to similar successes back in Mendip's caves. Meanwhile, the local men, 'Digger' Harris and Jack Duck, were busy forming the Wessex Cave Club; the nucleus of the pre-War support parties. Harris was destined to complete the very first successful exploratory dive in this country in Wookey Hole Caves the following summer when he entered the new Sixth Chamber beneath its water surface, whilst Jack Duck was to draw up maps for what is still regarded as the most definitive survey of the Great Cave as known then.

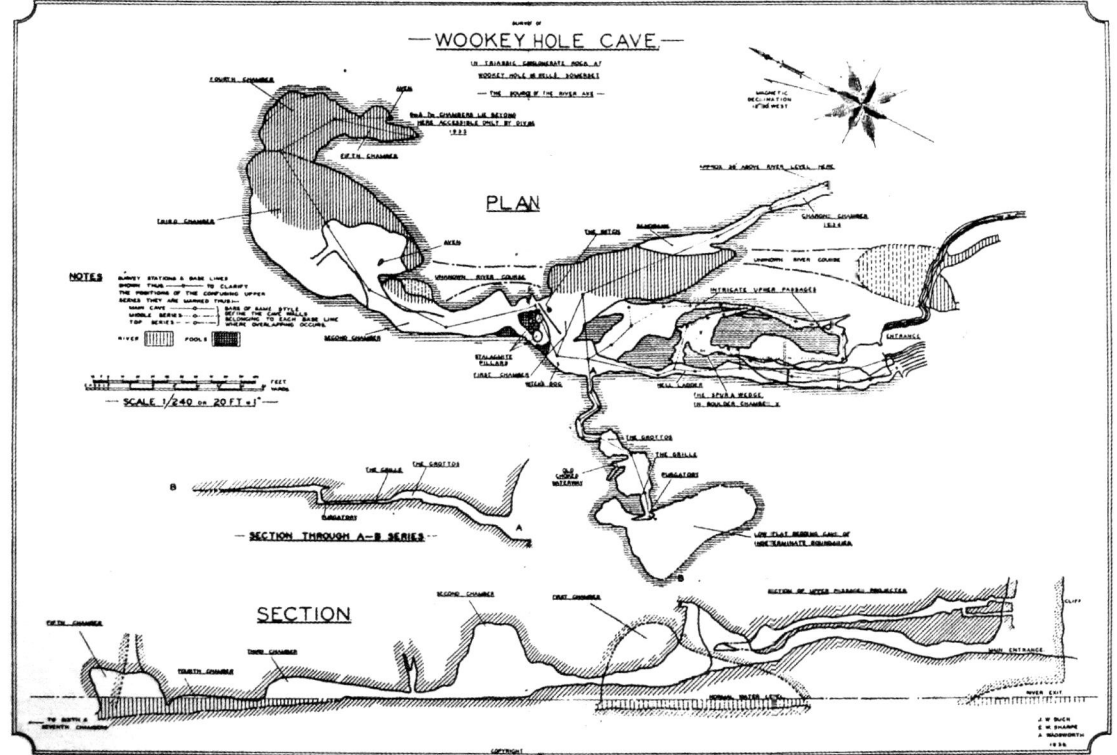

Fig. 1.16
The first definitive instrumental survey of the Great Cave made by Jack Duck and MNRC members in 1934 (see Chapter Fifteen)

By 1935, Jack Sheppard was engrossed in the task of building from scratch a home-made respirator that was pressure fed to work in deeper water. As we will see shortly this apparatus was not an adaptation as some have since suggested, but was brand new and designed very much with Sump 1 down Swildon's in mind. Jack named his diving suit 'Jimmy' and his own description of it in operation follows shortly. Meanwhile, Graham Balcombe took the direct step of seeking advice from professionals and contact was made with the diving experts Siebe, Gorman and Company Limited in London. Since he was living nearby, Jack Sheppard made an appointment with the company's owner, Sir Robert Davis, to find out if they had any equipment suitable for use in Swildon's Hole. Sir Robert showed great interest, but it took a long time to explain conditions down this cave and why lightweight gear was essential. Even the equipment for use in flooded mines was too large, and Jack recalls:

I made another visit and was shown their lightweight equipment as it was then. The dress was little lighter than the Standard Admiralty Diving Suit and the air pump was a curious wooden affair like a see-saw worked by one man at each end. From memory, I would say it was about six feet long and weighed around 70 lbs.

On a third visit, I tried once more to make clear why such gear could not be used in Swildon's and, in further very detailed discussions, I explained to Sir Robert my ideas for lightweight equipment to which I had been giving much thought. The availability of completely self-contained apparatus was also mentioned. He asked me to keep him informed on the progress of my suit and, during another visit, he questioned me closely about construction details, finally requesting me to send him a complete specification, which I did. In a written reply, Sir Robert offered to loan to Graham Balcombe and myself any equipment the company possessed, and also the services of their chief instructor, Charlie Burwood, whom I had met earlier.

Graham took up this very generous offer which included Standard Admiralty Suits, with all the accessories, and the 'see-saw' pump.

Since this bulky equipment would not have passed through the entrance of Swildon's Hole let alone the constricted passages inside, they were forced by circumstances to consider the larger caves and sumps at Wookey Hole. Thus, the known River Axe became the next objective. The little and large approaches to cave diving had arrived and were developed in parallel by Sheppard and Balcombe. As James Cobbett noted in his address to the Eighth Annual Congress of the old British Speleological Association (BSA *Journal* Vol VI, No 45, 1970):

Such equipment would be useless in Swildon's Hole, but down the road was Wookey Hole Cave where there was room for standard 'Hard Hat' diving gear. So it was, almost by accident, that the great Wookey Hole adventure began.

Both the home-made lightweight and manufactured standard equipment approaches were to prove successful in different ways: Sheppard's epitomises the economy of equipment outlook so crucial to later developments in cave diving down Swildon's Hole, and subsequently elsewhere, so that Jack has the distinction of being first in this field; Balcombe's, on the other hand, requires the organisational drive that is necessary for big expeditions, so that it may be claimed that Graham is the founding father of big diving operations. By general agreement, Balcombe led the subsequent exploration of Wookey Hole Caves by example; that rare mix of dash and determination. Yet Graham Balcombe sees it in a different light for, although not a conscious decision at the time, recently he told me that in retrospect he probably gave up climbing because the risks were becoming unacceptable and he feared the worst. Cave diving would be a safer pursuit! To both Jack Sheppard and Graham Balcombe, in fact, cave diving was not an end in itself, but simply the means to discover more passages beyond the sumps in Swildon's and Wookey Hole Caves. Their high expectations were rather different from what we know of these caves today, however, as Jack has described recently:

During the early days at Wookey Hole, our hope was to emerge beyond the upstream sumps into a dry passage sloping uphill, along which we might walk towards the source of the Axe. The reality of the chambers, deeps and high avens such as were actually discovered was completely outside our thoughts.

But high expectations are essential to make progress, and the prospect of passages connecting Wookey Hole with Swildon's strengthened their resolve. The challenge shared between Jack

Sheppard and Graham Balcombe reflects that presented by both caves. It is as if both these great but very different caves needed each other as much as the bringing together of the two pioneers. The ideal places had fired the imaginations of the right people at a ripe moment in the development of cave exploration. Phil Davies puts it nicely in the Cave Diving Group's pioneer *Somerset Sump Index*, published on 13th April 1957:

> If the Cave Diving Group was conceived in Swildon's Hole, then it was cradled and nurtured in Wookey Hole, where many have been the diving scenes enacted, from light-hearted water-romping to grim tragedy.

In 1957, the explorations of both streamways were said to be by far the greatest single achievement of the Cave Diving Group – and, another twenty years on, Ray Mansfield was able to add the following phrase to the revised *Index*:

> ...to triumphant discovery.

These glimpses ahead suffice to show how well Jack Sheppard and Graham Balcombe did their job at the start for their respective legacies still persist.

Because Swildon's is not a show cave and its streamway descends deep underground before reaching the sumps, it poses a considerable transport problem for bulky diving equipment; not least on the way out after an exhausting trip. The comparable sumps in Wookey Hole, on the other hand, can be reached in comfort and are much larger. The result is that, when technical problems have arisen during the exploration of one system, they have often been solved by experiments at the other one.

This is how we see the events of fifty years ago, with hindsight, of course. But, when you set out in the beginning, it is not even possible to anticipate nor easy to predict the exact form that any discoveries may take. To explore the underground River Axe, new and untried techniques would have to be mastered with no room for mistakes. Add to this intriguing doubts as to whether Swildon's drained to Wookey Hole or, perhaps, to the more distant risings at Cheddar. Even Herbert Balch latterly was inclined to think that Swildon's headed to Cheddar and that other swallets north of the Priddy Mineries also drained westwards, e.g. Waldegrave Swallet. Not only was the Swildon's streamway trending away from Wookey Hole, but new six-inch to one mile geological maps made by Dr. Francis Welch showed hitherto unsuspected folds and faults in the limestone which might guide groundwater to the northwest. Balch argued that Waldegrave Swallet probably went to Rodney Stoke because, after the nearby Wheel Pit pond burst on a dry summer's day, the only spring found polluted with suspended sediments was the one in this village. He wrote later:

> These facts were confirmed when the pond filled and burst again in 1935

This puzzle added an extra touch of uncertainty that intrigued Mendip cavers until well after Balch died in 1958. Its solution was a spur and even diverted Graham Balcombe into joining digging teams in search of the elusive 'Rodney Stoke River'; an element of Mendip's drainage that remains indeterminate to this day. Although Balcombe gave generously of both time and effort to excavate Waldegrave Swallet early in 1935, his heart was set on diving at Wookey Hole. He says so in his report on this unsuccessful dig:

Fig. 1.17
'Hordes of excursionists' at the Waldegrave Pond dam in the Priddy Mineries, 1935. The swallet dig was in a depression by the pine trees.
Photo from Graham Balcombe

At this stage the Wookey Hole work matured, and the Waldegrave Swallet had necessarily to be suspended awhile; this was no great sorrow for in the heat of the summer sun and surrounded by hordes of excursionists, the work was markedly distasteful... Central Mendip as a cave hunting ground is in my opinion poor, and the excavator faces very serious odds... the odds against success in this venture had been realised for some time and this realisation has helped in no small measure to soften the final blow.

And it was back to Wookey Hole Cave where the connection of the River Axe with the water that overflowed from St. Cuthbert's Minery at Priddy was definitely proven. For Graham Balcombe, this was the more positive way forward and he was right.

A dam was built across the mouth of the escaping River Axe in 1852 so that a higher level canal or leat could feed water to the Wookey Hole Paper Mill downstream. More water power was needed, for the local paper industry was booming after hard times. In digging the canal significant archæological remains were unearthed at what is now the Hyaena Den, the site where William Boyd Dawkins 'won his spurs' whilst still an undergraduate at Oxford. He was later to write the evocatively entitled *Cave Hunting* published in 1874; a book that lured many to study the caves.

Fig. 1.18
William Boyd Dawkins in later life and the Hyaena Den site he excavated when an undergraduate at Oxford University in 1859. Respectively from records at Buxton Museum and a contemporary wood-cut at Wells & Mendip Museum

Other fortuitous arrivals in the 1850s were the Hodgkinsons from London to take over ownership of the caves and mill. They rescued the paper making by winning vital orders for high quality banknote paper with delicate watermarks. The industry and the whole village were soon transformed by these go-ahead and generous owners. Future generations of cavers owe much to successive members of the Hodgkinson family and it is timely that some of the most notable discoveries by cave divers should have occurred just before the last owner, Olive Hodgkinson, finally sold the caves to Madame Tussaud's in 1973[5]. As with her husband Gerard, Olive encouraged cavers and particularly enjoyed celebrating their successes.

These early episodes were the seeds from which the exploring of Wookey Hole Cave took root. The canal raised the water level inside the cave as much as 6 feet to submerge the low arches to the formerly accessible Fourth and Fifth Chambers. As for Balch's exploration of these chambers by boat, already mentioned, access could be gained by lifting the sluices at the dam and lowering the water level. Thus, the Fourth and

[5] Now in the ownership of circus impresario, Gerry Cottle.

Fifth Chambers provided an unforseen opportunity for the pioneer divers to 'acclimatize' and make up the rules of their new craft in relatively reassuring surrounds.

The first Hodgkinson connection is celebrated for the outcome of the court case taken out by Gerard's grandfather, William, against Nicholas Ennor, an ebullient Cornishman then the owner and director of lead washing at the Priddy Minery. To prove that Ennor was responsible for polluting the River Axe at Wookey Hole, the courts required independent evidence and so elaborate water-tracing experiments were carried out in 1860. Ordinary red earth, copper sulphate, ochre dyes and even 29 gallons of common ink were poured into the swallet at Priddy Minery during October of that year. The venetian red ochre worked best and discoloured the rising at Wookey Hole 17 to 30 hours after it had been put into the swallet. It was enough to convince the court and, in 1863, the judgement went in favour of Hodgkinson. Since then, a perpetual injunction has existed to guarantee the right of the paper mills at Wookey Hole to 'a flow of water in a pure state' from the cave. The stream entering St. Cuthbert's Swallet at Priddy is thus protected. The novelty of the case aroused much public interest at the time and details of the experiments were published in the *Bath Chronicle* of 25th October 1866. Here was the sort of positive proof that gave added purpose to exploring the River Axe in Wookey Hole. It is an interesting coincidence that Graham Balcombe and the first cave divers carried out their open water training in the deeper Priddy Minery Pool in 1935; then, with nearby Eastwater Cavern, the only proven source of the underground Axe.

Fig. 1.19 Hodgkinson's Great Cave and Paper Mill in the Ravine of the River Axe at Wookey Hole: showing the sites of Hyaena Den (1), Badger Hole (2), and Rhinoceros Hole (3). Photo by Gavin Newman, 2004

The first 'acclimatization dips' in the cave itself took place on the weekend of 13th-14th July 1935 with a team of six: F.G. Balcombe, Mrs. P.M. Powell, F.W. Frost, C.W. Bufton, C.W. Harris and W.J. Tucknott. We have already met two of the team and Penelope Powell, who was always called 'Mossy', will become known a little more later in the book. Although her caving was confined to the pioneer hard-hat dives in Wookey Hole Caves during 1935, she became Graham Balcombe's 'second' in the exploration. As he recalls, when 'Mossy' first visited the Waldegrave

Swallet dig with F.R. Brown, she became involved straightaway, 'as an indispensable member of the team, for all recognised her mettle'. Frank, or rather 'Mac', Brown was then Manager of Wookey Hole Caves and did a bit of caving himself, including being on the historic pushes of the sump in Swildon's. Later, he was mistakenly referred to as 'P.R. Brown' to become Mendip's mystery caver of that era! A conundrum that was not solved until the research for this book fifty years later.

Frank Frost was another Post Office man who lived in Bristol all his life. Frank's reputation as a cave photographer was won during the Wookey Hole dives and his insight into peoples' caving interests led him to be the doyen of the Wessex Cave Club until he died in April 1981. Charles Wilfred Bufton, known as Bill, was an enthusiast from Exeter who travelled to Mendip every weekend on his motor cycle, loaded with caving and diving gear. Graham Balcombe's and Bill Bufton's parents were friends and Bill was yet another Post Office man attracted to caving by Graham's enthusiasm. He became a member of the Wessex Cave Club too, but R.A.F. duties during the war occupied most of his time. The closest he got to Mendip was a posting to the wartime airbase at Lulsgate, now Bristol Airport. It was here that he met his wife and, subsequently, both lived in Clacton-on-Sea in Essex, rather too far away from Mendip for regular visits. Happily, however, he has recently found the ideal excuse to return as his daughter and family have moved to Wedmore just down the road from Wookey Hole. What a coincidence that she should have returned to the Hunters' Lodge Inn, Priddy, to enquire about her father's caving days there exactly fifty years before. So it was that the third surviving member of the pioneer hard hat Wookey Hole cave divers made a timely comeback with local cavers. Bill Tucknott came from Wells where he later ran a road haulage business; a boon when heavy pumps, hoses and bulky divers' paraphernalia were shifted to and from the cave. Bill Tucknott also had a glass eye which, the story goes, popped out during swimming one day and prompted an exhaustive underwater search mounted by 'Digger' Harris. The glass eye was recovered!

Jack Sheppard was unable to join the team full time owing to his job and study commitments in London. Nevertheless, he vividly remembers the problems involved in transporting heavy diving equipment to Mendip when car transport was unavailable, a situation that was to persist well after the Second World War ended for many:

> Long train journeys became the only way to reach caving country. I have unpleasant memories of humping large suitcases loaded with diving gear from home to railway station and station to diving camp. One journey from my house half-a-mile to the local station was made in short stages of about 50 yards at a time, then pausing for a rest. I recall one occasion following a weekend at the Mineries Pond on Mendip when we packed all the gear into cases and left Mossy Powell to hire a horse-drawn cart to transport it to Wookey Hole. Faced with cases she could not lift, she unloaded them, put the cases into the cart, and then re-packed them!

Underwater, the problems were no less.

Progress beyond the Third Chamber into the already known Fourth Chamber was made rapidly on the first weekend, although initial impressions did not seem to tally with those found on Balch's earlier boat trip. Moreover, Balch had deduced that another spring downstream

Fig. 1.20
Jack Sheppard packs bags for the long train journey from London to Wells for a weekend of cave diving at Wookey Hole in 1935. Photo from Jack Sheppard

of the village at Glencot was fed by a leakage from the river bed within the cave, probably forming a deep pit in the floor of the Fifth Chamber. 'Holus Balchi' would pose a problem to divers weighted to walk along the bottom. What if they fell into the pit, or even got sucked down by the leaking current? Just as bad would be a 'blow-up' because an over-inflated suit would make the diver rise like an out-of-control balloon, maybe into a recess of the roof from which there would be no escape short of the highly dangerous procedure of deflating the suit of vital air. It was the accepted practice that divers must not try to control their own buoyancy like this; however, cave diving was different and the risk worth taking given that the necessary drill was practised. It became an emergency routine because cave divers soon concluded that they would have to rely upon themselves, for umbilical air lines and shot-ropes snaking around rocks underwater would be hopeless for hauling anyone out.

The following weekend saw the first totally new discoveries beyond the Fifth Chamber. Appropriately, in many ways, it fell to Wyndham Harris to pass through an underwater squeeze named 'The Letterbox' at a depth of 16 feet to enter a new part of the Great Cave for the first time, soon called the Sixth Chamber. An eventful three years had passed since his bold attempt to get through the terminal sump in Swildon's Hole. The hard work of line laying had taken up most of the Saturday and it was in the early hours of Sunday 21st July that Harris returned to the Third Chamber base to report his discovery of Six.

Now Wyndham Harris was a colourful local personality. As an active fire officer in the Wells Brigade, for example, he acquired the affectionate nickname 'Bunky' because of spectacular exits from his Market Place offices when the Fire Station siren sounded the alarm. The first to the station usually had the thrill of driving the engine! But he was also notably the gentleman, always fair, modest and not a media man. Not many know that Wyndham Harris helped to form the St. John Ambulance Brigade in Wells in the early thirties with first aiders in the old railway companies; also that he became the Government's Chief Legal Officer in Germany at the end of the Second World War in 1945, holding the rank of Colonel. He was responsible for ensuring that the defendants at the Nuremburg Trial had proper legal representation. In 1959 when Oliver Wells credited him with the discovery of Six, he replied:

Oh no!... Sorry... This is unfair to Balcombe... As leader of the expedition, most of the exhausting work of carrying forward the heavy concrete blocks had been performed by Balcombe with Powell as his chief underwater assistant. Pride of place had been accorded to them by the remainder of the team who had agreed to take their turns underwater only when these two pundits were resting. Well past midnight, Balcombe returned from placing his final weight at a narrow slit which, in his fatigued condition, appeared of doubtful accessibility. Thus it happened that I had the opportunity of going afresh to this point following the rope so securely laid. Lying on my back I slid through the slit feet first holding a 20 foot distance line shackled to the last block and had the joy of entering the Sixth Chamber [though still underwater and weighted to its floor]. The water was surprisingly clear, and looking to my right a chamber appeared to extend to considerable proportions... The mud started to rise and obscure vision and the air supply from the special small pump became reduced and I returned with no visibility, feeling my way along the heavy rope. This was the invariable condition for the return journey for all divers. Mossy Powell then followed, entered Six and spent 15 minutes there. On the way back she lost the shot rope and got her air hose foul on the rock... [On the following weekend] we entered Five by boat to see where the shot rope went and it was ascertained to go OUT of the Fifth Chamber. Powell and I were acquitted of the charge of 'fishermens' tales.

Charles Wyndham Harris died in Wells on 4th August 1978. The tributes to his memory were many, of course, yet one of the most eloquent came from Graham Balcombe on reading the above letter recently:

How like Digger! One of the finest men I ever met.

The next few weeks in the summer of 1935 were fully occupied with surveying the finds with Jack Duck, more publicity events in the cave and, importantly, practices with improved equipment. Wyndham Harris continued:

The small pump ... was replaced with a huge Admiralty pump with two outlets enabling two divers to be underwater at the same time. Such was the friction of the airpipe and lifeline that one diver could not drag it further forward without assistance.

Thus it was that Graham Balcombe and 'Mossy' Powell together were able to take the next historic step to find the Seventh Chamber at the very end of August as described in their 'Prologue'. For Balcombe especially this was a great reward for so much intense effort. A few days afterwards he wrote:

The Seventh Chamber, as then revealed, was a crowning discovery which removed all the regrets at having to conclude operations, and created a feeling of achievement, justifying our past efforts, but still leaving the prospect of fully exploring it as a glorious incentive for some future date when the water supply difficulties have been overcome. The Great Seventh, though only some twenty-five feet wide, transcends in architectural magnificence all other chambers in the cave... to find the end, to measure, and to photograph, are the goals that await us in future... So concludes the work of the Wookey Hole Exploration Expedition of 1935. Hats off to Sir Robert Davis, who has made it possible!

The full story of this expedition was privately printed and published as a unique book entitled *The Log of the Wookey Hole Exploration Expedition 1935* by 'The Divers'. The key sections of this prized book are reprinted in the next chapter as a tribute to Graham Balcombe and all his team.

But the scope of this chapter would be incomplete without details of the complementary push by Jack Sheppard the following year in Swildon's Hole. After his self-imposed exile whilst he worked for his degree during the summer of 1935, Jack made the first test dive with his own pump-fed diving dress specifically designed to crack the Swildon's sump. He now takes up the story:

On Wednesday October 30 in 1935, I wrote to Graham Balcombe saying: 'Diving suit is now practically complete. I spent 20 minutes in a pond at Borough Green on Sunday and am still alive. The suit is not quite according to theory, as even when deflated as far as possible, there is still too much bouyancy. Hope to make a further test next Sunday but it will probably be necessary to use weights of some sort ... Weight of gear has been kept to a minimum but a party of 9 will be required'.

My suit was built in two parts joined at the waist, a flexible rubber-twill helmet being permanently fitted to the upper part. The waist joint was formed on the same principle as the usual form of wrist joint. The lower part of the suit was jointed into a tight section of thin rubber at the waist forming a smooth surface over the underclothing and over which a short rubber diaphragm of small diameter (9 inches) fitted to the upper part of the suit made a water-tight joint. Incidentally, the waist and wrist seals were made from a material of amazing elasticity, so that the diver's head, shoulders and arms could get through without difficulty and without risk of tearing the diaphragm. The incoming air was discharged round the eye-pieces to reduce condensation. Ordinary nailed boots were worn.

On the question of the origin of the name 'Jimmy', I can only say that it results from the strong human tendency to call anything or anybody by a nick-name. 'Jimmy' just stuck: it has no special significance.

The suit was designed with a floating discharge valve, the idea being to prevent dangerous inflation, and to ensure that the water pressure on the lungs did not make breathing difficult. It worked perfectly.

In use, the pump was started before the suit was put on. As the diaphragms sealed, the suit inflated and breathing was normal. But also, the eye-pieces always misted over. The air blast quickly cleared this, and a check was always made to be certain that demisting was complete and did not recur.

On entering the water, the suit was deflated by water pressure without difficulty and synchronised breathing started automatically. Underwater, no difficulty was found in breathing, and no uncontrolled inflation took place in any position adopted by the diver. There was no way to breathe without the pump with the suit on.

Synchronised breathing was not in the original plan: I visualised a continuous supply of air. I tried, but failed to get a double-acting pump. For the first test dive, in a flooded quarry, the plan was for a slow down-stroke of the pump and a rapid return, to make the air supply as continuous as possible. When the pump man became tired, the return stroke slowed and suddenly the pump was roughly at the pump-man's breathing speed, and synchronised breathing by the diver followed. This was an absolute delight, the pump doing most of the work of the diver's lungs.

A practice dive was made at the Mineries pond on Mendip to familiarise the support party with the gear and to demonstrate the pump-action. This was followed the next weekend by a planned descent of the cave,

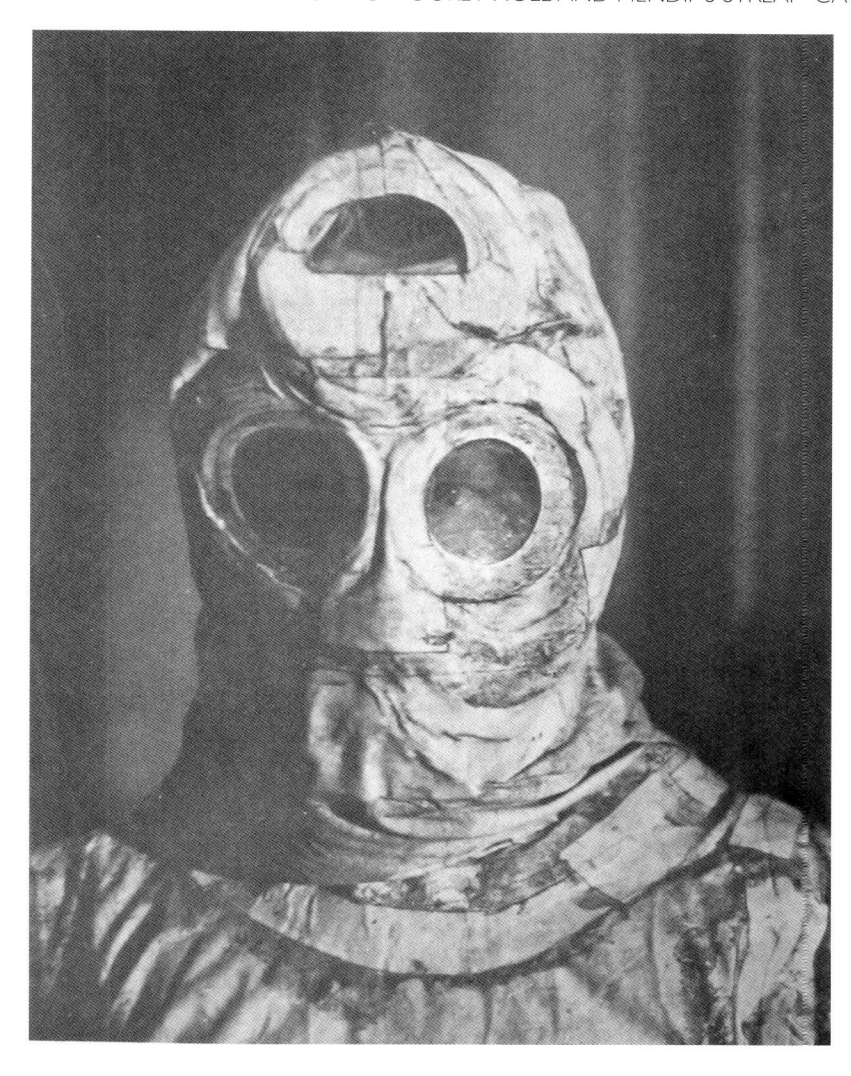

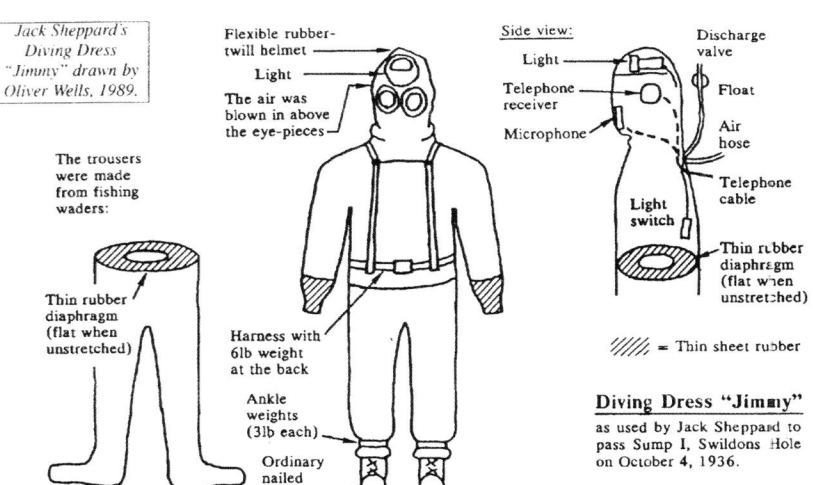

Fig. 1.21
a. Jack Sheppard's home-made rubber diving dress called 'Jimmy', 1936.
Photo from Jack Sheppard in Wells & Mendip Museum

b. Details of its construction drawn by Oliver Wells, 1989.

Jack Sheppard's Diving Dress "Jimmy" drawn by Oliver Wells, 1989.

Flexible rubber-twill helmet
Light
The air was blown in above the eye-pieces

The trousers were made from fishing waders:

Thin rubber diaphragm (flat when unstretched)

Harness with 6lb weight at the back
Ankle weights (3lb each)
Ordinary nailed boots

Side view:
Light
Telephone receiver
Microphone
Discharge valve
Float
Air hose
Telephone cable
Light switch
Thin rubber diaphragm (flat when unstretched)

////, = Thin sheet rubber

Diving Dress "Jimmy"
as used by Jack Sheppard to pass Sump I, Swildons Hole on October 4, 1936.

Based on CDG records, photographs and notes from Jack Sheppard.

(O.C.W. Jan. 1989.)

but the entrance was two feet underwater, and that trip had to be abandoned,

On being asked how it felt when he dived through Sump 1 on 4th October 1936, Jack replied:

My feelings were strongly influenced by the fact that I was a very experienced (competition) swimmer, had spent hours underwater in the local swimming baths diving for coins, and so was very familiar with the feel of being submerged. But the occasion of the dive was a great event in my life and left very vivid impressions which still persist, and these I shall try to describe. Before the actual dive in Sump 1, I had been underwater three times in 'Jimmy' and so had complete confidence in the safety of the suit. Donning it in preparation for the dive produced more reaction from the assistants than fom me, as most of them had not seen 'Jimmy' before. However, the lower half was already on, the top was unpacked, the pipe, pump and telephone connected, and I was still one of the party. Then with the starting of the pump, I was suddenly in another world, completely cut off, and viewing the cavers through my eye-pieces rather as one might view a film or play, without being part of it.

This feeling persisted during the routine testing of telephone communication, headlamp, demisting of the eye-pieces, and a curious 'still' until a tray of flash-powder was persuaded to ignite. I then sank slowly into the water while the suit deflated. The thrill came, as divers will know, when the water rose above the level of my nose without affecting breathing, then above eye-level and suddenly, no longer watching a party of cavers as in a film, I was completely alone inside the small world of the diving suit. I had no feelings of danger, but did experience a supreme pleasure in being underwater and in using the suit at last for the purpose for which it had been designed. Now for the result of so much thought and work.

I was in a clean rocky tunnel brightly illuminated by my headlamp, proceeding cautiously in case of loose rock. Suddenly the beam from my lamp found a water surface instead of the rock roof. I stood up and waded out of the water. The helmet and upper suit were removed. I then experienced to the full the delight of true exploration, alone in a place where no man had ever been before. Just for a moment it was 'my' cave, even if only a dark and cold stream passage.

I called for the pump to stop... and had a look around. The bed of the stream appeared to be composed of black pebbles where it formed a dam that held the water back. I spent some time trying to remove these to lower the water level in the sump, but without success. Then, deciding to explore my cave, I followed the stream passage until reaching a point where further diving appeared necessary. I returned to 'Jimmy'.

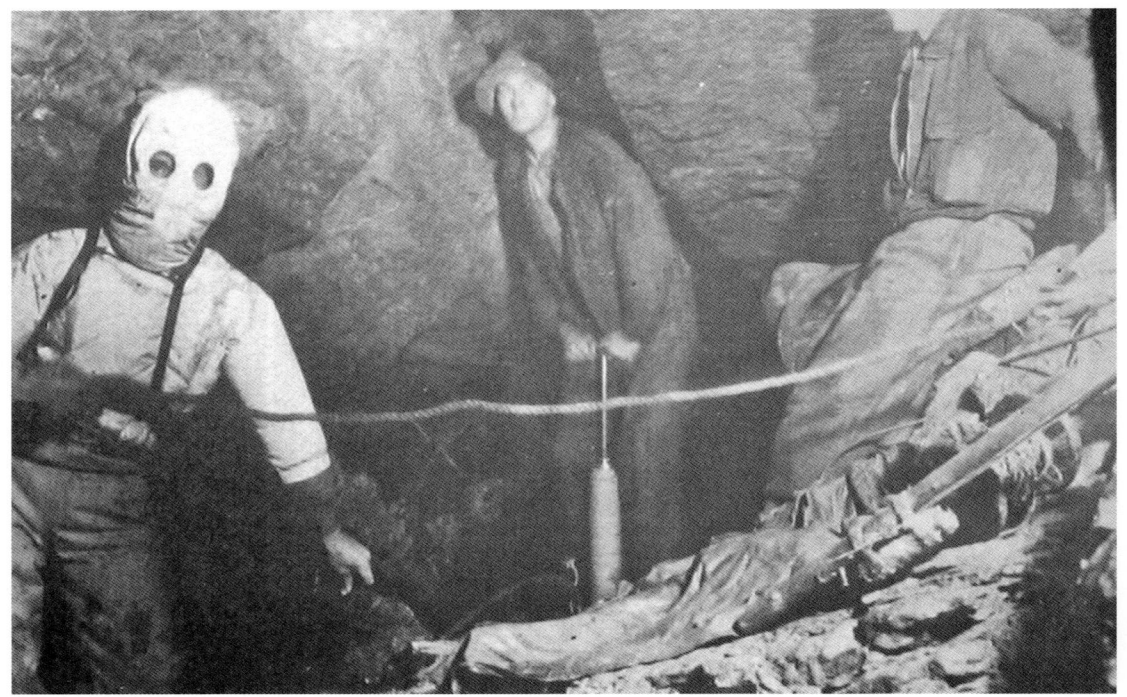

Fig. 1.22 Jack Sheppard on the first successful cave dive through Sump 1 down Swildon's Hole on 4 October 1936, wearing 'Jimmy' with air pumped by 'Digger' Harris. Photo by Hywell Murrell from 'Digger' Harris in Wells & Mendip Museum

By that time my head was beginning to ache, due no doubt to spent carbide fumes in the pump intake. I could get no reply on the telephone, only a weird echo from the cave itself. I did feel somewhat lonely sitting alone in a dark cave. Finally the voice of Graham Balcombe came through, followed by the thud and hiss of the pump. I replaced the suit and returned, my head now splitting from the fumes.

Back in Swildon's One, I thought 'never again'. This feeling lasted for nearly a week before thoughts turned to passing the sump by a direct swim, and I wondered why I had not left a rope in position.

Only a fortnight afterwards, Jack Sheppard free-dived the sump and laid a rope so that Graham Balcombe could follow. They explored the streamway together past two very low 'ducks' to Sump 2. 'Digger' Harris was the lucky third man through to help survey the new passages:

Two weeks later they took me with them – the first terrified 'tripper' to Swildon's Two. At Sump 2 there were poles floating from the former 'bomb' activities and Balcombe decided to push on underwater holding breath and hanging on to one end of a jointed rod. Sheppard held the other end. It disappeared further and further under the arch. Jack reached under to the full length of his arm as well and then remarked 'What do I do now? Let go I suppose'. The next thing I saw was Balcombe swimming back under water with a surprising over-arm stroke, going hell for leather. We laughed until we cried.

Balcombe's telephone call to me after his first trip to Swildon's Two through the sump remains in my mind: 'We got through,' he said, 'and it goes a long long way. It is just like dying. You feel the cold waters close over you and you come-to in a new silent world. And there is a spike coming down from the arch ... I got caught on it and stuck there under water'. Despite this I agreed to go with them on their next trip.

Imagination played hell with me ... until, of course, the day came and all went merry as a marriage bell. 'I'm depending on you, Shep,' I said as I entered the water of Sump 1 to follow Balcombe. 'I'll come and look for you after five minutes,' he replied.

Do you know Orwell's Nineteen Eighty-Four? *I am sure the special chamber of horrors reserved for me would be a sump with hooks on it to catch on clothing.*

These recollections will be enriched further when Jack Sheppard's own essay is published in *The Last Adventure*[6]. They help to redress an imbalance in contemporary and later records which happen to highlight the earlier work at Wookey Hole more than those equally momentous moments in Swildon's; understandably, perhaps, for Wookey Hole is very much in the public eye and Gerard Hodgkinson had made the most of the free publicity accompanying the 1935 dives in his cave. Jack Sheppard's dive through Sump 1 in Swildon's in 1936, on the other hand, was a unique event of significance among a tiny circle of caving enthusiasts at the time. Very soon, too, others were swimming through, simply holding their breath, yet, like 'Digger' Harris, inwardly in awe of Jack's pioneering work. It only becomes easy when you know, of course!

As has happened several times since, however, some remained unprepared for the ordeal of 'sump swimming' when the moment of truth arrived. Sheppard recalls:

A further trip was made with a larger party, although many cavers contacted found that they had other engagements for the weekend. The trip was a success but must have been terrifying to cavers with less underwater experience than myself. On the return, two of the party flatly refused to go underwater to enter the tunnel and it took them a long time to overcome their feeling of panic. I had serious thoughts that it was going to be necessary to knock out the most badly affected of the pair and drag him through unconscious! Shortly after this, by request, a 'Ladies' Day' was organised, and there were two takers – Ruth Johnston, who later married Hywel Murrell, and Molly Hall. In pleasant contrast to the previous party, they showed no signs of fear and the trip went smoothly. It was a great experience.

It almost seems appropriate that 'Jimmy' was never used again. However, he did have a final public appearance at the British Speleological Association Conference held at Bristol in July 1937. The suit was stuffed with newspapers and exhibited with its pump in attendance. Jack should have the final word on his brainchild:

He was yet the prime attraction among the maps, pictures and models on display. After this, he just

---

[6] Jack Sheppard died on 14 July 2001, aged 92. *The Last Adventure* compiled by Alan Thomas was published in 1989.

disappeared and I do not know what was his final fate.

Meanwhile, the prospects for further finds down the Swildon's streamway looked good. Cave diving with breathing apparatus was vindicated and Jack Sheppard's discoveries beyond Sump 1 are rather special in the honours of caving achievments.

On 22nd November, Graham Balcombe returned with the old 'bicycle respirator', connected it to an oxygen cylinder and dived the larger Sump 2 to find the Little and Big or Great Bell Chambers. Apart from a life-line, he was on his own as a self-contained diver. Balcombe's account of this historic event was published later in a *Letter to Members* of the Cave Diving Group dated 29th March 1947:

The 'Bicycle Respirator'

The 'bicycle respirator' was born at Broadstairs in 1934. It was the first known attempt at cave diving equipment, unsound in design but careful in execution. It came into being to meet the demand of the now famous Trap I in Swildon's Hole, where already the urge to conquer the water barrier was becoming irresistible and the happy ultimate consequences were to have such a great effect on the technique of cave exploration. It is a simple device of the type toyed with throughout the centuries by experimenters who probably should have known better; the diver by his own lung-power drawing in air through a tube from the surface and blowing it to waste into the water through a system of valves. Looking back on such a proposal it would seem that anyone with the most meagre knowledge of hydrostatics should have known that the lungs would be incapable of working against the pressure of normally expected depths of water. Yet it is remarkable how many people I met about that time who knew somebody who had used such a rig for diving six feet and more below the surface. I should have known better, but it is probably just as well I did not, for the endeavours with the simple gear stimulated interest which might not have withstood otherwise the trials of making more complicated equipment.

The basis of the assembly, Fig 000, is a piece from the seat-tube of a derelict lady's bicycle. The air-pipe (1), a 40" length of ½" garden hose, is attached to a tap connexion (2) modified to fit a flange brazed on the end of the tube (3). This tube ends in a spigot with a flange (4), see Fig.000, which carries the inlet valve (5) and is bolted against a flange (6) brazed on the T-tube (7). The T-tube is flattened at (8) to accommodate the inlet valve which is of the flutter type. The breathing tube section of the respirator is the usual corrugated tubing (9) strengthened with curtain rings and terminates in a bite-on mouthpiece (10). Exhaled air passes out into the water through a second flutter valve under the sheet-metal housing (11).

Valve Arrangement in Part-Section.

The valves are made of bicycle inner tubing, see Fig. 000; to deform the tube from its natural shape it is stretched into a flat mouth by a wire (12) and as that would cause the mouth to gape by the thickness of the wire, the edges are solutioned down and then clamped under a lead edge (13). The respirator was strapped to the diver's chest and used with a nose clip, swim-goggles and headlamp, but with no protection against the coldness of the water. In the event it failed for two reasons (apart from the occasion when the hose-pipe fell off when Sheppard was well and truly under the sump), the diver has to submerge to a depth approaching two feet in Trap I so the breathing strain became serious, and the exposure to the cold (coupled with the nervous tension associated with the undertaking) soon rendered the diver incapable of further work. In 1936 the respirator was dragged out from its home among the workshop junk and an oxygen cylinder coupled up in place of the hose-pipe. The diver wore the cylinder dangling between his knees and gripped it with his thighs, turned on the gas while he took a breath, turned it off again, and blew the gas out to water as he breathed out. Did we carefully note the gas pressure before starting? Good gracious, no! We left the cylinder parked at the top of the Forty-Foot for a week or more, but did we test it for gas seal? Good gracious, no! Anyway, there was quite enough gas left after the dive to brighten up a candle quite distinctly. This crazy contraption should long ago have been thrown out into the rubbish bin, but no it lies, or what is left of it does, in the workshop junk box, and woe betide the bod who thinks a part of it might be useful for some job or other!

FGB.

Without waterproof clothing, Graham Balcombe was hampered by cold and unable to plunge further into Sump 3 as his life-line limited him to 40 feet out from the base back in Swildon's Two. He could see the way on underwater. It is lucky that he could not go further, however, for, on returning, the tiny oxygen cylinder had almost run out as he surfaced to safety. Extremely cold and, no doubt, anxious to get out of the cave after such a near miss, Graham Balcombe set off alone ahead of his support party. They caught him up just upstream of Sump 1, virtually unconscious and huddled over the pitiful flame of his carbide lamp for precious warmth. He must have been hovering around the fatal stages of hypothermia by then and it must have taken an immense effort to defeat the cold and climb up the streamway.

After Graham Balcombe's tantalizing dive on 22nd November 1936 there was a lull in the great streamway pushes on Mendip, and then the Second World War cruelly intervened. It would be another ten years before diving began again at either Wookey Hole Caves or Swildon's Hole. There was more than enough time to reflect upon past achievements and future promises. Balcombe and Sheppard had pointed the way ahead in their use of self-contained breathing apparatus in Swildon's, for both Sump 3 there and the submerged River Axe beyond the Seventh Chamber in Wookey were wide open.

In their day, the exploratory dives reviewed here were no less remarkable than those that have been achieved since. For the older men such as Herbert Balch, they represented the ultimate possible. His contribution to the *Somerset Year-Book* for 1936 recounts the work of 'our young members' in pioneering cave diving in the so-called 'flooded caves of Mendip.' Of the helmet dives in Wookey Hole Caves he writes:

They safely returned, minus certain tackle, which is still there, possibly to remain for ever, for it is not likely that this venture will again be made.

He was even more concerned over the risks taken with home-made respirators down Swildon's Hole. Relieved that Graham Balcombe had emerged safely with such a 'narrow margin of safety', he concludes:

There the matter at present rests, and I do not think the advantages to be gained by pursuing the effort warrant the risk of valuable lives.

Whilst writing his last book shortly after the discovery of Swildon's Two and Balcombe's penetration to the Great Bell, Balch commented in the MNRC *Report* for 1936:

He stood shoulder deep in the icy water, and looked for the first (and I expect the last) time, on these furthest confines of the Mendip underworld.

The rest of the present book is written by those prepared to challenge the 'blind alley' recognised by Balch but beyond his reach. Herbert Balch's last essay about his own vintage appeared, appropriately, in the first volume of the magazine *Caves and Caving* for November 1938, the forerunner of the modern 'glossies' now read widely by cavers about their pastime. Its cover features a contemporary advertisement for Wookey Hole Caves and Balch's contribution is aptly headed 'Fifty Years of Caving on Mendip'. In it he reflects upon his own introduction to caving:

My earliest memories of the call of the caves belong to 1885 when, as a youth of 16, I listened to Professor (later Sir William) Boyd Dawkins as he stood in the valley of Wookey Hole and told a small party of his experiences in the Hyaena Den where he had worked 25 years before.

By a happy coincidence, then, the first Fifty Years of early cave exploration on Mendip dominated by Balch spanned the period 1885 to 1935. And then a new era began with the success of the pioneer cave dives in Wookey Hole Caves during 1935 and down Swildon's Hole in 1936. How uncanny it is that One Hundred Years of Mendip cave exploration since 1885 should be precisely halved by the second Fifty Years commemorated here.

One wonders what Herbert Balch and his fellow explorers might have foreseen fifty years on from their own discoveries in 1935. Happily Balch lived to learn of the discovery of St. Cuthbert's Swallet; but, since his death in 1958, the length of explored passages in Swildon's Hole and Eastwater Cavern alone has almost doubled. Less than two thousand feet of the old show cave at Wookey Hole was known to him, but, nowadays cave divers go over five times as far under Mendip.

And it still goes on, of course. Those with a sense of history and knowledge of cave divers will not be surprised that our story here begins as that of the older men comes to an end.

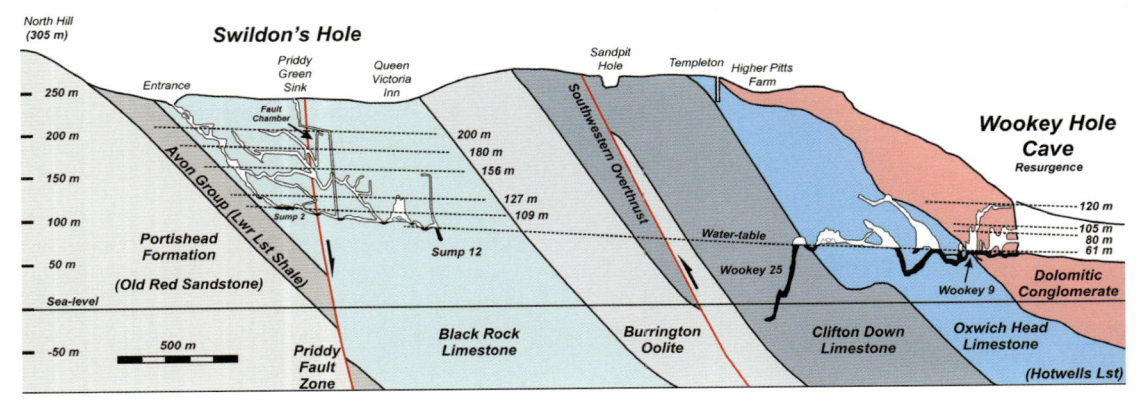

Fig.1.23
A review of today's hard-won knowledge of the links between Swildon's Hole and Wookey Hole Cave by Andy Farrant, in Swildon's Hole: 100 years of exploration pp. 264-5 (Wessex Cave Club, 2007)

Fig.1.24
Herbert Balch in later years at his Badger Hole dig, 1950s. Photo by Luke Devenish.

Chapter Two
Foreword, Discovery and Divers' Observations
by Herbert Balch, Graham Balcombe and Penelope Powell

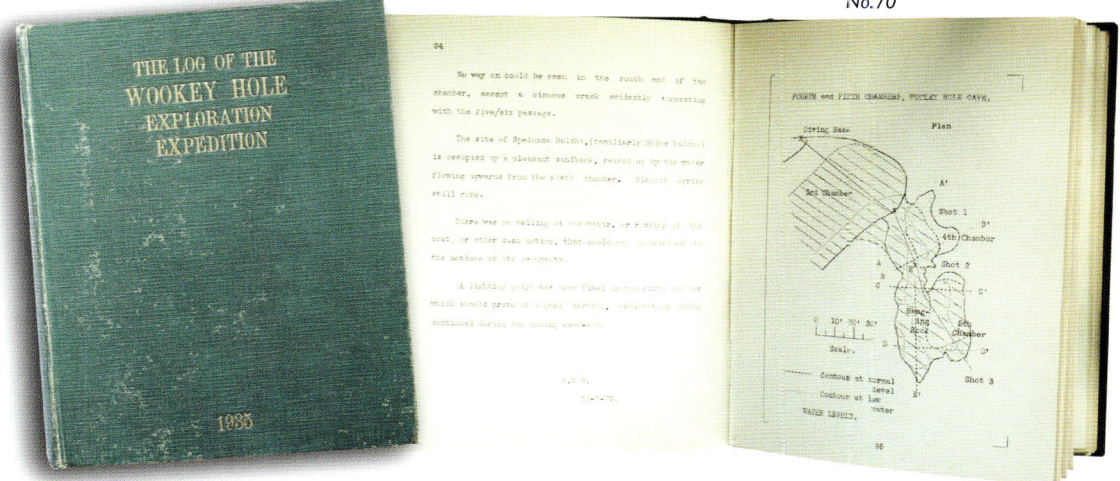

Fig. 2.1
Jim Hanwell's personal copy of *The Log of the Wookey Hole Exploration Expedition (1935)*, No. 70

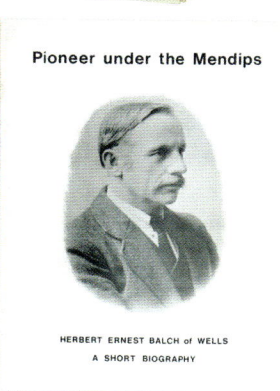

Fig. 2.2
Herbert Balch in the early 1930s, when he went to live in Wells Museum (now Wells & Mendip Museum).
From the cover picture of *Pioneer Under Mendip* by William Stanton, Wessex Cave Club, Occasional Publication Series 1 Number 1 (Oct 1969)

The following accounts are taken from *The Log of the Wookey Hole Exploration Expedition 1935*: a book written by 'The Divers' and produced by Graham Balcombe who did all the typing, printing, stitching and distributing during 1936 with the help of a local schoolboy in Ascot and a firm who did the blocking and binding in covers. There were one hundred and seventy-five numbered copies and the publicity consisted simply of the notes to fellow cavers reproduced here. The 'Log of the Wookey Hole Divers' is, therefore, a rare prize to those who appreciate its significance to the history of caving. Here are parts of that prize.

29

Foreword

by H.E. Balch, M.A., F.S.A.

That eternal law of progress, in exploration as in other matters, will always find new ways by which man will find an outlet for his energies and his insatiable curiosity. So it came about that the blind alley, in which we, the passing generation of cave explorers, found ourselves at Wookey Hole, when our raft, or boat, was brought to a standstill by submerged passages, has been challenged by the writers of this Log, and their companions in adventure, who, undeterred by dangers greater than those faced in ordinary diving attempts, have faced perils by water unknown to my generation.

The Great Cave of Wookey Hole, formed by countless aeons of the activities of the subterraneous Axe, has exercised a peculiar fascination for mankind for ages past, and never more than at the present time, when many thousands, who have never given a thought to Nature's marvellous works, now revel in the beauties of the underworld there revealed.

This place has not only been the objective of visitors who for century after century have left records of their impressions, often weird and wonderful, detailed at length in my various books on the cave, but, as proved in our five-years-long digging, was the actual home of a most interesting community for many hundreds of years, just as pre-history was passing into history. The whole valley outside the cave told its older story, too, under the spade of Sir William Boyd Dawkins, who dug there many years ago, when its denizens proved to have been Mammoth and Rhinoceros, Lions, and Bears, Hyaenas and Wolves, and the other great beasts of 50,000 years ago; and Man was here even so long ago as that.

The Great Cave has thus been known as far as its three great chambers are concerned, for centuries, and in our explorations the two distant chambers, no by water, had been visited several times to the limit of further progress being possible.

Fig. 2.3
Standard Diving Equipment
provided by Siebe, Gorman & Co.
Ltd in June 1935.
The 'Frontispiece' of The Log.
Photo by Frank Frost

It was thus that the matter stood when Mr. Balcombe conceived the plan of diving, by which the water-barrier, which we had measured 14 feet in depth, might be passed, and the absolutely unknown might be revealed.

Such a hazardous enterprise was not to be undertaken lightly, and long and arduous experimental work was necessary, carried on in deep pools in the open air before they felt themselves capable of making the venture. In this way six volunteers became proficient divers.

Taking advantage of low water conditions, former expeditions by boat had found advantages which were denied to those plucky adventurers. The reason for this was that some of the passages are low and the water shallow except under flood conditions. Therefore, as flood conditions were essential in order that the divers should be submerged, no boat could follow them in case of need, or feed them with supplies. Hence no boat could ever reach them even in dire necessity.

So, foot by foot, yard by yard, with extending air lines and shot-rope (a weighted guide-line) the submerged floor was explored. Stirred by the passing feet and dragging lines, sediments swirled about their heads, obscuring their vision, or limiting it to a very short distance.

It was under conditions such as these that measurements, or if not actual measurements, close approximations of distance must be taken and compass read. Possibilities of error are numerous and are to be expected.

Our great hope is that the final result of these plucky efforts will be to show us what direction we may attempt surface excavations which may, by ways at present unknown, lead us to the same objective. There are two miles of caves between Wookey Hole and the Priddy swallets. Much of this must be submerged. There are other miles stretching eastwards towards Hillgrove Sanatorium, and so I conclude that neither this generation, nor the next, are likely to reach finality in this 'land of caves, whose palaces of fantastic beauty, still adorn the mysterious underworld, where murmuring rivers first see the light.'

The distance traversed in this venture may appear insignificant, and not to be measured by ordinary rules, but by the almost insuperable difficulties encountered. In our experience on many occasions progress of five yards has involved strenuous work for as many days, and thus it may well be in the present instance, alternated perhaps, by sudden leaps forward into the unknown.

The Wookey Hole Exploration Expedition

by F.G. Balcombe and P.M. Powell

Is it a press stunt? A pertinent question raised by a member of the caving and climbing world, well known for impressive description, perhaps not so well for accuracy of detail.

The answer. No. Publicity is the unavoidable accompaniment of work in a commercialized cavern.

The divers

F.G. Balcombe	*Ascot*
'Mossy' Powell	*The Base Camp*
'Digger' Harris	*Wells*
Frank Frost	*Bristol*
Bill Bufton	*Exeter*
Bill Tucknott	*Wells*

The work of the 1935 Expedition has been brought to a close.

The possibility of exploration at Wookey Hole has for a long time been evident, but work elsewhere, lack of experience, and a certain diffidence about working in a commercially operated cavern, have all combined to deter until 1934, the decision to start an expedition. Arising out of a magnificent offer from Sir Robert H. Davis, Managing Director of Siebe, Gorman & Co., Ltd., submarine and safety engineers of world-wide repute, the major obstacle, that of equipment, was removed from the path.

By late June this present year, our apparatus was ready, and diving lessons began. Here again we are indebted to Sir Robert Davis, for we had his firm's technical representative, Mr.C. H. Burwood, to give us our instruction. Burwood is past-master in diving and the allied arts, and his diving, anti-gas, and anti-smoke proselytes number thousands.

Fig. 2.4
'The Leader Survives First Dip'. Graham Balcombe thanks Siebe, Gorman's Diving Instructor, Charlie Burwood, on completing training in the St. Cuthbert Minery Pond at Priddy, on the last weekend of June 1935, watched by Bill Bufton. Photo in The Log facing p.44 by Frank Frost

It was some weeks before we essayed the first dip in the grim surroundings of the River Axe. Diving is no easy art to acquire, and it is only now on concluding this year's work, that we feel really capable and masters of our job. Our technique of progress under the unusual diving conditions has been almost perfected, and our mentalities properly modelled to the new surroundings: we are ready now to challenge the unknown.

Fig. 2.5
Charlie Burwood supervises 'Digger' Harris (left) and Graham Balcombe (right). Photo in The Log facing p.24 by Frank Frost

VALVE ADJUSTMENTS.

Our first dip in the River Axe on July 13th was not expected to produce any exploratory results, but only to give us some idea of the conditions awaiting us, but since then and until September 1st, every week-end has been spent in the serious work of fixing underwater plant, pushing forward into the unknown, examining every possible outlet and inlet, mapping and recording.

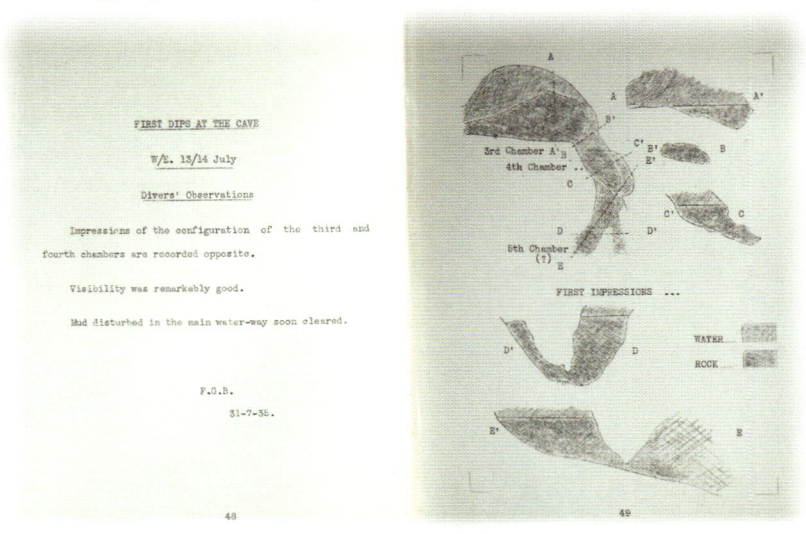

Fig. 2.6
The 'first dips' into the Third, Fourth and Fifth Chambers take place 13/14 July 1935. From The Log pp.48-9

From week-end to week-end the magnitude of the undertaking has been further pressed home to us, but so far each time we have been able to rise to the occasion and find the solution to the problem. The first trip up the bed of the River Axe is a revelation of the beauties of this underwater world. It is almost impossible to describe the feeling as, leaving the surface and the dazzling glare of the powerful lights, slipping down from the enveloping brown atmosphere, one suddenly enters an utterly different world, a world of green, where the waters are as clear as crystal.

Imagine now a green jelly, where even the shadows cast by the pale green boulders are of the same hue, only deeper; advancing, light green mud rises knee-high and falls in deadly silence softly and gently into the profound greenness behind. So still, so silent, unmarked by the foot of man since the river was brought into being, awe inspiring, tho' not terrifying, it is like being in some mighty and invisible Presence, whose only indication is this saturating greenness.

Travelling along a gully about twelve feet deep, against a wall of rock on the farther side of the river, passing beneath the archway into the fourth chamber, the rock above appears to be a warm pinkish brown, while every depression in its surface contains a silver bubble, created by the slow escaping air. The archway passed, we enter the Fourth Chamber. Overhead, the surface of the water is moving away with the rising air, and a myriad silver flakes dance with it.

Here the floor becomes steeper, and slopes down to the right. In front is a climbable bank of sand reaching to the surface, where it is suddenly cut across by a line of inky blackness. Over the descent, pale and mighty, looms a huge projection of rock, like the opened lid of some mysterious casket, hiding in the baffling greenness of its interior, secrets too deep for human eyes to see.

Ahead is an interminable distance, becoming more and more intense as it increases, broken occasionally by some ghostly shape, as the boulders insinuate themselves into the channel.

Turning around, and facing the archway whence we came, the water is lit up with fairy-like sunshine, the attenuated rays of the immensely powerful lamps on the shore, and on the floor a series of great foot-prints ever changing shape with the current of the river, until as if dissolving they become round dimples in the mud and then gently, oh! so gently, they disappear.

Four pulls on the life-line, the signal to come up, and the diver leaves those majestic depths, to be unceremoniously hauled ashore by a couple of cut-throats, who wrest his belongings from him in a very few minutes, and with his wits still in fairyland, he is sent to take his spell on the pump.

That first was the most edifying night's work, and we returned to our stronghold in the Mendips with a great feeling of exhilaration, having at last got our teeth into the job, and made acquaintance with the outskirts of this great field for adventure.

It was intended that all six divers of the party should go down, for then in our inexperience we thought that in one or two bids we would walk right through to our destination, open unsubmerged cave, or alternatively, would have run out our full four-hundred feet of pipes, and have realised the impossibility of reaching our goal.

Conditions soon proved to us that the task was no child's play, and that the original intention stood condemned by the time required to send down relays of relatively inexperienced men. Continuous concentration on one man and a competent assistant was obviously demanded, but in a party of volunteers, such discrimination is a difficult problem. The choice of No. 1 Diver was easy, it was Balcombe's job, as inaugurator of the expedition; it was with No. 2 that difficulty would arise, since all the divers were about equally competent.

It was finally decided that the best way was to give the place to the woman of the party, and royally has the choice been justified. Cool, collected, knowing no fear, she has carried out her task with an assurance and reliability that none could better. There is only one criticism which can be levelled at the choice; time has been lost owing to her physical handicap of small wrists.

Relatively small that is. The diving suit was made to fit a seaman's wrists, and stoppage of the circulation from the amount of tight packing required almost always ensued. This, however, is a difficulty which should have been removed by the substitution of more suitable cuffs or bands, and the trouble not allowed to persist. Various experiments were in fact made, but up to the closing of the present work, no satisfactory solution to the problem obtained. Before the opening of the next attack this will certainly have been remedied.

The particular obstacles to our progress, of which we had been warned in no uncertain terms, were a hole in the floor of the Fifth Chamber, whence Glencot Spring, a little farther down the valley, was suspected to draw its water, and which, owing to the volume flowing, would present a danger indeed; the second consideration was that the river had been said to rise up into the same Fifth Chamber, (the farthest point previously reached), with such force of current as to rock a boat.

The point of exit for Glencot Spring has not yet been met, wherever the spring has its source, it is not in the Fifth Chamber, and the hole in the floor is now buried by a long sand-bank, presumably silted up during the last thirty years, that is, since the original observations were made; the water from the presumed, but certainly not legendary, Sixth Chamber flowed up so slowly that the drift of the disturbed sediment could scarcely be detected. This was encouraging, although a faster current would be of tremendous assistance to the divers, for the greatest handicap is the persistence of disturbed sediment. The onward march progressed with unsuppressed excitement. Deep down under the overhanging wall at the far end of the Fifth Chamber, a full twenty feet below the surface, a narrow squeeze was discovered, which seemed to lead up beyond into a further chamber.

Wyndham Harris was the first man through this tight hole, and returned to tell of his discovery, of a great water-filled space, similar in construction to the underwater portion of the Third Chamber, in which we have our diving base. Later, another way through was found, to the left of the tight part of the entrance, which gave comfortable access to the cave beyond. A surface was discovered; here indeed was the Sixth Chamber, the first milestone in our progress.

```
            WOOKEY HOLE EXPLORATION EXPEDITION
                      DIVER'S LOG

Logged by..      W.G.Tucknott.
Diver's name..   C.W.Harris.
Venue..          Wookey Hole Cave.
Date..           21st July, 1935.
Purpose of dip.. Acclimatisation.

       CHRONOLOGICAL RECORD   (Use 24 hour system)
```

Time	Depth	Record
03.18		Left surface.
03.49	18'	At the low portion of passage, fixed distance line to the shot-rope near to the shot.
		Passed under the arch, and into a large chamber.
		Diver unwell, coming up.

Fig.2.7
'Digger' Harris is first to enter the Sixth Chamber. Recorded in The Log p.71

Meanwhile, as there was some doubt about our point of exit from the Fifth Chamber, a boating party was arranged, and with the water lowered to the maximum extent by means of the mill sluice-gates below the cave-mouth, the leaking boat was propelled under the two low arches into the far chamber, and behold! there was the guide shot-rope, streaming away in the jade green water below, down and under the overhang in the south east corner, and away into the green infinity!

This shot-rope is the divers' best friend, anchored firmly at the diving base, and at intervals along its length to heavy concrete and iron weights, it serves both as a means of progress and as a guide when the water has been disturbed by his passage.

The task of the first man is difficult and arduous, for he has to carry the shot-weights, over half a hundredweight each, and drag along the rope, besides having to haul his own air-hose and telephone cable behind him. Under these conditions, the ascent of a steep bank taxes his ability to the uttermost. Indeed, one problematic pitch at first seemed unsurmountable. Across the entrance to the Sixth Chamber, a yard or so from The Squeeze, lay a huge fallen block, many yards in extent, half buried in the sandy floor, and blocking the path ahead; the top of the block was smooth and slippery with fine silt.

Three times the diver lost his hold and fell back into the pit at its foot. The next effort was crowned with success, his cragsman's training stood him in good stead, even in those clumsy lead-soled boots, and a hand grasped the incut back of the rock, and the task was accomplished. A weight was attached to the shot-rope, and dropped over the back of the block, and now its ascent and descent is a mere bagatelle.

All these efforts after the first were in total darkness, for lights are unavailing once the diver's passage has sent up those beautiful yet fatally handicapping clouds of red and white silt, which covers floors and ledges, and clings thinly to walls and roof: it is better to save the batteries, than to clamour for its useless but undeniably comforting glow.

Forward is the order, and forward they go, through the Sixth Chamber, to a point where the low roof shoots suddenly upwards, and the diver's discharged air can be seen sending out great radiating waves, which lose themselves in the darkness beyond. Yet another chamber, the seventh, and this time of great magnitude. Excitement runs high as the components of the floating shot-rope are dragged through from the base and assembled.

The need for this device will be evident when it is explained that although a diver can float himself up to the surface, and then sink again, he has little or no control over the speed of this manoeuvre. Imagine our diver rushing up to the surface, out of control, the air in his suit expanding in obedience to Boyle's Law faster than he can get rid of it, and shooting out of the water like a porpoise; with a spile, or solid roof an inch or two above, waiting to greet him! Not a pleasant meeting.

Or again, imagine the diver sinking out of control, and call to mind that a fall from the surface to a depth of some 30 feet or so is sufficient to kill him, or at least that he will be the victim of severe internal injuries from the squeeze to which he is subjected. The depth at this point is some twenty feet, and a yard or so away yawns a pit of unknown depth, waiting like a death-trap to catch the unwary.

Therefore a rope to control the speed of ascent and descent is needed. The device used here is probably unique in the annals of diving history, and consists in its elements of a float, attached to which is a rope, with an anchor at the lower end. In practice, two oil drums were used, filled with air from the diver's helmet, and connected by an iron bar, with a thick Y-spliced tail rope tied down to two heavy weights. The diver ascends the rope as slowly, or fast, as he pleases, and rises between the drums, and there leans across the bar to view the surface.

The point had been reached where the diver could no longer pull his trailing ropes and pipe, and the assistance of a second diver to help him do this was found essential. Here at the point where the roof rises so steeply, was found a flake of parent rock, handily offering its security as a tie-on point.

In optimism it was christened Belay 1, and here Diver No. 2 was belayed with a loop of rope and a karabiner, a handy device used by mountaineers abroad, and probably having for ancestor the little spring hook on a dog's leash, and from here No. 2 sees to it that the pioneer has no further trouble from this source.

Tense with anticipation, the divers went aloft in turn, the first ever to set eyes upon this vast and gloomy recess of Pluto's Kingdom.

Well may they have risen in the River Styx itself, and be looking towards the gloom of Hades. Above, the vertical and parallel walls shot upwards towards the surface of the hills, and the feeble lights were unable to penetrate their immense height; away up stream they ran still parallel, till swallowed in the darkness. No more than a stone's throw could be discerned, though this makes it the greatest chamber yet, and within this compass the red-brown walls revealed no beauty of draping stalagmite, they are too steep, no drip of water from the roof would ever strike them; just at one spot is a patch of this white substance, dribbled over a ledge, where perhaps a stream once flowed.

It is a world detached, where, were it not for the steady hahrrh hahrrh of the pumps, and the the occasional call on the telephone, which keep the lone pilgrim tethered to the real world, now far behind, the spirit would wish to obey the command and slip quietly out and over the still river, into the darkness, where Charon would await it to speed its journey into the land of the lost.

Thus, in a series of seven efforts extending over a period of more than two months of continuous week-end work, this point has been reached, some 170 feet from the base, and with a great chamber overhead.

This is but the threshold of the realms which will now be accessible to us, the work has been stopped for the moment, though 1936 should see another expedition launched, this time knowing what has to be faced, and prepared for it in consequence. In comparison, the 1935 Expedition will be classed as little more than a preliminary investigation to ascertain what might be expected.

The River Axe is the source of drinking water for part of Wookey Hole village, as well as being the supply to the old-established paper mills in the valley below the cave-mouth. The operations cannot, therefore, be over-welcomed by these parties. The repeated disturbances obviously cannot be allowed to continue, greatly indeed is appreciated the tolerance on the part of the water-users, which has permitted work in the river each week-end for over two months. Next time some arrangements must be made to overcome this difficulty.

With a fresh expedition established by attention to this source of difficulty, greater progress is expected. The divers will be experienced and competent in this unusual type of work; every forethought will be concentrated on having available at a moments notice all equipment likely to be needed. A sectioned raft will be built, ready to be taken through and assembled underwater where required, and with a search-light and universal camera, this for both under- and above-water photography.

The size of the party necessary will be reduced considerably by introducing a motor-driven pump. This is an important factor with a volunteer party; it is very difficult to find a couple of dozen enthusiasts, who are in a position to spend night after night at the arduous tasks of the diving base.

On reaching that part of the cave where the floor again rises above the water, the type of diving equipment must be changed. Having proved by exploration with a pipe-fed suit that the passage can safely be made without great delay, self-contained apparatus must be employed, and on reaching the far bank, the amphibian divers will crawl out of the water like some lesser saurians of pre-historic days, to continue the exploration, upwards towards the source of the Axe. Maybe they will be cut off from the world for days at a time, examining, measuring, sketching, and photographing.

This then is the ambition; its successful execution is a mighty task, in which many factors play their part. Barring accident, we, the divers, feel confident that we can carry it through; finance will present one of the most serious problems, and any subscriptions to aid the work will be welcomed by our society.

Preparations are already well in hand, and the zero hour creeps slowly on. The day when next we venture into the green and chilly waters of the Axe is awaited with unsuppressible keenness.

What will it reveal of the hidden wonders in the Great Cave of Wookey Hole? This mighty cavern, which throughout history has drawn men from all parts of the world to view its present known magnificent vaults and chambers, and been the dread of many as the home of old Pen-Palach, the Witch of Wookey, is about to reveal some more secrets, which would otherwise lie hidden, until countless ages hence, when the cracking frosts and ceaseless dripping of water will have laid open the cave as a yawning gorge, giving access for the casual glance of a race of future creatures.

Or will the Witch of Wookey, whose spirit still seems to haunt the cave, decree otherwise?

F.G.Balcombe,

P.M.Powell.

16-9-35

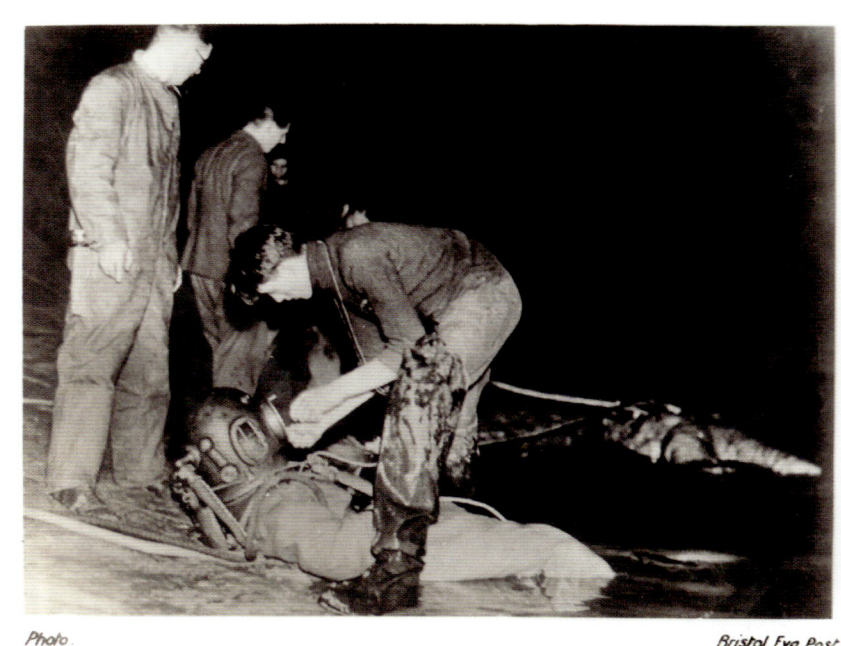

Fig. 2.8
The BBC and Press arrive to record the next push into the Sixth Chamber.
Photo in The Log facing p.148 by Bristol Evening Post

Discovery

Saturday, July 20th, and a lorry flying the colours of the House of Tucknott thundered into camp. Signal practice was willingly abandoned, and the gear hurled aboard, then off again at such a speed that a whole Midgetful of forgotten odds and ends was forced to follow in the rear.

Signal practice, or The Attendant's Revenge, is a great game, the victim wears helmet, corselet, breast-rope and air hose, and in some cases the bull's-eye, since no one can spit or swear through half an inch of plate glass. He is then guided into every possible obstacle by his attendant, and forced to climb, top-heavy and sweating, piles of loosely stacked timber, and tripped up on heaps of ropes, and finally, if the attendant is skilful, can be caused to blunder into his own tent, completely wrecking that structure, and drag the remains after him all tangled up with his various 'blow-pipes', as they were heard to be called on one occasion.

The diver, in his turn, can retaliate by sending constant demands for ropes, slates, and more air; he can too, should he be cunning enough, get possession of the full length of rope and hose by dint of ringing a continual series of 'four bells', and then, squatting out of sight behind a bush or pile of tarpaulins, with one well-timed jerk, throw his tormentor to the ground.

But to return to the night's activities. We arrived about 9 o'clock at our destination, the Third Chamber of the Home of the Witch; where the BBC was in attendance with coils and coils and coils of wire everywhere, myriads of microphones, wreaths of cigar smoke, a wealth of gents' natty suitings, fortunes in cuff-links, in fact the only thing missing was adhesive tape, which Mossy provided off an Oxo tin, and a sock to put in the loudspeaker.

The Western Electric Company were well represented by an amply cut motoring coat, containing a not so amply cut Engineer, who provided the public address system for the benefit of the general mob.

Through the smoke, one caught occasional glimpses of the ample starn-pieces of the BBC, more coils of wire, pipe and rope, sometimes even a diver, and on rare occasions, the River Axe itself.

There was some trouble with the telephone at first, but the experts soon settled that, and after sitting fully dressed, barring the weights, for quite half an hour, Balcombe, the man of the moment, was allowed to enter the water, where the stress of that last half hour was immediately forgotten in those serene depths.

At short intervals he reported his progress and what he saw, as one by one he lugged his colossal monuments, the concrete weights, along with him, and attached them to the shot-rope, there to remain until dissolved by the etching waters.

The loud-speaker, as has been mentioned before, had forgotten its sock, and the result was terrible, one long and awful blare of voice, and occasionally an intelligible word. The cave guides will have us believe that the accoustical properties of this chamber are perfect; if accoustical perfection includes a ten-second echo, they are probably right!

Signalling has been difficult, for the lines catch on the rocks, and it is next to impossible to get a signal through by them, but the new telephone was perfect.

It had been raining during the week, and the water was not so clear as on the last occasion.

After Balcombe came up, Harris made a trip, on which he descended to the low archway, the limit of Balcombe's forward march, and crawled through to the space beyond, where he discovered a chamber of large dimensions. Still holding on to the distance line, he stood peering round in this newly found wonder-world, until the clouds of billowing sediment arose to obscure the view.

On his return, a third diver went below, for the gang was still fit, and this time, myself. The divers had agreed to record their impressions separately, but it is doubtful whether those lazy blighters Balcombe and Harris will ever submit theirs. Here at any rate is my story:-

> 'I did not think I was to have a chance that night, it was getting awfully late, well after four, and when the gang said they were O.K. and in spite of the fact that the remaining woollens were being used as a bed for someone, they succeeded in dressing me in record time.'
>
> 'My wrists are an awful nuisance, they're so puny I have to have rings, and rings, and rings,' (rubber rings used for packing) 'and they are not too comfortable then, but after fixing them up in a new way, and testing them in a bucket, everything seemed alright, and I was soon slithering down into the water. That way in over the mud is an absolute gift.'
>
> 'The river was not so clear as the first time I went down, it was sort of thundery brownish foggy colour, if you know what I mean, and instead of enticing you like the fairy green of the week before, it kind of hated you, and said "get out!", as if it couldn't tolerate a third diver that night. Anyway, I went on, wallowing in the colossal boots, like a slow-motion footballer, holding on to the shot-line with one hand, and flashing the torch about with the other, everywhere was this baffling fog, the rocks only came into view when the torch nearly touched them, and they glowed back with a sort of reddish-brown.'
>
> 'I travelled along the rocky and muddy terrace to Harris's low archway, and secured the distance-line

round my right wrist, and waited for a few moments for the water to clear, before I inserted my cumbersome bulk into that depressing little orifice. It was the first one I had navigated, and I dared not lie right down and wriggle like a lobster, for fear of blowing up, so I proceeded very carefully, as some of the stones seemed a bit loose, on one hip and shoulder, with my helmet bumping and scraping at intervals on the roof.'

'Flat slabs of a sort of tufa stuff kept falling past the bull's eye, ringed round with little silver bubbles, and finally I came into the new chamber. I saw a huge boss of stalagmite on the floor, and went across to it to rest, the mud I'd kicked up rose above me, and curled down again round my helmet like heavy smoke clouds, and finally dispersed.'

'On the far side of the chamber, opposite the place by which I came in, I could see what looked like a long dark archway, low but very tempting, unfortunately the distance-line wasn't long enough for me to get close to it to examine it, so I sat a bit longer, hitting one or two edges off the boss with the torch, in true tripper style, then I rang up the shore and announced my intention of returning.'

'Again that beastly squeeze, but much less difficult the second time, as it slopes up, and you don't have the feeling that you might suddenly go up faster than you want to. The mud was very thick by this time, and my hands were getting cold, and the rubber rings making them numb, but once I'd let go the distance-line, and got a grip of the shot-rope, progress was easy despite the thick fog all round me.'

'Suddenly I found myself pulled up tight, and to take another step, however hard I tugged, was impossible. It requires a good steady pull to lug the lines along, but this was no ordinary resistance; it then dawned on me that, quintessence of bad diver-craft, I was not on the shot-rope at all, I was gaily using my own breast-rope, which had somehow become hitched up good and hearty behind me, and I had doubled back on it.'

'The telephone was just the limit, I could hear nothing they said, and apparently they were quite deafened by my silvery voice, and could get very little of what I said. It was lovely to hear the steady hahrrh-hahrrh-hahrrh of the pumps, and to know that never mind how long you stayed there, or what predicament you were in, it would continue; the good old gang breathing for you! So I sat there with a huge bank of mud looming up beside me in the thick still water, waiting for it to clear a bit. After a while my hands got too numb to use my fingers, and I could only do my useless best with my two wrists and one knee, lifting and gently jerking, I dared not do it too hard, in case something got loose and fell on me, then waiting again for the water to clear a bit to see if it had been any good. Gradually the air pipe became more tractable, and by dint of first my pulling a yard, and then the crew on shore pulling a yard, it got loose; oh! the joy as it slowly but surely floated past my bull's eye!'

'With a little encouragement, the breast-rope came too, and I shall never know what they were hooked by owing to that fearful mud. Then I 'bout turned, and slunk home; I felt simply awful, I knew I'd been a fool again, and had broken one of the most vital rules, by letting go the shot-rope, and I crept out of the water wishing I'd never been born. If only I'd discovered something wonderful, or got an awful cut, or anything but no, they didn't even rate me, which was awful, my teeth chattered, though I wasn't cold, and I was afraid they'd hear them. My rings were taken off, my hands chafed, and I was undressed in stony silence, broken only by a curt 'stand up!' or 'sit down' as the occasion arose.'

'Still, I've learnt something, I've learnt the consequence of not ascertaining that it is the shot-rope, before letting go the distance line, I've gained an enormous amount of confidence underwater, and further I've learnt to sit quietly down and to think a thing like that out, instead of getting a vertical breeze, so in face of the fact that I always shall be a fool, I do feel that in one direction at least I'm a wiser fool.'

Mossy.

Dear Mossy,

If you wish to know why there was a stony silence, it was because we had been scared stiff; you had been trapped in a place whence there was no rescue.

Graham.

On Sunday evening, having partially recovered from the return to camp at 8 o'clock that morning, semi-respectable clothes were unearthed, and a small expedition set off to Wookey Hole to report progress, return utensils borrowed from various departments the night before, and incidentally to gather in any news likely to be useful to us of the gang. We were greeted with such kind words, and so many promises, that we were beginning to feel quite popular, in spite of the fact that we have been certified as mad, but delightfully so, by the BBC!

We were informed that we had entered the Sixth Chamber, but told to keep it under our hats.

Mossy.

12-9-35.

Divers' observations

Week-end 31st August-1st September.

Permission for a final effort had been obtained, thanks to the splendid efforts of Brown, secretary to the cave management, and now a firmly established and prominent member of the Expedition. This permission had been given in order that the underwater tackle might be recovered.

It was felt that the opportunity would be much more profitably used for a final bid at exploration, at the cost of abandoning the gear.

The programme was therefore:-

(a)   An early descent to Six, and if circumstances justified, better to explore the northern end of this part of the system.

(b)   To go to the surface at Belay 1.

(c)   To make a supreme effort, after waiting for the water to clear, to press on upstream at least a hundred feet.

The start was unfortunately delayed owing to a new member misconnecting a telephone cable, and damaging the fitting. This temporarily remedied, Diver Balcombe descended to Shot 3 with two 40 lb. shot weights, and leaving them there, proceeded to Belay 1 with the 80 lb. weight left at that point on August 16th.

Meantime Diver Powell took a 75' shot-rope extension to Belay 1.

The oval type karabiner is proving an appliance par excellence for this work.

Hope of any attempt to explore the northern end of Six was abandoned. Here Powell confirmed Balcombe's observation of a surface as shown in report of Week-end 3/5 August. This confirmation must therefore reinstate the surface then reported, irrespective of the conclusion drawn in the General Observations of Week-end 24/25 August, where it was stated that it could now 'be fairly definitely said not to exist'.

Shot 4 was replaced so as to avoid the sharp corner round the big block between Shot 3 and Shot 4, and the slack taken up at Belay 1 to prevent it from settling back.

At Belay 1, looking up from under the overhang, surface was clearly evident and the waves caused by the divers' discharged air could be seen flowing away until lost in the distance, in this case about fifty feet. This therefore is the seventh chamber.

The trapeze was assembled and sent up, the process being interrupted by an unpremeditated return to the base. Powell's wrist joints, ever troublesome, became disarranged, and it was found impossible to remedy below.

Balcombe had to accompany Powell, since she had no telephone receiver, and had to rely on rope signals from him, who preceded her and was in full communication with the base.

Fig. 2.9
The trapeze enabling weighted bottom-walking divers to ascend to the water surface in the Seventh Chamber is launched. Photo in The Log facing p.121 by Bristol Evening Post

Fig. 2.10
The trapeze, used successfully on the last weekend of August 1935, is illustrated in The Log p.117

OFF WITH THE TRAPEZE

Fig. 2.11
Standard lead-weighted boots used by bottom-walking cave divers in 1935.
Photo by Mark 'Gonzo' Lumley

Returning to Seven, the trapeze was sent up, and both divers went aloft in turn.

The seventh chamber, as then revealed, was a crowning discovery which removed all the regrets at having to conclude operations, and created a feeling of achievement, justifying out past efforts, but still leaving the prospect of fully exploring it as a glorious incentive for some future date when the water supply difficulties had been overcome.

The Great Seventh, though only some twenty-five feet wide, transcends in architectural magnificence all the other chambers of the cave, rising with sheer and parallel walls, straight up into the blackness impenetrable by either the 100-watt lamp or the diver's torch, and running as a mighty rift, away into the unknown, the end, like the roof, swallowed in the gloom.

A patch of stalagmite on the North Wall marks the one spot where the drips from the roof have found lodgement, elsewhere is just stark brown conglomerate. Doubtless, if the roof were visible, pendant masses of stalactite would be revealed to compensate for the paucity below.

To see these, to find the end, to measure, and to photograph, are the goals which await us in the future.

So concludes the work of the Wookey Hole Exploration Expedition of 1935. Hats off to Sir Robert Davis, who has made it possible!

F.G.B.

4-9-35.

Going through from chamber to chamber in a meditative manner, you cannot help being struck by the strangely different atmosphere in each. Four is rather a scrambling place, all big untidy stones with very little beauty, except it be contrast to Three. Then Five, where you have the illusion that you are a giant standing miles above earth, and looking down onto a vast expanse of desert, where no sign of man is visible, not a mark disfigures the even ripples of the sandbank, and where the light fails a little in the distance, the deep shadows cast by the overhanging rocks creep together to make dark mysterious forests, silent on the desert's edge. There is not the slightest sensation of being under water, and the whole illusion must be caused by the position of the lamp, high up like a sun, above the fleecy clouds of the surface.

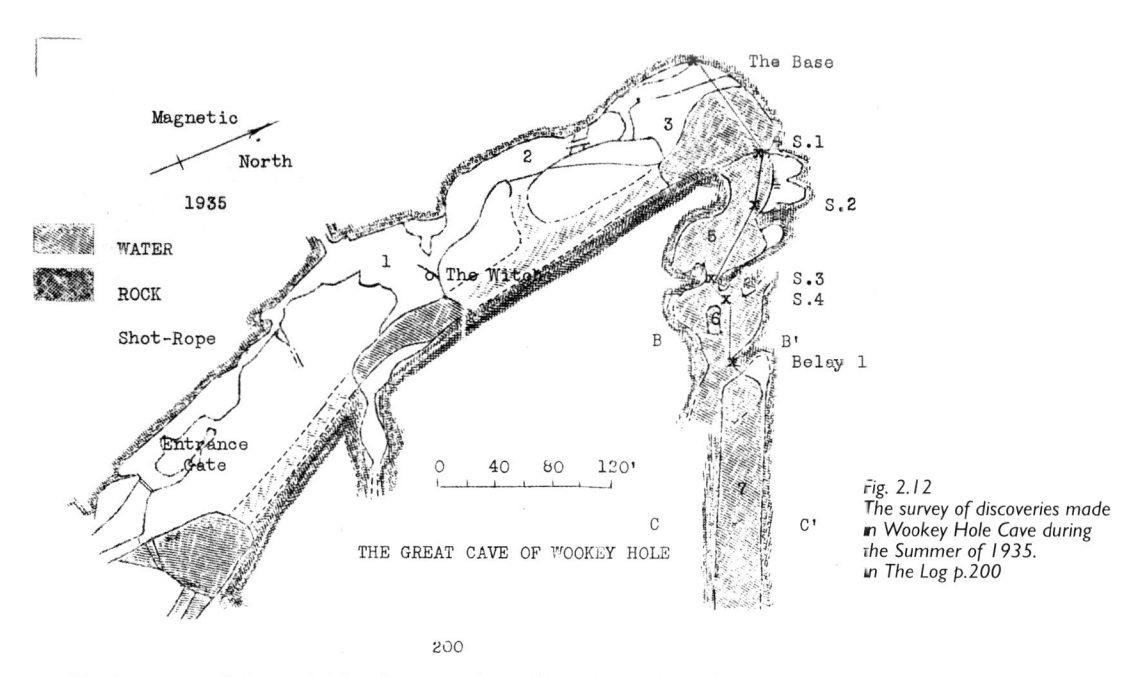

Fig. 2.12
The survey of discoveries made in Wookey Hole Cave during the Summer of 1935.
In The Log p.200

Having passed through The Squeeze into Six, where there is no lamp, you suddenly find yourself a tiny awestricken creature, standing in a colossal world, half dark and rising round you like the sides of a bowl; pale clouds billow and fall away on either side, the floor is tumbled rocks again, like that of Four, and in front an enormous sandbank looms. Away into the distance the shot-rope trails over its ridge, and out of sight, so you know that it is not new country, but somehow it is new to you. It is the charm, the spell the Witch had cast upon us, which makes us wander, for ever seeing anew the things with which we are already familiar, like the spell-bound folk of old, who wandered in an enchanted forest, which was really their own orchard, planted by their own hands.

Above the sandbank is an aperture, a window shaped like two stars, a large one and a small one holding hands. Behind you, a little to the right, and far, far above is a crazy moon, expanding and contracting, with each movement shaking off great flakes of itself, which writhe and turn, travelling for ever away from their parent, disappearing as suddenly as they came. It is the air space in Six. The moon is the light of your torch shining on the escaping bubbles from your helmet.

As you begin to ascend the sandbank, and navigate the rock in the middle travelling towards the stars in front, reality again descends and before many minutes have passed an ordinary human being has arrived at Belay 1, between Scylla and Charybdis, and contemplated a trip up the trapeze.

There was a lot of hand tapping, signalling on bull's eye, and receiverless No. 2 was again belayed to the rockside, and No. 1 on his way to the surface. His suit began to swell, he waved his hand, grasped the rope, and up he went slowly and in the most dignified manner possible; the last that Diver No. 2 saw was those fearful brass toecaps slowly disappearing in a sulphur cloud, like a person going up to Heaven. Diver 2 is earthbound, so was more or less prepared when, with a sudden crash, down came Diver 1 on her helmet! Diver 1 then did some curious antics, whether from pleasure or from rage it was difficult to tell. Diver 2 promptly put through a request to go up also, granted by No. 1, who proceeded to dekarabine (or should it be entkarabinen?) the coil of rope, spare shot-rope, so that the ascent would be less encumbered. Then up, up went Diver No. 2, and when her head popped out, and she saw Seven in all its glory, Three echoed and re-echoed with squeals of delight, and she hung, swinging and wobbling, on the iron bar of the trapeze, bathed in the orange, red, brown, and gold reflections on the water.

Here was Seven, winding away as far as the light could reach, towering above as far as the eye could see, two gigantic walls of clean rosy conglomerate; how could it be so many million years old? It looked as fresh and new as the day it left the Hand of its Maker.

This cathedral of peace is guarded not by an angel with a flaming sword, but by a huge and pointed boulder, ready to destroy all who are not fit to enter, and for whom it is waiting, so keen and sharp, so watchful. La Guillotine!

Along one side, the river's left, is a huge overhanging square-cut ridge, like the one in Five; the opposite wall is practically vertical, though high above is a smaller ridge that has at some time caught a drip of water and made a tiny cascade of white stalagmite, it gleams with a pearly radiance and in that lofty place it is as a sleeping soul, perhaps the little white keeper of the cavern, resting, head on arms, against the rosy grandeur of the wall.

Upon return it was described as a place like a shag's nest.

When the divers were reunited on the river-bed below, they executed a regular war-dance, hand in hand until their helmets crashed together, and finally Diver 1 pushed his companion over, which ended the performance, and sorrowfully, but triumphant, they wended their way home, stooping, climbing, crawling and at times shooting along face downwards as the attendant and coilers waxed more and more energetic and tugged with all their might.

The Wookey Hole Exploration Expedition

The thanks of the Expedition are specially due to the following:

Sir Robert H. Davis, to whom it is endebted for the loan of the diving apparatus, and for tuition in diving.
Captain G.W. Hodgkinson, M.C., for permission to work in the cave, and other facilities provided.
C.H. Burwood, Esq., for the pains he took in teaching us to dive.
H.E. Balch, Esq., M.A., F.S.A., for negotiations with Wookey Hole Caves, Ltd., for permission to operate.
W.G. Tucknott, Esq., for haulage facilities at unreasonable hours.

With Circular No. 11.

> THE LOG
> of the
> WOOKEY HOLE
> EXPEDITION
> The record of the findings
> and the story of the explorer-divers
> who for over two months
> laid siege to the water-chambers of Wookey Hole Cave,
> will soon be available.

Although issued primarily for the benefit of the caving world, it is hoped that some profit will be made on the sale to assist the continuation of the work. The expenses of the Expedition are heavy, and the amount of work outstanding great.

> LEND A HAND
> by purchasing a copy, its success
> depends on your support.

The Log will contain some two hundred pages, sixteen photographs, and numerous sketches and diagrams.

> It will cost 7/6d. a copy.

Clubmen, we are counting on your support, do not fail us. Send a post-card now to Graham Balcombe, at High Street, Sunninghill, Ascot, Berks., and a copy will be reserved for you.

---oooOOooo---

> Supported by
> members of
> The British Speleological Association
> The Wells N.H. & Arch. Soc.
> The Wessex Cave Club
> The Bristol University Speleological Society
> The Northern Cavern and Fell Club.
>
> --ooOoo--

Fig. 2.13 The Wessex Cave Club publicises The Log in Circular No.11 (December 1935)

DIVING PLANT

1	...	1-Double-Diver Pump. 2-Cyl. D.A.
2	...	1-"Baby" Pump. 2-Cyl. S.A.
3	...	1-No.1 Diving Suit.
5	...	2-No.2 Diving Suits.
5	...	2-Lengths Breast Rope.
6	...	2-Pairs Divers' Boots.
7	...	3-Corselets.
8	...	3-Telephone Helmets.
9	...	3-Corselet Nut Spanners.
10	...	3-Collars.
11	...	2-Pairs Weights.
12	...	2-Divers' Belts.
13	...	2-Divers' Knives.
14	...	1-Diver's Torch.
15	...	5-Pairs Trunks.
16	...	5-Sweaters.
17	...	9-Pairs Stockings.
18	...	4-Woollen Helmets.
19	...	12-Grey Wrist Rings.
20	...	12-Red Wrist Rings.
21	...	1-Single Diver Telephone.
22	...	1-Double-Diver Telephone.
23	...	6-D.E. Hose Spanners.
24	...	Sundry Air Hose Washers.
25	...	Sundry Telephone Cable Washers.
26	...	1-Male/Male Hose Union.
27	...	1-Under-water Lighting Fitting, for 2,000 c.p. bulb.
28	...	4-Lengths of Male/Female Sinking Hose.
29	...	1-Length of Female/Female Sinking Hose.
30	...	5-Lengths of Male/Male Floating Hose.
31	...	3-Lengths of Female/Female Floating Hose.
230	...	1-Length of Male/Female Floating Hose.
34	...	5-Large Chests.
35	...	2-Small Chests.
36	...	1-Tea Chest.
37	...	1-Diving Ladder.

All the above are the property of Messrs. Siebe, Gorman, & Co. Ltd., loaned to us by their Managing Director, Sir Robert H. Davis.

#		Item
38	...	1-Diver's Compass.
39	...	1-Diving Stage.
40	...	1-250' Length of 2½" Shot Rope.
41	...	3-75' Lengths of 2½" Shot Rope.
42	...	20 lbs. 1¼" Manilla Rope.
43	...	6-Pear-shaped Karabiner.
44	...	6-Oval Karibiner.
45	...	1-½ cwt. Iron Weight.
46	...	12-½ cwt. Concrete Weights.
47	...	12-14 lb. Concrete Weights.
48	...	12-Pitons, Rock Pattern.
49	...	1-300' Length of Female/Female Tele. Cable.
50	...	2-150' Lengths of Male/Male Tele. Cable.
51	...	1-Wooden Base for "Baby" Pump.
52	...	1-Canvas Sheet.

These the property of the Expedition.

#		Item
53	...	2-Lighting Fittings.
54	...	1-250' (approx.) Length Lighting Cable.
55	...	1-100' (approx.) Length Lighting Cable.

These the property of Wookey Hole Caves. Ltd.

THE COST OF THE JOB

	£	s.	d.
Telecommunication (except postage.)	1	10	3
Divers' Medical Certificates		17	0
Rope & Associated Miscellany	4	13	6
Locomotion (Bus fares, and a nominal charge of 1d. per mile towards the running expenses of SH.P. car; major trips only.)	4	14	6½
Pitons & Karabiner	2	5	9
Timber & Associated Miscellany	5	0	3½
Press Cuttings, &c.	1	6	0
Stationery, Postage, &c.	1	7	5½
Railway and Cartage Charges	8	4	7
Wages	11	18	4
Repairs to Apparatus	1	12	6
Diving Apparatus, including Rail Charges where received carriage paid	31	9	7
Miscellaneous		19	4
	£75	19	1½

Subscriptions &c.			
The British Broadcasting Corporation	9	3	9
"The Observer"	5	0	0
Wookey Hole Caves, Lt.	20	0	0
	34	3	9

THE WOOKEY HOLE EXPLORATION EXPEDITION

The thanks of the Expedition are specially due to the

Sir Robert H. Davis, to whom it is endebted for the loan of the diving apparatus, and for tuition in diving.

Captain G.W.Hodgkinson, M.C., for permission to work in the cave, and other facilities provided.

C.H.Burwood, Esq., for the pains he took in teaching us to dive.

H.E.Balch, Esq., M.A., F.S.A., for negotiations with Wookey Hole Caves, Ltd., for permission to operate.

W.G.Tucknott, Esq., for haulage facilities at unreasonable hours.

... ...

F.G.B.

Chapter Three

'Mossy' Powell and the Base Camp

by Frank 'Mac' Brown

'Mossy' Powell came to Wookey Hole in late 1933, the same time that Graham Balcombe first arrived in Somerset. She was appointed as Manageress of the Shop and Museum adjoining the Great Cave which Gerard Hodgkinson had developed as a tourist attraction to rival the Gough's and Cox's Caves in Cheddar Gorge.

THE BASE CAMP

Fig. 3.1
At the Priddy Mineries by Fair Lady Well.
Photo by Frank Frost
facing p.121

Apart from the occasional exploring of the dry upper levels of Wookey Hole Caves and a, to me, memorable foray in Cow Hole with Gerard Platten's party, I do not remember 'Mossy' being involved in any serious caving until she appeared at the Plantation Swallet where Graham Balcombe's team used to camp to become 'head cook and bottle washer'. This site is by Fair Lady Well, appropriately, within easy reach of the old Waldegrave Swallet Dig and near the St. Cuthbert's Minery Pool where Graham Balcombe and his party of divers were being put through their training by Burwood, the Siebe, Gorman Instructor in helmet or Standard Diving Equipment. She must have had a go at diving early in the programme because she became accepted as the Number 2 diver to Graham. And this was entirely on merit for such was the training for the task that lay ahead in Wookey Hole.

We have since learnt from her daughter Jane that Penelope Margaret was born on the 14th October 1904 at Penryn near Falmouth in Cornwall. Her father was the local doctor. When she was eighteen years old, Penelope married a rubber planter by the name of Powell and went to Malaya. It was a 'runaway marriage' that eventually came to grief. She returned alone to take up the appointment at Wookey Hole; an energetic and delightful personality, always helpful and full of fun. Jane remembers her mother's zest fondly as do all involved in the first cave dives. She also recalls that 'Mossy' was intolerant of fools, which is why Penelope Powell got on so well with Graham Balcombe and his companions. How came she by the nickname 'Mossy' is less certain, except that she loved small animals and, later in life, found much happiness running a small-holding on the lovely unspoilt island of Bryher in the Scillies. She died there in 1965.

Fig. 3.2
The Wookey Hole 'Pushball Team' at the newly opened swimming pool. Left to right: 'Mossy' Powell; Miles, a waiter; 'Mac' Brown, the Manager; Gerard Hodgkinson, the owner; Jerry, head cave guide, and Hudson, the chauffeur. Hodgkinson's daughter 'Doodie' sits on the ball, holding 'Ting', 'Mossy's Jack Russell.
Photo from 'Mac' Brown, Easter 1935

I knew 'Mossy' personally as I was the Manager of Wookey Hole Caves at the time she arrived here. From time to time we conducted some rather amateurish cave digs together in Ebbor Gorge. Much later, she was to recall these times with great affection to her daughter Jane; especially of the time that she found the tooth of a 'sabre tooth tiger'. With responsibility for the little Museum, 'Mossy' knew about the dispute between Gerard Hodgkinson and Herbert Balch over the Great Cave's archaeological treasures being displayed at Wells Museum rather than at Wookey Hole. So, maybe she had it in mind to increase the exhibits under her care when first setting out to excavate likely rock shelters in Ebbor accompanied only by her dog Ting. But digging caves alone was not rewarding and just a little alarming it seems, for this is how 'Messrs Brown & Powell, Excavators, Wookey Hole' came into being. It is nice to know that most of the finds are still to be seen at Wookey Hole in the new Caves Museum and not lost. Our caving and digging equipment was simple: boiler suit, excellently shaped hand pick, and Cornish tin miners' hard hat. The hats were quite useless really in a cave as the brims kept catching on rock faces in tight squeezes and tilted over one's eyes. Quite why I took photographs of these, I don't know. It wasn't in any sense for the record because I usually appear wearing a trilby on such historic pictures as the first party to attempt diving the Swildon's sump early in 1934! We got our carbide lamps from a Cornish tin mines friend of mine and had them inscribed: 'Brown, Powell and Ting'. 'Whoomp No 1' belonged to me and 'Mossy's was 'Whoomp No 2'. I still have both inscribed lamps as keepsakes.

Fig. 3.3
'Whoomps 1 and 2' together at Wookey Hole. Photo by 'Mac' Brown, 1935

Fig. 3.4
A self-portrait of 'Mossy' Powell in The Log p.163

'Mossy', who was quite an artist, made a caricature wash drawing of me with my pick through a pottery mug at one cave shelter in Ebbor; but, if the truth is to be told, I trod on it as it came away from under the roots of an overhanging nut bush at the entrance! The very mug is still in the Caves Museum somewhere. Several of her drawings illustrate chapters in *The Log of the Wookey Hole Exploration Expedition 1935* and she also wrote most of this classic book, including the chapter on the first successful dives called 'Discovery'. 'Mossy' Powell's accounts are stylish and capture the enthusiasm and fun of the whole enterprise. She also included a whole story about a hound called 'Diligence', which wandered into camp from the Mendip Farmers' Hunt Kennels and caused havoc. Her own Ting was a dear Russell Terrier who was always present and is recorded as being part of the divers' team. On one occasion, Ting was having puppies at the camp by the Mineries pond and got into difficulties. 'Mossy' put her in a basket, made her way at night over to a car parked near the Hunters' Lodge Inn, 'knocked up' a loving couple who, bless them, motored 'Mossy' and Ting to the vet' in Cheddar. As recorded in 'the log', Ting did not survive for very long afterwards. It was probably the only sad event in 1935 for everything else proved to be a success.

Fig. 3.5
'Mossy' Powell's cartoon of 'Mac' Brown discovering a pot at a rock shelter in Ebbor Gorge.
From a copy made by 'Mac' Brown, 1935

Fig. 3.6
'Mossy' Powell, Diver No. 2, assists Bill Bufton on his first training dive in the St. Cuthbert's Minery pond.
Photo by Frank Frost from the Wessex Cave Club, June 1935

I have re-read Graham Balcombe's and 'Mossy's accounts and particularly remember the occasion when she lost the shot rope, then both her breast rope and air line became fouled behind a protruding rock. For some reason she could not hear us over the telephone line and so, unable to exchange effective messages as to what to do to help, we could only wait on the shore. The usual way of sending signals by pulls on the breast rope was impossible because this was jammed. All we could do was to continue pumping air to her and trust that she could sort things out alone.

Imagine our relief when, after what seemed hours, the breast rope suddenly went slack and the lovely gurgle of escaping air bubbles could be heard and seen coming to the surface at the entrance to the Fourth Chamber. When we unscrewed her glass face piece, she seemed remarkably cool and collected: I seem to remember that, if not at once, then certainly that night, she went down again for a quick trip under the first arch, rather like a car driver getting back to the wheel to restore confidence following an accident. She told her daughter about this near miss much later and Jane thinks that her mother's lack of fear about death stemmed from the time she might well have died on the first cave dives in Wookey Hole. It was, I recollect, the following weekend that the whole gang sat down one night and spliced 400 feet of breast rope with built-in telephone lines to the pipe line. After the previous week's near catastrophe, it was considered far too risky to rely on any vital links between shore party and diver being separate. Cave diving over rocky terrain was clearly prone to different lines getting snagged.

There was another difficulty we had with 'Mossy' apart from the loss of circulation in her hands because of the tightness of her wrist bands, for she wasn't very tall. When the telephone connection was made, the divers rang to the shore by pressing down their chins on a bell push situated just inside the helmet. 'Mossy' being small inside her diving suit, spent a lot of her time, quite involuntarily, with her chin resting on this bell push! This resulted in those on shore suffering an almost perpetual ring, and we quite fairly got fed up with answering nothing. All we got in reply was the gurgling of bubbles and, at times, certain rude remarks about the difficulties she was encountering. Of course we forgave her.

'Mossy' was obviously helpful and resourceful other than when diving. At the beginning, I mentioned a trip to Cow Hole with her; another cave dug open by Gerard Platten, Hywel Murrell, 'Digger' Harris and his team in 1935 between Nordrach at Charterhouse-on-Mendip and the village of Ubley. Gerard Platten had intended doing Swildon's that night but the entrance when we arrived was like a sink full of water with the tap still running and the plug pulled out Quite impossible! So, Cow Hole was suggested and 'Mossy' came with me in my 1926 six cylinder A.C. But we failed to find where the others had parked and finished up off the road some two to three hundred yards across the fields from the entrance to the cave.

One moment on the trip down Cow Hole brings back memories. I was standing on the floor of the cave with a small man who was holding the bottom of the double life line which was threaded through a pulley block belayed at the head of the main pitch in the roof of the chamber. A massive fifteen stone man trying to squeeze himself through the boulders at the top before disengaging his end of the line suddenly 'came off' and, with a strangled cry, swung into space. My companion fortunately held tight on the rope, but weight for weight was no match and he slowly ascended into the dim darkness of the roof of the cavern. At least he didn't get very far because I grabbed his feet as they rose past my face and, by our combined tactics, the man that had fallen managed to get onto the ladder again.

That was, of course, by the way. The most serious thing happened to me later when, on coming out of the cave, I made my way in pitch darkness to the car which had been parked well away from the others. It was freezing cold and a real hoar frost settled on my saturated clothing. With only a Cornish miners' acetylene lamp to see by, I missed the way and got hopelessly lost. I went back to the cave entrance and the dying embers of a fire someone had lit some four or five hours earlier to keep the tail end of the party warm whilst the leader rigged the pitches underground. Slightly warmed up, I started off for the car again with no better result. But 'Mossy' who had come out of the cave earlier had found her way to the car and, bless her, had the sense to flash the headlights at intervals. Quite frankly, even in those days being young and fit, sodden frozen clothing and no protection might have been serious otherwise. Fortunately, having got to the safety of the car, I had a change of dry clothes. Interesting! Fifty years ago there were no car heaters or, at least, not in my

Tune:
"The Lincolnshire Poacher."

THE DIVING GANG.

On Saturday nights the DIVING GANG,
   A wild and lawless crew,
With pumps and ropes and scarlet hats,
   And shirts of navy blue,
Come roaring down from Mendip's heights
   In Wookey Hole to pitch -
So call the Captain quickly, boys,
   To chaperone the Witch.

Begad!   They are the toughest crowd.
   That ever filled the cave.
The Celties and Romano-Brits
   Lie shaking in their graves.
They'd use a pound of gelignite
   To open any niche -
So call the Captain ....

The Diver takes some holding down,
   It's done with leaden weights,
His frightful boots are made of brass,
   As safety first dictates.
His range is quite four hundred feet,
   Before there comes ahitch -
So call the Captain ....

Before they go, on Sunday morn,
   They take a last look round,
And anything they may have missed,
   Will now be surely found.
The mermaids from the River Axe,
   Lie swooning in the ditch.
So call the Captain ....

Pen-palach sits and listens to
   The noises in her cave,
And wonders if it's worth the fag,
   To keep her virtue safe.
She quivers as they thunder past
   And shout "Wotcher, old bitch"
So call the Captain quickly, boys,
   To chaperone the Witch.

Fig. 3.7
'Mossy' Powell's celebrated song 'The Diving Gang', with John Hassall's portrayal of the 'Witch of Wookey' on display in Wells & Mendip Museum, and Pump Gauges at Wookey Hole Cave museum.
Montage by Mark 'Gonzo' Lumley.
Verses of the song from British Speleological Association archives (20 March 1942), and Potholers' Songs collected by Bob Leakey (c.1947)

Tune:
"The Lincolnshire Poacher."

PUMP .... PUMP.

Oh, Balcombe, as you know by now,
   Is siezed with notion queer,
He's diving on the Mendips,
   And there ain't no water there.
He called his troops together
   On the Waldegravian dump,
And announced his new intentions,
   Shouting -
        "Pump, you ..... Pump"

He covered up his box of tricks
   With canvas pure and pale,
Then tootled down to Ceddar,
   And got Mossy out on bail.
"Now you and Ting must guard my store,
   Or you'll have cause to jump,
So keep the frogs and lizards
   Out of
        "Pump, you ...... Pump"

He won a lovely diving suit,
   From distant London Town.
He tried to catch the tadpoles
   As they wriggled up and down.
Then moved he off to Wookey Hole,
   Where Captain got the hump,
For Balcombe bust the telephone
   With
        "Pump, you ..... Pump"

Some fat men came to B.B.C.
   What Balcombe meant to do,
And brought their wire entanglements,
   And left them there on view.
The Gang produced the diving gear
   And stacked it in a lump,
Then Balcombe promptly shattered mike
   With
        "Pump, you ..... Pump"

So now he leaves us on the shore,
   Beside his muddy tracks,
While he dallies with the mermaids
   In the green and limpid Axe.
But all night long we listen in,
   And many a heart goes thump,
When he rudely brings us back to earth,
   With
        "Pump, you ...... Pump."

                    Mossy   Powell.

open two-seater. We drove home to Wookey Hole passing the Mineries at Priddy and never dreamt that the pools and camp site there would soon see the start of the helmet divers' great adventure in the cave I managed. Or that I would be asked to remember some of the incidents, perhaps not recorded before, fifty years later because diving in the Great Cave would continue.

Running all the tourist attractions at Wookey Hole kept everyone very busy, and our 'Base Camp' was working usually up to 9 p.m. for seven days a week, especially during the summer season. These duties gave little time for long caving trips; but, it was great fun in our time off. When the tourists were not around, games of pushball were played in the swimming pool between teams made up of staff employed by Gerard Hodgkinson and locals, sometimes joined by cavers. Our team included Miles, a waiter in the restaurant, Hudson, who was Hodgkinson's chauffeur, Terry, a cave guide, 'Mossy', of course, and myself, Mac. On Thursdays until midnight we held dinner dances in the pool surrounds and, after then paying out the weekly wages, I rarely got to bed before about 3 a.m. This went on without a break, including Sundays and Bank Holidays, all through the summer. So, as the historic divers' log sheets show, when the first dives took place during July and August operations could not get going until very late at night. It was usually after midnight before anyone got underwater. Even on the broadcast night for the BBC on 17th August 1935, Graham Balcombe did not set off before 10.45 p.m. They were long days and nights. Approval had been given by Gerard Hodgkinson, of course, and the staff did all they could to help.

I must pay tribute, and I am sure that the cave diving group of the times in the thirties if they were still alive would want it so, to Eric Lawrence who was at that time a guide in Wookey Hole Caves. He spent hours helping the divers and crew fetch and carry the tackle from the cave entrance to the Third Chamber, and out again in the early hours of the morning before opening up for yet another day of visitors. I recollect his helping out on the pumps when we were short-handed. Also, I spent hours with him on many caving trips: Swildon's, Eastwater and so on. With 'Digger' Harris, we tried to unblock the chimney in Charon's Chamber, across the river in the First Chamber, for this was thought to connect with unexplored upper levels above the main show cave. Subsequent discoveries show that we were on the right lines. Eric Lawrence was a great man to have beside one on a caving trip.

My own time with the Wookey Hole Divers was very limited and the 'Log Book' gives me more credit for the help I gave them than I deserve. However, I must have been about quite a lot and well remember the broadcast night in August. Balcombe had lost his man for the telephone and I had been pressed into doing the job during the previous weekends. On the night, however, Gerard Hodgkinson had organised a large party for the BBC, including dancing at the swimming pool surrounds with loud speakers to relay to his guests the radio broadcast direct from the cave. As I always undertook the job of being master of ceremonies at the Thursday night dances, Hodgkinson expected me to M.C. his party too. Graham Balcombe stood up to him, however, and said: 'No Mac on telephone, no broadcast!' For once Hodgkinson gave in and we were glad not to have to defend our stance.

Anyway I was allowed to operate in the cave with the divers. For some reason, Graham Balcombe had faith in my ability to keep communications open between shore and diver, even though at times he must have heard me tell the crew to stop pumping air to him when he became a bit impossible! I hope he will forgive me mentioning this, if ever these memories of mine reach print. Unwittingly, Balcombe enhanced the commercialism and publicity for the caves at the very end of the broadcast. After he had been down for some time, Balcombe was wont to open his air exhaust valve wide, clear his suit of stale air and then ask you for an extra effort on the pumps to fill his suit again. Just before 11 p.m. when the live broadcast was due to finish, Balcombe emptied his suit and called: 'Air! More air!! I must have more AIR, you so and so's'. And at that very moment the BBC ended the programme and switched over to dance music. Perhaps they didn't like the flow of language; but, whatever happened, they could not have timed a better climax. The BBC switchboard was jammed for some time to follow by anxious listeners asking about the diver's safety!

Hardly true to her nickname, 'Mossy' Powell left Wookey Hole soon after the diving operations,

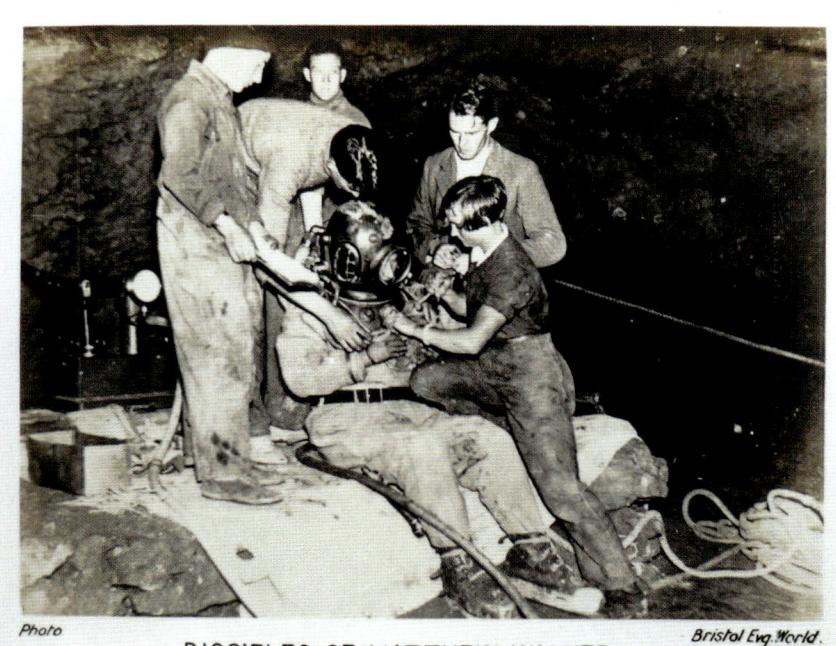

Fig. 3.8
'Mossy' Powell secures Graham Balcombe's helmet and chest weight, Frank Frost (left) keeps the logbook and 'Digger' Harris (right) holds the upstream guide rope. Photo in The Log facing p.172

DISCIPLES OF MATTHEW WALKER

worked at Cheddar for a while and then rolled on to Bristol. Here she remarried to a Clifton housemaster by the name of J.B. Wyllie. During the War, they set up their own boarding school in Hampshire, complete with animals, and made a final move to the Scilly Isles sometime afterwards. When the Cave Diving Group formed in 1946, however, Graham Balcombe acknowledged 'Mossy's achievements in 1935 by consistently listing her as a 'Section Honorary Member' right to the end of 1949. It is only in his last Letter to Members at the end of 1950 that 'Mossy's name finally disappears and the inevitable break with a remarkable pioneer cave diver is recognised by Mendip cavers.

'Mossy' was not on the original diving attempt at the sump in Swildon's in 1934, but it seems appropriate to close these recollections of early cave diving with some first hand memories of where it all started deep beneath Priddy with Jack Sheppard, 'Jumbo' Baker, Graham Balcombe, 'Digger' Harris, Bill Offer, Bill Tucknott, Frank Frost, and myself, Mac Brown. It is worth mentioning that Graham Balcombe, having spent twenty minutes or so under the sump, connected only to his shore party by a hose pipe through which he breathed, and which must have been a most hazardous undertaking, duly came out and helped Jack Sheppard to kit up for his go. He soon submerged but, when well under the arch, somehow pulled the hose connecter from his mask and inhaled a plentiful supply of muddy water. Fortunately his legs still protruded from our side of the sump and he was rapidly pulled out. Having recovered his breath sufficiently to travel, Balcombe took him up to the surface immediately, which I should have thought must have been an exhausting trip, especially getting Jack up the Twenty Foot and Forty Foot ladders. But upon seeing Jack in safe hands on the surface, Balcombe returned all the way to the sump to help in bringing the equipment out. The remaining half dozen of us needed his help as it was bulky stuff to handle in the narrow passages and we had, I suppose, been down for five or six hours. His was true leadership which was also seen many times during the Wookey Hole diving events.

I am pretty sure, too, that whilst Frank Frost clearly took the picture of the first sump diving party in Swildon's, he was using my camera at the time. In those days I had my films specially developed and printed in sepia by a firm in Cheshire. The original print appears in my collection with the photographs I took of 'Mossy' Powell's and my own Ebbor digging gear mentioned earlier. Sometimes it was for the record and often not. But then, fifty years on I am still taking photos, mainly natural history, birds in particular, and people frequently say: 'Whatever did you take that picture for?'

Fig. 3.9
'Mac' Brown's historic photo of the first cave diving party at Sump 1 in Swildon's Hole. Standing (left to right): Jack Shepparrd; 'Jumbo' Baker; Graham Balcombe; 'Digger' Harris, and Bill Offer. Below: 'Mac' Brown and Bill Tucknott.
Photo taken by Frank Frost, 17 February 1934

Chapter Four

'D-Day', then 'Bung' to 'Innominate'

by Jim Hanwell

Fig. 4.1
The Cave Diving Group forms in South Wales over Easter 1946.
Bill Weaver dives in Ffynnon Ddu.
Photo by Frank Frost from the Wessex Cave Club

Fig.4.2
Pioneer cave divers meet again after the Second World War.
Jack Sheppard, Mavis and Graham Balcombe, with Frank Frost outside the Ancient Briton in South Wales, Easter 1946.
Photo by Frank Frost from the Wessex Cave Club

There is a saying among Mendip cavers that events happen in threes and things last for three years. On a third trip you may be in luck, and new ground is gained! After three years of probing sumps, for example, the first one was successfully dived at Wookey Hole in 1935 and then in Swildon's Hole in 1936. During the Second World War, there would be another three years of trials with new equipment to the dawn of the post-war cave diving. As Graham Balcombe noted at the time, Easter 1945 was 'our D-Day'. The actual revival of operations at Wookey Hole Caves would span a similar period from Whitsun 1946 to the spring of 1949; starting optimistically with Operation Bung and ending tragically with Operation Innominate. Since these operations stem from the thinking and experiences of Balcombe's 'D-Day' campaign, we had better review these too. History is about continuity.

The War may have cheated Graham Balcombe's generation out of time but, as if in compensation, it also stimulated overdue developments of diving equipment that might otherwise not have happened just then. During this period, Balcombe had the foresight and the skills to build his own waterproof suit, an ingenious contraption for underwater navigation, and, of course, a breathing apparatus. The suit consisted of chest waders and jerkin joined by a rubber cummerbund and sealed with normal cuffs and a tight neckband; maybe not very elegant by comparison with the contemporary 'frogman's suit' and the modern 'wet suit', but moderately effective in cold cave water. The resourceful navigation gear was a substantial all-purpose device made up of cannibalised motor car headlamps, alkali battery, line reel, depth gauge, compass and a current-direction tell-tale. They were assembled with Wookey Hole Caves in mind and made to size on the cat's whisker principle. This 'Apparatus For Laying Out Line And Underwater Navigation' was called an 'AFLOLAUN' or, in the manner of the times, simply 'Aflo'; a contraction that subsequently lent itself to many versions, the most bizarre name being the 'Aflohonk' because of its built-in hooter for signalling underwater! The inventor, Bob Davies, recalls that the intended clang became more of a tinkle, but was nevertheless a highly successful underwater signalling device up to 600 feet in distance. Balcombe's vital breathing apparatus may have looked like a plumber's nightmare to the uninitiated but was, in fact, a skilfully constructed and delicate piece of equipment whose antecedents had done honourable service in mines rescue and underwater salvage since 1880. Graham Balcombe described the ensemble as:

> The best that I could put together at the time.

It certainly was, and it did the job.

Fig. 4.3
'The Grandfather of all Aflos', at Keld Head, Kingsdale, Yorkshire, 6 August 1945. Ray Nunwick (left) and Graham Balcombe with AFLO Mk.I.
Photo from Bob Davies album in Wells & Mendip Museum

A glance even further back in time is appropriate here. Dreams of being free to breathe and swim around underwater must be as old as Man, but most ideas of how self-contained diving might be done wisely ended on the drawing board. Even Charles Condert, the likely pioneer of self-contained underwater breathing apparatus, called SCUBA, drowned in the East River, New York, when a tube to his compressed air suit broke in August 1832. Details of this apparatus and Condert's demise were widely published at the time, as noted by Howard Larson in *A History of Self-contained Diving and Underwater Swimming* (1959). Diving bells and suits with helmets fed by air lines to the surface had no serious contenders until after the International Exhibition of 1855 in Paris. Ten years later, Frenchmen Benoit Rouquayrol and Auguste Denayrouze designed an air reservoir from which divers could breath briefly whilst detached from the pumped supply. Their so-called 'aerophore' supplied air at the right pressure, for the need to regulate breathing with depth was well known. Jules Verne took aerophores into the realms of fantasy in *Twenty Thousand Leagues Under the Sea*, written in 1869.

The first fully self-contained breathing apparatus with a useful duration was invented by Henry Fleuss in 1878. As the main safety requirement was to ensure the longest possible duration of the breathing gas, any system which enabled one's exhaled gases to be recycled or rebreathed was more attractive than one where the air is breathed once and then blown to waste. Thus, the continuing competition between closed-circuit rebreathers and rival open-circuit respirators began in favour of the former. Although the latter were essentially simpler, gas economy took priority. Fleuss was the wiry son of a Marlborough College schoolmaster and his ingenious reputation was made when the device worked successfully during a mining disaster in 1880. The same year, the railway tunnel under the River Severn flooded and it was impossible to drag air lines down the 200 foot shaft then along the submerged passage to close an iron flood door 1000 feet away. So, the celebrated diver, Alexander Lambert, used the new self-contained respirator to carry out this task successfully. He was supported, in total darkness, by no less than Fleuss himself who had never dived before but obviously possessed great nerve and extreme confidence in his invention. The difference between this Severn Tunnel epic and cave diving is, of course, that the way was known. Finding the way poses additional problems in caves. Where will the passage go?

The secret of the Fleuss respirator was its use of pure oxygen rather than air. Exhaled gases were not wasted, for the carbon dioxide was removed by the absorbing chemical, soda-lime, so that the oxygen breathed out was purified. This oxygen was then stored in a bag for re-use, so prolonging its useful life. Apparently, Fleuss was unaware at the time that Paul Bert in France had conducted experiments to show that pure oxygen becomes unsafe to breathe on diving below 30 feet, because its effects are poisonous when normal atmospheric pressure doubles at 33 feet deep. In any case, the oxygen 're-breathers' were intended chiefly for mines rescue work such as fire-fighting and foul air problems and, perhaps, shallow water work. Depth limitations were not a handicap. Plenty of oxygen rebreathers were designed for various tasks during the First and Second World Wars. They were the easiest respirators to lay hands on, particularly during the 1940s. So, when Graham Balcombe and other cave divers adopted the available re-breathers during and after the Second World War, they knew of the restriction but did not anticipate that sumps would go deeper; a perfectly reasonable assumption at the time. Most negotiable sumps subsequently explored bear this out. They could not have known that beyond the conglomerate the tilted beds of limestone on Mendip would cause the underground River Axe to go down and up in great steps, one after another. This was to be the main problem in the explorations of Wookey Hole Caves, and it brings this chapter to an end with the need for deeper dives in 1949.

Young Henry Fleuss worked at the time for the famous diving dress and air pump firm established by the late Augustus Siebe and inherited by Henry Siebe and his brother-in-law William Gorman. The new owners moved their premises to Lambeth in London, and it was here that eleven year old Robert Davis joined the firm as a 'factory boy' in 1882. He assisted Fleuss in his laboratory. Davis's promise was clear and the more eccentric Gorman encouraged and moulded him for management after the untimely death of Siebe in 1885. When Gorman himself died in 1904, Robert Davis was appointed Managing Director of Siebe, Gorman and Company Limited when only 33 years old. 'Everything for safety everywhere' became his slogan and he consulted old Fleuss on likely developments. By 1906, the Fleuss-Davis patent respirator won an international competition for a reliable self-contained mines recovery breathing apparatus. It was aptly called 'Proto'.

Proto was adopted by mines rescue teams around the world for firefighting and foul air work. During the First World War, Davis produced a smaller version called the 'Salvus' for trench warfare and use under gas attacks. Perhaps the best vindication of their design and durability is the successful use of both to rescue all 116 survivors from the horrendous Knockshinnock Colliery disaster as late as 1950. It has been said that they have protected and saved more lives than any other such apparatus in the world. The longer duration Proto could also be adapted for use underwater and, with rescue

*Fig. 4.4
An introduction to the use of closed circuit oxygen re-breathers by Siebe, Gorman and Company, Limited, from a war-time Handbook of Instructions for using the SALVUS A.N.S. apparatus.*

SIEBE, GORMAN AND COMPANY, LIMITED

I. EXPLANATION OF THE SELF-CONTAINED PRINCIPLE

Before dealing with the apparatus in detail it will be found helpful if a general idea is obtained of the simple principle upon which the apparatus functions.

The air in which we live contains, approximately, the following:

Inhale	Nitrogen	79%
	Oxygen	21%
	Carbon Dioxide	trace

After inhaling this atmosphere we exhale, approximately, under normal working conditions, the following:

Exhale	Nitrogen	79%
	Oxygen	17%
	Carbon Dioxide	4%

On studying the above approximate figures it will be seen that, to bring the exhaled atmosphere back to normal, it is necessary to replace the 4% of Oxygen lost and to absorb the 4% of carbon dioxide.

The "Salvus A.N.S." apparatus carries out the above requirements by means of an automatic supply of oxygen and a medium for the absorption of the carbon dioxide.

The apparatus meets all the demands of the human body, which, as is well known, vary in accordance with the work in hand.

For instance, if a man is required only to search about to find a smouldering fire, his respiratory exchange will probably be normal. If, on the other hand, he is called

"NEPTUNE" WORKS, LONDON, S.E.1

2

from submarines in mind, Siebe, Gorman produced the Davis Submerged Escape Apparatus, or D.S.E.A. Its reputation was proven when H.M.S. Poseidon sank to 125 feet in the China Sea in 1931. Of the 27 crew trapped below, eight managed to leave the boat using the D.S.E.A., although two failed failed to reach the surface and one died later. Shortly afterwards Robert Henry Davis received his Knighthood. This may be why the D.S.E.A. kit was dubbed as the Davis Submarine Escape Apparatus in subsequent accounts, notably by those cavers who adapted it for diving sumps.

Professor John Scott Haldane who wrote the classic work on *Respiration* with J.G. Priestley in 1920 collaborated with Siebe, Gorman and was a life-long friend of Davis until Haldane's death in 1936. Appropriately, his son Professor J.B.S. Haldane continued both his research and the association with the company and cave divers. The J.B.S. Haldane Fund was to be a boon to the Cave Diving Group. So it was that the ideal blend between theory and practice led to the Admiralty Experimental Diving Unit being based at Siebe, Gorman's headquarters during the Second World War. Audacious raids by Italian divers riding human torpedos called 'pigs' had eventually succeeded in sinking Allied ships actually at anchor in Gibraltar and Alexandria harbours in 1941. Prime Minister Winston Churchill demanded retaliation from the Chiefs of Staff and the task of making a bewildering variety of amphibious equipment became urgent. Integral breathing sets and suits were made for every combative tactic possible. All used the proven oxygen re-breathing system in one form or another. Apart from duration, these closed-circuit sets did not emit streams of tell-tale bubbles and could be used in secret. Examples were: the Siebe, Gorman Amphibian Mark One and its improved Mark Two (called SGAMTU or even SGAM2), the tiny Amphibian Tank Escape Apparatus (ATEA), and the versatile Port Party (P-Party) designed for mines clearance in harbours and general salvage operations. The Dunlop Rubber Company made many of the suits for the so-called frogmen, and it is worth noting that the first serious use of fins for swimming was in 1943. Most divers walked on the bottom or rode on 'torpedoes' and 'submersible canoes' known as Special Boats. All are illustrated in post-War editions of Sir Robert Davis' own book *Deep Diving and Submarine Operations* first published in 1935.

In pre-War France, Yves le Prieur had built a simple 'air-lung' in 1933 and founded the first diving for sport group the following year, the 'Club des Sous-l'Eau'. They were open-water enthusiasts, not cavers, so there was no great reason for prolonged dives. During the War, when France was occupied, Georges Comheines developed a semi-automatic valve to regulate the breathing of compressed air when required, and Jacques Cousteau with Emile Gagnan built the first practical 'aqualung' in June 1943. Understandably, all this was done in great secrecy and remained unknown to would-be cave divers in Britain. But aqualungs will have to wait until later in this story.

And what of Graham Balcombe during the War? Well, as so often happens, chance favours the prepared. He was posted to work in Harrogate, Yorkshire, and found time to make up the equipment for his next exploratory cave dives at his new home in Bramhope. The opportunity to try out his home-made equipment came when he was was able to return to old haunts in the Yorkshire Dales during time off from Post Office duties; the very place where he had been first won over from climbing to caving ten years before. Old friends and young ones were there to help too, notably Reginald Hainsworth and Raymond Nunwick. Reg Hainsworth ran a garage business in Ingleton which was a handy depot, workshop and welcome refuge. More to the point, as an old member of the Gritstone Club and founder of the Cave Rescue Organisation in 1935, Reg Hainsworth was an ideal mentor and companion. It is worth noting that the C.R.O. shared its Fiftieth Anniversary with that of the Wookey Hole Divers. Such are the close connections among British cavers. To Graham Balcombe, Reg Hainsworth was also:

> Pioneer of cave diving in this country...a great man who had the misfortune to choose (as we have since proven) the worst possible sumps for his attempts (Keld Head and Alum Pot).

Now, imagine the excitement throughout the Dales when the irrepressible Bob Leakey succeeded in free-diving through a short sump to open up Disappointment Pot as yet another, and perhaps the finest, way into the mighty Gaping Gill system. It was January 1944 and the impact of this fillip during the war years is very evident in *Underground Adventure* written by Arthur Gemmell and Jack

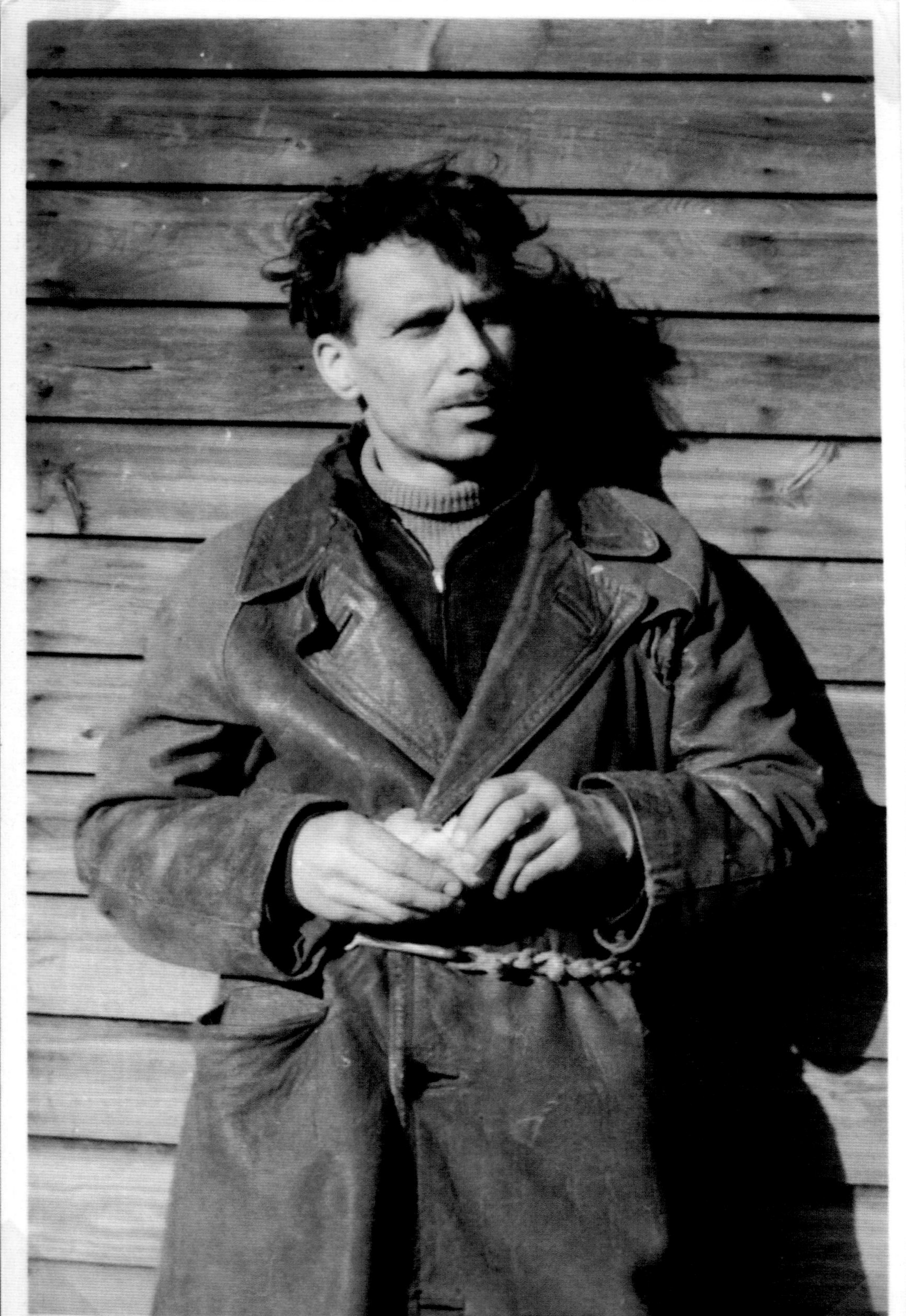

Myers, the first post-war classic to revive the sporting appeal of caving in 1952. It was during 1944 that Graham Balcombe planned his comeback to cave diving; a period shared with the momentous return of the Allied Forces to the Normandy beaches, coded 'D-Day'. Like them, he had reached his goal within the following year!

To those who may wonder why on earth Balcombe's 'D-Day' in Yorkshire should be part of the Wookey Hole story, the following extracts from his notebooks of the day should suffice. Obviously delighted by the news that divers had found major new chambers in Wookey Hole during January 1970, he published these 'practically verbatim' in issues 142 and 144 of *Journal* Volume 12 (1972) of the Wessex Cave Club. His notes tell the full story of how he tried and tested the first closed-circuit respirator used in cave diving. They also reveal the sense of fun and sudden fears, the necessary commitment and the need for caution. There is also a hankering to get back to grips with the underground River Axe in Somerset, for his home-made suit was called the 'Wookey Hole Divers' Dress' and the respirator the 'Wookey Hole Divers' Breathing Apparatus'. He referred to the kit as 'Whodd-Whodba'. The transport was a tandem, behind which they towed a trailer made by Raymond Nunwick, then an apprentice engineer. Balcombe described Nunwick as having an 'insatiable appetite' and as an 'amiable butt'. No doubt he had to be, for the sight of Balcombe, his wife Mavis, Pete the family dog and the diving kit festooning the trailer was very comical and must have attracted dry humour from leg-pulling folk in the Dales!

It is the weekend before Easter 1945 and the start of a week's trials of Whodd-Whodba:

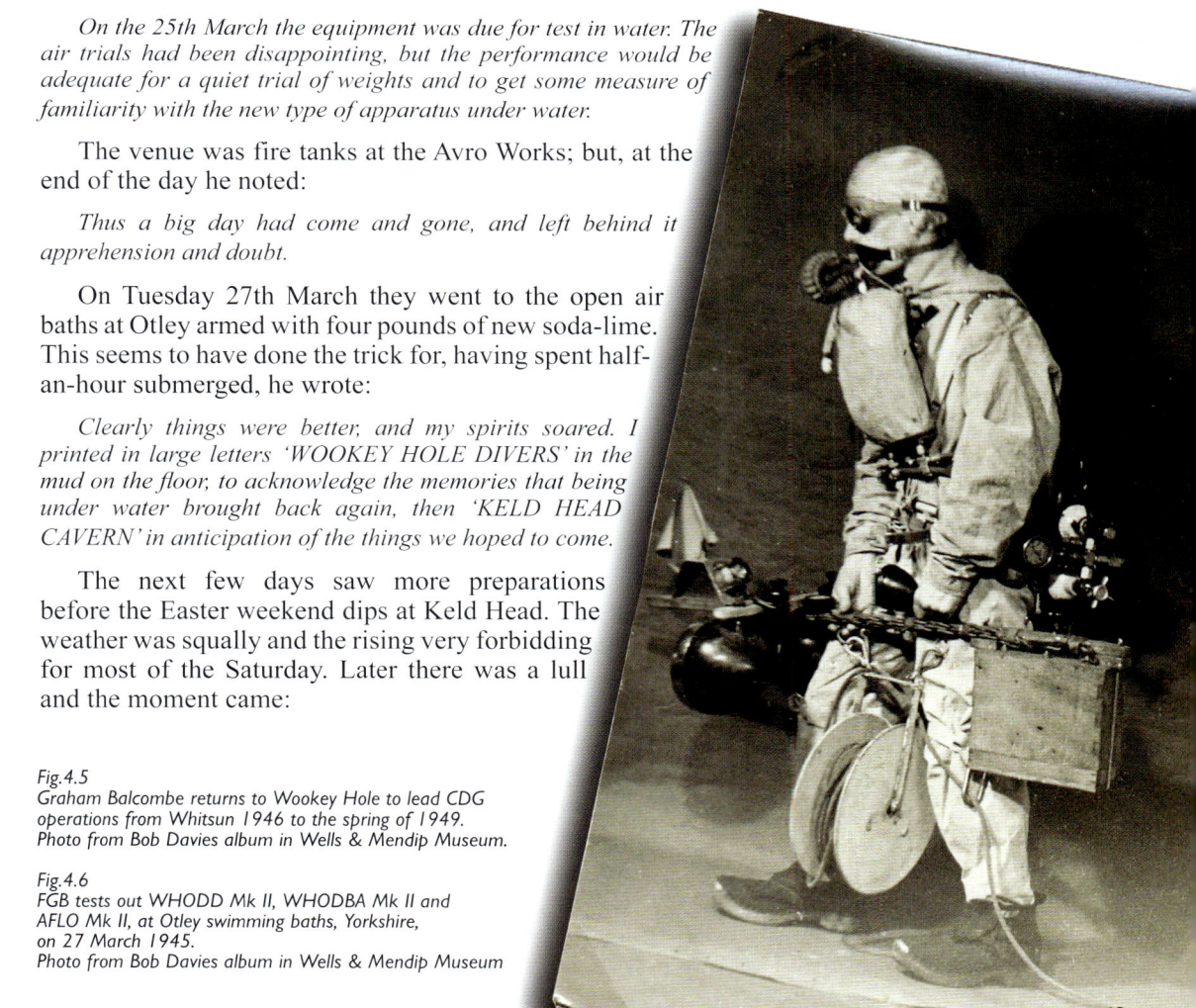

On the 25th March the equipment was due for test in water. The air trials had been disappointing, but the performance would be adequate for a quiet trial of weights and to get some measure of familiarity with the new type of apparatus under water.

The venue was fire tanks at the Avro Works; but, at the end of the day he noted:

Thus a big day had come and gone, and left behind it apprehension and doubt.

On Tuesday 27th March they went to the open air baths at Otley armed with four pounds of new soda-lime. This seems to have done the trick for, having spent half-an-hour submerged, he wrote:

Clearly things were better, and my spirits soared. I printed in large letters 'WOOKEY HOLE DIVERS' in the mud on the floor, to acknowledge the memories that being under water brought back again, then 'KELD HEAD CAVERN' in anticipation of the things we hoped to come.

The next few days saw more preparations before the Easter weekend dips at Keld Head. The weather was squally and the rising very forbidding for most of the Saturday. Later there was a lull and the moment came:

Fig.4.5
Graham Balcombe returns to Wookey Hole to lead CDG operations from Whitsun 1946 to the spring of 1949.
Photo from Bob Davies album in Wells & Mendip Museum.

Fig.4.6
FGB tests out WHODD Mk II, WHODBA Mk II and AFLO Mk II, at Otley swimming baths, Yorkshire, on 27 March 1945.
Photo from Bob Davies album in Wells & Mendip Museum.

At last I was dressed (it is a lengthy business), launched and set about the job of trying out the new idea. After some time wallowing about and checking previous observations I called for the light, but it was then time to return. So, having undressed, the kit was stowed in the tent for the night and we went down for tea. Meantime I can tell you first hand that the black abysses gaping under the rock walls looked very fearsome and suggestive of the lair of savage monsters! Visibility was two to three feet and likely to be much less with artificial light, so it was not entirely with regret I left the place to exchange for a mighty feed and the fun of the local dance hall.

On Easter Sunday, the 1st April no less, he had a scare about his gas supply:

Clutching my drum of rope, I slowly groped my way downwards. Visibility quickly dropped from two feet to one, then about six inches from the light. Each stone on the slope seemed to be the last, yet after each stone there appeared another. The slope was very steep, and the stones about 6 inches in diameter were perched very precariously. When disturbed they rolled down out of sight with a queer high-pitched clink! clink! Soon the daylight disappeared, last seen as a dim orange foggy patch almost overhead; the sound of breathing and of the inrushing gas disappeared, a deadly quiet ensued. It was disquieting, and after another few feet I left my drum and climbed back with the aid of the rope anchored on the bank; the dim patch appeared again, grew lighter and was daylight at last. The noise of breathing came back, and on opening the gas valve, yes, there was the rush of gas, all was well.

Going to a depth of about 20 feet without a hard hat was a new experience. The increasing pressure had probably caused a simple ear blockage; the sudden and unaccustomed silence that followed must have been both eerie and unnerving after the friendly noises heard earlier at shallower depths. Understandably, in such a forbidding environment he just had to turn back to sort things out. Keyed up, no doubt, he made two more dives only for more troubles to pile up:

Fig. 4.7
FGB dives at Keld Head, Kingsdale, Yorkshire, watched by Ray Nunwick, 6 August 1945. Photo from Bob Davies album in Wells & Mendip Museum

Going down I thought that all was not well with the respirator, it was by no means free of trouble, but I knew it was not far to go and I would soon be back. By now I was sure that the respirator was going badly, so quickly opened up the [water] sampling bottle, I heard it gurgle as I sat there half-dozing, sealed it off mechanically, and slowly started to climb. It seemed ages before that patch of light above broke through and the difficulty of the last few feet broke irritatingly through to my dulled consciousness; at last I surfaced, but there was still the stretch to the bank to cover. I started to walk it and came to a shallows; in temper I threw myself down into deeper water the other side and crawled to the bank.

This time it seemed that he suffered from an excess of carbon dioxide and almost blacked out. There was also the nagging possibility that the trouble may have been caused by a far more dangerous oxygen-lack, however. It was the uncertainty that led Graham Balcombe to be so hard on himself, being frustrated as much with his own performance as that of the kit. But, then, when learning to do something new, entirely alone and without instruction, you have to drive yourself for no one else will. Future cave divers would have the benefit of training with experienced instructors, but someone has to make a start.

Early in June 1945, Balcombe and his team went back to Kingsdale and he dived the Keld Head rising to reach two air bells beyond the 20 foot deep. He reported to the BSA that he had penetrated upstream at least 165 feet and, maybe, as far as 210 feet. And still the sump went on. Two further trips in mid-June and early July had to be called off, however, as poor visibility had been caused by heavy rains on the fells. Water churned brown with peat and suspended sediments could not be dived. Over a long weekend from 4th to 6th August the water conditions were kinder. With clear visibility and a good light, he soon reached his previous limit and was able to survey the distance dived more accurately. As a result, his earlier estimate was revised to 180 feet; still a significant achievement. But, in the event it turned out to be a rather sweet and sour weekend. On his second dive, Graham Balcombe accidentally knocked off his gas supply somehow, an incident that he recalls as being:

A real hair raiser.

No further progress was made in the Keld Head rising by Graham Balcombe for later trips in mid-September and early October had to be abandoned because of more floods. So he gave up the struggle there and has recently made a telling point that could so easily be overlooked by those who now find their way around the Dales by car:

The ... effort with tandem and trailer from Bramhope to Keld Head was just about killing us.

A round trip of over 100 miles with camping, caving and diving gear was enterprising enough then, to say nothing of the progress underwater. Maybe good fortune was at work, too, for Keld Head has since become the longest continuous underwater cave passage yet dived in Britain (1979) and ranks highly in world records. Success was well beyond Balcombe's grasp anyway. His 'D-Day' had not been rewarded by great new discoveries and he did not return to dive in Kingsdale above Ingleton.

But Graham Balcombe did not give up and was drawn to other sites posing different problems. In June 1945, for example, he investigated the pool at the bottom of Alum Pot, that splendid abyss which overlooks Ribblesdale. This, too, proved unrewarding for, as will be seen shortly, the submerged shaft simply went down and down again.

One sump had gone in too far, the second was too deep, so what would the third be like? Well, it proved just right and nearer home. Undeterred, Graham Balcombe turned to the Goyden Pot sumps in Nidderdale during October 1945 and, although failing to get beyond the lower passage, he succeeded in making the return trip underwater to the lower river; the learning process was more convincing and confidence was restored. After all, things happen in threes remember and Goyden Pot is regarded with affection by Graham Balcombe for bringing him luck at last. But good luck has to be won, and it was exactly nine years

Fig. 4.8
FGB pushes his tandem and trailer up to Alum Pot, Yorkshire, assisted by Mary and Ray O'Neill, 24 June 1945. Photo from Bob Davies album in Wells & Mendip Museum.

since Graham Balcombe had used the 'crazy' bicycle respirator to make the very first self-contained cave dive to the Bell Chambers beyond Swildon's Two. The approach worked and the old enthusiasm was rekindled for future cave divers could use self-contained oxygen re-breathers; what else! They would be on their own and free to explore without umbilical air lines to base. What is more, the War was over and one could begin to spend time on one's own interests. Without Graham Balcombe's 'D-Day' campaign the revival of diving in Wookey Hole Caves would not have happened so soon after the War.

Fig. 4.9
Bulky cave diving gear heads south from Yorkshire to London, after the War.
Photo from Bob Davies album in Wells & Mendip Museum

Shortly after his Goyden Pot dive, Graham and Mavis Balcombe returned to London for a new life there. But, before leaving wartime Yorkshire, we should hear from Graham Balcombe's companion there, Raymond Nunwick. In October 1985 he wrote the following recollections; a cameo of the sort that only comes from shared interests and firm friendship:

I joined the BSA, I think, in 1940, when in addition to caving as a very junior partner to some of the tigers of the day, I read quite a bit of their library material. The exploits of Graham Balcombe in Wookey Hole, which I pored over in great detail, created a deep and lasting impression. I attended a meet of the BSA at Horton-in-Ribblesdale at Whitsuntide 1941. The more affluent members of the party were resident in The Crown Hotel, but having little money myself, I had to be content with camping in the field behind the hotel. It soon became apparent that other tents in the field were occupied by members of the Northern Cavern and Fell Club. Imagine my incredible surprise therefore, when I heard someone shout out the name of Balcombe – surely there could only be one man with that name!

I contrived to have a chat with him. I found him abrupt, not exactly friendly, a man of very few words, and gained an immediate and very clear view that he was a man who would not suffer fools gladly. I was amazed to find that he was living about four miles from my home, so some weeks later, with considerable trepidation, I called on him. (I don't think either of us had a telephone at that time.) So commenced a friendship which has lasted through to the present day, and where, over the years, we have been involved in all kinds of exploits, which have unfortunately been curtailed since the tragic death of Mavis, which was a very severe blow to him, and changed his life completely.

FGB was an engineer of great talent. I was studying engineering at that time, and it was not long before Graham was sorting out some of my problems, for which I was eternally grateful. He had, it seemed to me at the time, a fully equipped workshop, with tools galore, and even a power driven lathe and drill items beyond my wildest dreams. He was always making something, and as time passed, I deemed it a great honour to be allowed to assist. I think Mavis must have found me a bit tiresome, as I was always at the Balcombe house, and in due course travelling the limestone areas of the Dales on the tandem looking at the odd cave, but concentrating in particular on sites of resurgence, and plotting where there might be a sump.

'D-DAY', THEN 'BUNG' TO 'INNOMINATE'

I cannot recall the actual time when a positive move towards diving was made but I know we had spent many hours sitting on the bank at Keld Head, and walking over Kingsdale. I and my friends had a hut on a farm at Westhouse near Ingleton. It was only about three miles from Keld Head, and so formed a focal point in the area. The workshop was thus soon in full production making the Mark I version of 'Whodd', and I was introduced to the mysteries of soda lime, absorption rates, flow rates, valves and gauges.

Progress on the diving apparatus was slow, but it must be remembered that the complete assembly was constructed by hand, during a period when sources of material were nil. Much of what was used came from Balcombe's pre-war collection. FGB took infinite pains with regard to the technicalities of his construction work, and it was a source of amazement to me at the time that he would spend hours and possibly days on sheets of calculations to prove one small point in the design.

There was a field of about three acres behind his house, where we commenced dry testing of the completed diving dress. The drill was that FGB would get kitted up, a lengthy job in itself. We would then jog round the field boundary, FGB gasping away through his respirator, and me trailing along behind, armed with a clip board. I was supposed to try and interpret what Balcombe was thinking and saying totally muffled by the breathing apparatus, take the time with a stop watch at various intervals, and note it all down. This I found difficult whilst trying to run at the same time.

It was now 1944. There followed a period of madness, when FGB indicated that the only way into Keld Head was to lower the water level, and so create air spaces where a diver could surface. I think I lost a number of friends on that work. We would spend hours pulling stones out of the stream bed, where it empties out of Keld Head pond, without any apparent effect on the water level, then the rains would come, and the next time we went, all the stones had been re-deposited. Needless to say, this effort produced nothing of any significance. And 1944 was a bad year for weather.

We pottered around looking at sumps, including a fairly detailed examination of Manchester Hole and Goyden Pot river systems, which I remember as a particularly good caving trip. Copious notes were made by FGB. I learned a lot about what to look for, so a good time was had by all. We even went down the newly opened Disappointment Pot, just the two of us with presumably grand ideas of finding the sump system which must lead towards Clapham Cave. We had used a very old and rather short rope ladder on the bottom pitch. When we returned cold, wet and weary, the ladder, which had been hanging in a stream of water, had shrunk, and was quite out of reach. It took some time, and much energy to get out of there and we were late back.

I think the first wet trial of the diving kit took place in what we would now call a pleasure park Golden Acre where there was a lake. As I recall, this was a dismal failure. The edges of the lake were soft and muddy, and the water shallow, with the result that the diver soon became totally immobilized by mud, and quite unable to see in the water. A better solution had to be found, the result of which was permission to use the public swimming baths at Otley, it being out of season, but they were still full of water. I don't know whether any money changed hands for this exercise, which as far as I could tell, was a success.

Clean water meant that FGB could see to make adjustments. It was easy to come ashore for modifications, and re-enter the water for further tests. We were enclosed by the surrounding fence, so the whole thing was quite private, and we had no spectators.

Serious trials then began with the diving dress at Keld Head. The permanent problems were cold, rain, wind, and discoloured water. Most of the time it was thoroughly miserable. One rather horrifying sight was FGB, stark naked in late March, swimming in the pond, and groping about at the base of the cliff, looking for alternative ways of entry.

The breathing apparatus, and the dress seemed to work well, and it was during Easter 1945 that we laid siege to Keld Head. Much hard work was put in, most of it in very poor conditions, and in the light of what is now known about the place, there is no wonder little headway was made. I had a few nasty scares when the diver was overdue. We had no backup system, but I guess one did not worry unduly about such things at the time.

The attempt on the sump of Alum Pot in June 1945 was almost a holiday compared with what we had been doing at Keld Head – at least that is how I saw it. There were seven of us in the party, and we had struggled to Ribblesdale with the tandem pulling a trailer, and various other cycles loaded down with equipment. Getting the lot up the field to Long Churn was the last straw! We had no long ropes, so everything was hauled in via Long Churn passage, it must have been quite a procedure. This was done on a June Saturday, and we were the only party in the hole all day – no sherpas to assist. Imagine that situation today, when the only time you could have Alum Pot to yourself would be in the middle of the night! The diving was a success in that all went well, but of course with the sump going straight down, instead of horizontal, it was impossible to make any progress.

There followed a number of further visits to Keld Head, again with much effort, but little progress, but by this time, I was in the Army, and it was not long before FGB returned to London. He pursued development of his apparatus, and whilst I was stationed in London, we had an evening of testing at Willesden swimming baths. I didn't return to the scene until the early 1950s, when once again I was resident only a few miles from Graham.

The work at Swildon's Hole and Wookey Hole Caves was well organised, and we were down there at frequent intervals in the company of considerable groups of people, a total change from the Balcombe-Nunwick combination of earlier years. My sole job was to act as diver's assistant, which I remember doing with almost mechanical action weekend after weekend, and I have little recollection of who was there apart from my immediate circle. I don't remember ever eating much, or even sleeping during that period, let alone ever having a drink – I would never contemplate it now!

All things change, and gradually the interest in diving declined. (I think Mavis had a lot to do with that.) We spent some time entering car rallies together, and even won some awards.

As time passed, I had to pay more attention to my work, and crazy weekend plans became few and far between. Visits to the Balcombes became gastronomic, and much caving was done from armchairs. I always count myself fortunate that I was camping in Horton-in-Ribblesdale in 1941.

So it became clear that future cave diving operations would need considerable back-up and well-laid plans. Like-minded companions and an organisation that ran smoothly were called for. It was back on Mendip among old West Country friends that Graham Balcombe found both. There were enough cavers on Mendip not only keen to tackle the sumps in Wookey Hole Caves, but also to take on virgin sites in South Wales such as Ffynnon Ddu in the Swansea Valley. Two teams called the Wrington Group and Downend Group, respectively led by Peter Harvey and Paul Dolphin, helped

Fig. 4.10 Graham Balcombe dives at Ffynnon Ddu, South Wales, Easter 1946. Photo by Frank Frost from Wessex Cave Club

to get caving going in South Wales. Peter subsequently formed the South Wales Caving Club and helped to discover Ogof Ffynnon Ddu, now one of Britain's longest and deepest systems[7]. Paul had led the Wessex Cave Club party that opened the Dolphin Pot route in Eastwater Cavern in 1940, before digging at sites in South Wales. After many years overseas he came to live near Burrington Combe back on Mendip[8].

Fig. 4.11
Early members of the Wessex Cave Club who helped to form the CDG and South Wales Caving Club after the War. Seen here at the Priddy Minery testing a pump-fed gas mask in the late-1930s. Showing Jack Lander (on the pump), Frances and Geoff Tudgay (centre) and Paul Dolphin (wearing the mask). Mrs Kidd and Mrs Dolphin rest on the blanket.
Photo by Frank Frost from the Wessex Cave Club

Fig. 4.12
Harry Stanbury, founder of the Bristol Exploration Club in 1935, follows tradition by undertaking his open-water training in the Priddy Minery during 1946, supervised by Don Coase.
Photo from Bob Davies album in Wells & Mendip Museum

The Somerset cave divers naturally divided into a South Wales contingent and those more concerned with Mendip in which Wookey Hole was the main prize. Many of the latter came from the Bristol area and were members of the Bristol Exploration Club. Indeed, the immediate post-War birth of the Cave Diving Group in 1946 and the welcome revival of the BEC coincided. The Club also shared its Fiftieth Anniversary celebration with the old Wookey Hole divers in October 1985.

The connections are such that the same characters play parts in both, and Don Coase in particular was the young lion who came to lead most of the advances up the subterranean River Axe in Wookey Hole. When the BEC finally succeeded in digging open St. Cuthbert's Swallet in 1953, it was Coase who organised and led most exploration parties. Sadly, he died following surgery in 1958. Today down St. Cuthbert's Swallet is a memorial plaque in the Cerberus Hall to remind cavers that Don Coase discovered major stretches of the underground river from both ends. Countless tourists to Wookey Hole Caves are now able to admire the majestic Ninth Chamber found by Don Coase and Graham Balcombe on 24th April 1948; a fitting tribute to the pupil and his teacher, of course.

Back in the War, whilst Balcombe was in Yorkshire, Coase was living in London and involved in ordnance work. Thus, he had occasion to visit the explosives factory near Bridgwater and every opportunity to cave on Mendip at weekends was taken. The rigours of train travel and cycling with all caving kit were considerable but acceptable. Bicycles were carried in the Guard's Van and, as carriages were invariably packed-out with troops and other travellers, a seat was a luxury; otherwise you found enough space to sleep in the corridor, or even stand throughout

[7] Peter Harvey died in 2009.
[8] Paul Dolphin died in 2005.

Fig. 4.13
Harry Stanbury undertakes his first dive in Wookey Hole Cave on 4 October 1947, supervised by Don Coase and logged by A.C. Johnson
Log from Cave Diving Group records

Fig. 4.14
Don Coase (above) spreads out his own P-Party breathing apparatus, flanked by boots, weights and Aflo in 1946.
Photos from Bob Davies album in Wells & Mendip Museum

the entire journey. And it took almost five hours from Paddington Station to the little halt on the Great Western Railway at Haybridge near Wookey Hole! Don Coase describes a typical weekend at Priddy in his third personal 'Log Book'. The War in Europe is almost over for it is March 1945:

> Friday 9th March. Caught 6.00 train from Paddington to Wells. Changed at Witham & arrived Wells 10.30 p.m. Started to cycle to Priddy via Wookey Hole, and passing Mr. Alpins House saw a light, so paid him a visit. Jawed caving etc. Wife got me some supper. Left about 11.45 & took it easy up Easton Hill [now Deer Leap]. Arrived at Main's Barn approx 1 o'clock a.m. Main had left wood, milk, bread & eggs so made tea, fried egg on toast, & pancakes. Then sat & waited for the others, wondering whether there was enough wood. They arrived about 2.45, Shorty, Betty & Charles. So fed them & then we retired to the straw.

They were up at 9.00 a.m., after a short sleep, to cycle over to Charterhouse-on-Mendip and complete trips to both G.B. Cave and nearby Read's Grotto on the Saturday. After cycling back to Priddy in damp caving kit, they changed, spent some time at the Queen Victoria Inn and then cooked a stew in the barn before settling down for the night at 1.00 a.m. On Sunday 11th, they walked over to fire an explosive charge and continue an old dig at Cross Swallet before returning to tidy up themselves and the barn. After an evening yarning with the locals in the Queen Victoria, Coase sets off for Bath on his bicycle:

> Sunday 11th March. At 10 went back to Barn settled up with Main & set off for Bath. Dry, slight tail wind & dark. Arrived Bath 12.20. Took 1¾ hrs, a record for night time I think. Train didn't arrive till 2.00. Bloody cold waiting on platform as had no mac or coat. Couldn't lie down in corridor as too wet with condensation off windows. Arrived Pad' 5.30 & cycled home.

This was a fairly typical entry which gives some insight into the mettle of the man who cannot tell his own story with the rest. It is also worth remembering that such transport was customary throughout the period to 1949 being reviewed here, for, until the fifties, these were the austerity years of rationing and queues.

Another account from Don Coase's log book describes a rarer but obviously memorable clandestine trip into Wookey Hole Caves the following month:

> Sat 28th April. We set off to Wookey about 11.30 p.m., Nick, Dave, Ron & Self. Cycled to top of Easton Hill, left bikes and walked over hills through Ebbor to top of Wookey ravine. It was still early for the others from B'water so I had a snooze. Woke up & found other three disappeared, so went to see if could see rest of party. Heard herd of elephants crashing through hedge & found it was party. Freda, Frank, Charles, Gordon Hall, two other females & several others. A Hell of a crowd. We managed to get down the ravine without too much noise & it was fairly light, the moon only just being on the wane. Got inside & surprised to find boat was not in 1st Chamber. It was actually in 3rd with canoe, so launched canoe. Freda, Frank & self got in & found it was definitely unstable, the other two jumped out & tipped me out into deep water where I went right under. I had intended to have a swim, so got out, stripped & went in again. Water was pleasantly warm...It was bloody cold putting on wet things after.

Although this was not the first time that Coase and his crew had entered the show cave via the constricted upper entrance on the cliff face, it was almost certainly his 'baptism' in the underground River Axe. Once again, light-hearted, if not exactly legal, play would eventually reap rewards for the owners. Don Coase's explorations would repay his wartime stealth.

To celebrate the end of the War, Don Coase and his close friend Dave Broadbent put their bikes aboard a very full train leaving St. Pancras at 9.30 p.m. for the North on Friday 22nd June 1945. They reached Horton-in-Ribblesdale station the following morning, some twelve weary hours after leaving London. It was pouring in the Dales and all the streams were in spate. They were unable to get down a pothole until Sunday 24th June and went to Alum Pot, descending it through Long Churn. On reaching the bottom, Don Coase takes up the story in his log book:

> More scrambling past several deepish pools brought us to the end chamber with Diccan Waterfall falling 100 ft from the roof. At the end the party who were in front of us were standing round Balcombe who was going to explore the final pool in his diving suit. After some time another party arrived & Tony Humphreys greeted us. Balcombe climbed into the pool & down the ladder. His light receded under the water and disappeared. After a minute he returned as he was at the end of the ladder. Another ladder was tied on & after some

difficulty with his goggles he went down again. He came up & reported the pool went down again about 20 ft, then on a ledge & then down again.

This chance meeting with Graham Balcombe in action led Don Coase to take up cave diving at the first opportunity. June 1945 in Yorkshire had been 'a perfect holiday'.

It is said that opportunities are only worth to a man exactly what his antecedents have enabled him to make of them. So, too, Graham Balcombe, Don Coase, and other would-be cave divers for years to come, once again turned to Sir Robert Davis. He is the third thread running throughout the early years of this story. If wars can be thought of as being at all fruitful, then Siebe, Gorman's wartime drive to manufacture breathing apparatus provided a bumper harvest. What would young cavers in the austerity years have done without the spoils of the old War Department. Ex-W.D. surpluses were a boon in these times. Those who, in better days, argue that these held back the development of good gear made especially for caving forget that cavers simply wanted to make up for the halt in exploration during the war years. Anyway, kit had to be cheap! Experienced cave divers such as Graham Balcombe and Jack Sheppard had both the ideas and the skills to adapt equipment for work underground.

Full details of his equipment and recommended training for cave divers were written up by Graham Balcombe and eventually published in *British Caving*. This influential book edited by Cecil Cullingford for the old Cave Research Group of Great Britain first appeared in 1953 and is the most authoritative account of how cave diving developed after the War. Further illustrations of the many ex-W.D. components that were put to good use appear in the *Pictorial History of Swildon's Hole* edited by Phil Davies for the Wessex Cave Club in 1975. But these accounts are backward looks and the different purpose here is to share some of the action as it appeared to those involved at the time. From the scarce, records that have survived, the 'Operation Orders' tell us most. Unlike the *Letters to Members*, and the *Reports* of the Cave Diving Group, the 'Op. Orders' reveal the hopes rather than the bare accomplishments, notable though these were. All the operations and publications of the CDG were master-minded by Graham Balcombe who had come to live in London. Number 6 Temple Gardens, London N.W.11, was the HQ and Balcombe, naturally, was 'Acting Honorary Secretary and Treasurer'. His energy was prodigious and his paperwork prolific. And we must not forget that he was turning 40 years old when it all re-started at Wookey Hole.

Cave divers returned to Wookey Hole Caves over the Whitsun weekend in 1946. 'Operation Bung' was arranged at short notice and began on 9th June. Another three years of exploration followed and then, on the very day that further progress was halted for want of breathing apparatus to go below 30 feet deep, the first cave diving tragedy struck. Who would have thought that one of the country's most experienced wartime Royal Marines frogmen would be the first to die in a diving accident underground. It is ironic that this operation on 9th April 1949 should have been called 'Innominate': the unnamed had seen the most unwanted happen. The event had a profound effect upon Graham Balcombe in particular as we shall see.

Over two dozen operations were mounted during 1946-49 and a great deal achieved in the circumstances. The efforts of the second generation of Wookey Hole Divers should not be belittled by the ease with which tourists now see almost all of what was discovered. Rather, the trip to the Ninth Chamber, opened by tunnelling in 1975, is a privilege made possible by the likes of Don Coase and Graham Balcombe. Their discovery of the Ninth Chamber was their greatest prize and moment.

Not only was the River Axe thoroughly explored and surveyed from the rising or resurgence outside to the deep lip in the submerged Eleventh Chamber, but the new equipment was mastered by many. There were important archaeological finds to record and such work cannot be rushed. Busy Summer holidays had to be avoided and, because tourist parties visiting the caves in those days went as far as the Third Chamber and returned the same way, the choice of diving sites was limited to avoid congestion and too much distraction. The owner, Wing Commander Gerard Hodgkinson, insisted on each dive being carefully managed on the day. No diver will admit to smoking there 'after hours', and those who wanted a fag learnt to dive into an air bell beyond for the rule was: 'No smoking except in trapped chambers'! More seriously, too much sediment stirred up at the wrong times would spoil the expected view of the underwater lights in the classic First Chamber that appeared on all the posters designed to attract tourists to the Great Cave. Mud would also get into the working paper mills downstream and there was a legal injunction to deter this. So, in his third *Letter to Members* of 2nd September 1946 Balcombe warns:

> Work at Wookey Hole must avoid infringing the paper mill's right to clean water. This appears to involve working only at those week-ends when the canal is normally cleaned or, if necessary, could be cleaned by special arrangement. We would also cause inconvenience to other water-users unless we restrict operations to normal canal-cleaning week-ends. We propose to do this and therefore will not be any burden on water-users for at such times their water will be cut off in any case. A further restriction is that the water flowing must be sufficient to flush out the cave within a reasonable period. Thus we are normally limited to one week-end in four during wet periods.

Diving into the rising was rarely undertaken because of the sediment hazard. Also, boulders could be seen guarding the obvious route under the archway and these would need 'gardening' somehow. When the canal was constructed back in 1852, the river bed here had been carefully puddled with clay to stop leakages, then covered with boulders to slow the current and create a stilling pool or mud trap. Like most naval divers and salvage contractors of the time, cave divers were weighted to walk along the bottom and this quickly disturbed settled muds into suspension. So, the first dives took place upstream in the lake of the First Chamber on 9th June 1946. The 'excuse' given to mount the operation seems blatant but was accepted by Hodgkinson or, more likely, simply tolerated. Balcombe wrote afterwards:

> The nominal object was to recover the bung of the barrel which rolled down the river during the work by H.E. Balch and party in the vicinity of Charon's Chamber. The real purpose was practice for both diver and attendants, and simultaneously such exploration of the lower submerged passages as might be practicable.

Now Gerard Hodgkinson was not someone so easily fooled, for he had been around a bit to say the least; a cavalry officer in the Boer War, an Army Captain in the First World War, and a Wing Commander in the R.A.F. throughout the Second World War. He was highly decorated for valour in action, and continued his passion for horse riding as Master of the Mendip Hunt. The 'Wing Co' was used to command and getting things done. It is certain that he, too, saw the potential of new discoveries for the business. He even put a wooden hut in the lower car park at the disposal of the divers. They called it Crooks' Rest for the pre-War 'gang' had returned as post-War 'crooks'; begging, borrowing and making-do in order to get along. Often the Hodgkinsons' chauffeur ferried both men and gear from the nearby railway stations at Wells and Haybridge. The 'Wing Co' and his last wife Olive had an eye for winners. He was pleased to accept Honorary Membership of the CDG along with Sir Robert Davis and Professor J. B. S. Haldane. Maybe, too, the spirit of the divers reminded him of the time when, as a young flier, he had flown his biplane through Cheddar railway station!

Two more stories about the 'Wing Co' seem appropriate. Although he had fallen out with the equally strong-willed Herbert Balch over the housing of Wookey Hole's archaeological relics at Wells Museum, apparently he did not object to the old man starting the Badger Hole dig on his doorstep in 1938 on land belonging to Captain Guy Hodgkinson, who lived at Glencot, for such was his influence in the village and the family. Captain Guy managed the old paper mill and so could have objected to any caving activities that might make the river water dirty. One day a young local caver paid a visit to the caves in the manner described earlier by Don Coase, but was caught in the act, hauled up before the 'Wing Co' and treated to the necessary reprimand. Again, however, the 'Wing Co's good nature prevailed and he even allowed the culprit to arrange an official visit to the upper levels of Wookey Hole Caves. Could it be that these passages led to the way on beyond the Great Cave? It was even possible that Balch and his young helpers would find another way to the unknown caves beyond through Badger Hole? Just think of the attractions that cavers could make available if given the chance.

*Fig. 4.15
Wing Commander Gerard Hodgkinson, Master of the Mendip Hunt, outside the Great Cave, leads the hounds towards the old Gallopers' Restaurant to join the popular Annual Meet held at Wookey Hole, 1948. Photo given by Olive Hodgkinson to Jim Hanwell*

There were seven 'Bung' operations strung out from June 1946 to October 1948 and coded B1 to B7. As they were confined to the river from the First Chamber, or Witch's Kitchen, to the rising outside, this stretch of the cave became known as B-Reach. After putting on a show for the BBC and making a pretty poor sound recording on 'Operation Scratch' in mid-summer 1946, more serious work began. During the following year floods shifted some sediment and it was hoped that the throat of the resurgence might have cleared naturally. Don Coase was granted leave to investigate on 15th November 1947 but considered it too dangerous to pass, and so B2 was inconclusive. He found problems in buoyancy because of minimizing his weights to reduce the effects of trampling the floor and did not fancy getting his back snagged on the roof whilst squeezing beneath. This dilemma is something that cave divers must come to terms with. It was not until B6 on 25th September 1948 that Coase finally dived from outside to arrive safely in the First Chamber. On B7 the 'Bung Ops' wound up when someone found the remains of Balch's barrel; it was 2nd October 1948. Whilst they did not find the bung, at least they had retrieved its stave and honour was satisfied. Who kept the bung stave as a souvenir will probably never be known!

The middle three B-Reach operations in the spring of 1948 involved survey work and recording important archaeological finds that the winter floods had turned up on the bed of the Lake in the First Chamber. Underwater cave archaeology was to become another 'first' for Wookey Hole divers.

The probable source of the finds lay upstream, for the stretch of submerged river from the First to the Third Chamber was even richer in remains. Imagine the fanciful ideas of the more credulous visitors when it became known that divers had discovered human remains, including skulls, in the Witch's Scullery between her Kitchen (First Chamber) and Parlour (Third Chamber)! As the searches through the mud became 'Operation Muckment', the submerged river from the First Chamber to the Third Chamber was called the M-Reach. It is to the credit of the early divers that the prospect of new exploration did not overcome their respect for the Great Cave's archaeological importance. In fact, Graham Balcombe's first dives over Whitsun 1946 began as 'Operation Prehistory'; a task coded as 'PH'. But the nature and significance of his finds must have been unexpected and highly rewarding. Just upstream of the First Chamber in a side bay to the submerged Scullery, he found three skulls and other human bones with a fine Romano-British food bowl. They were partially buried and so the need for systematic 'muck-raking' everywhere became evident. As the wealth of finds grew during the 'Muckment Ops', the entire M-Reach and B-Reach stretches of the river were scheduled as the archaeologist's preserve. However, the honour of supporting Graham Balcombe on the first through trips fell to Don Coase. Together they dived from the First to the Second Chamber on 5th July 1947, and then right through to the Third Chamber that autumn on 4th October.

Edmund J. Mason was appointed to be the archaeological adviser to the Somerset Section of the CDG. Ted Mason organised and led the underwater searches, producing detailed plans to locate every find and compiling lists of each object for CDG publications at the time. Many of the objects were independently identified by Romano-British experts at the National Museum of Wales, Cardiff, and at the City Museum in Bristol. His 'Report on Human Remains and Material Recovered from the River Axe in the Great Cave of Wookey Hole during Diving Operations from October 1947 to January 1949' was published by the Somerset Archaeological and Natural History Society in 1951 (Volume 96 of their *Proceedings*).

Mason, who was once gently chivied by Graham Balcombe for his 'tardy pen', in fact did more than most to popularize the long history of the Great Cave and let the general public know what cave diving was all about. He up-dated the old guide books written by A an Bell (1928) and L.B. Thornycroft (1948), who had barely time to squeeze cave diving in as an appendix, and the first of three delightful editions of *The Story of Wookey Hole* by E. J. Mason appeared in 1963. Ted Mason was also a co-author of perhaps one of the best sellers on the area: a book simply called *The Mendips* by A.W. Coysh, E. J. Mason and V. Waite (1954). He tells the 'inside story' of the archaeological 'ops' in the next chapter. The 'muck-raking' was over by the end of 1948 and a pump was used to try and unearth more deeply buried relics the following January. But tests for 'Operation Sandblast', coded 'SB', were more hilarious than effective as Ted Mason describes later[9]. Happily, his suggestions that the cave had been used as a cemetery were sounder as proven twenty-five years after when, with the canal out of action and the water level lower, Professor E.K. Tratman and his diggers found the site in the Fourth Chamber during 1973-76. Their account of the 'Romano-British Cemetery in the Fourth Chamber of Wookey Hole Cave, Somerset' for the University of Bristol Spelæological Society (1978) is a vindication of how it all began with the divers over thirty years before.

Now cave passages can go in any direction and it is sometimes difficult to decide exactly where the way on may be. In flooded passages, water even flows uphill because of the head of pressure upstream. And in Wookey Hole Caves the River Axe wells-up from lower levels under Mendip. Underwater, twists and turns with ups and downs in poor visibility present divers with buoyancy problems. Going upstream into clearer water could help the outward journey but returning through one's own mud cloud would be dangerous. Guide lines are essential in cave diving, even if definitely frowned upon in ordinary caving. But getting entangled with underwater lines could be lethal and so, all in all, it seemed sensible to be weighted enough to keep one's feet well and truly on the bottom. Alternatively, if no weights were used on the feet and the only ones were on a waist belt, the buoyant breathing bag like a neck collar pulled the head up a bit which enabled the diver to glide along the bottom, lightly crawling and partly swimming. Remember Don Coase's uncertainty over disturbing the bottom and threading his way around boulders when first entering the rising; and it

---

[9] Ted Mason died in 1993.

is open to daylight. On fresh exploration you must take 'everything in' and to 'look both ways' is a wise maxim. So it was that the new river bottom beyond 'The Gateway to the Seventh Chamber' found in 1935 became the Janus route (Janus was the Roman god of gateways, usually depicted with two faces so that he looks both forwards and backwards). Later, the whole section from the Third to the Ninth Chamber was referred to as 'J-Reach'. The more exciting 'Janus Ops' were coded simply 'J' in Graham Balcombe's operation orders. There were seven such pushes ending with the twice postponed J7 on 15th January 1949, the year when luck ran out.

It all started on 12th October 1946. The detailed operation orders circulated in advance to all taking part show that Gordon Ingram-Marriott, the an ex-Royal Marines frogman, had been invited to see whether or not swimming with fins had advantages over bottom-walking. This idea was Wing Commander Hodgkinson's own brainwave for Marriott was then employed by Butlin's as an exhibition diver, and any contacts with the thriving holiday camps would be good for business at Wookey Hole! Billy Butlin himself was well known to the 'Wing Co' and a fellow spirit with big expansion plans at Minehead in West Somerset. The cave divers did not share such enthusiasm for the venture, however, since it broke their rule that Cave Diving Group members should be established cavers first. But it was early days and they had to go along with the 'Wing Co'. Balcombe's plans for the ambitious 'Operation Flippers Magnum' tell their own story. It is worth highlighting that the 'controller of operations' was Denis Hasell; a grand Somerset character – always called 'Dan' by cavers, although even he does not know why. Dan Hasell became pillar and President of the Cave Diving Group[10]. Dan was an early member the Bristol Exploration Club – only Harry Stanbury, one of the divers, could pull rank on Dan for Harry founded the BEC back in 1935[11].

But the best laid plans sometimes do not work out. Marriott was unable to get along on the day and the show was curtailed as 'Operation Magnum'. One can wonder for evermore how different the history of early cave diving might have been, to say nothing of Gordon Ingram-Marriott's life, had this ill-fated diver been able to demonstrate the superiority of swimming with fins on 12th October 1946! Rather, let Graham Balcombe tell the story of the first advance along the Janus Reach as he wrote it as a possible press release after J1 on 22nd February 1947. But this is its first appearance in print:

ANOTHER HUNDRED FEET INTO WOOKEY HOLE

During the weekend in Wookey Hole Cave, Somerset, a party of amateur divers made another attempt to reach an inner series of caverns cut off by submerged passages. One of the divers succeeded in reaching a point a hundred feet further upriver than on their last attempt four months ago.

Slowly and resisting every inch of the way, the Great Cave of Wookey Hole is giving up its secrets. In 1935 my party of divers starting from a base near the far end of the known system of caverns succeeded in getting through to two new chambers, the Sixth and Seventh Chambers of the cave. The Sixth was entirely submerged, but the Seventh is a lofty rift standing in some ten to fifteen feet of water. The distance gained, however, was quite small, probably less than 100 feet of hitherto unknown ground. During 1946 we returned to the attack, and in October reached a point some 30 feet farther on, not in the Seventh Chamber, but by diving steeply under its left hand wall. It was a short advance, but a significant one, for in the first place we had overcome our worst fears of this aggressive-looking, boulder-strewn place, and in the second we had located the way up-river into the core of the hill.

On Saturday night we tried again, spurred on by this achievement and the hope of reaching the above-water series of caverns which must be farther on. We arrived at the cave mouth at seven o'clock and by the light of our torches carried in the diving kit to our dressing base in the Third Chamber, about 500 feet from the entrance and a hundred feet or more underground.

Two divers were dressed and when ready went on to the next chamber, the Fourth, which is the last shallow water chamber. The water flowing out from the deeper inner chamber was beautifully clear, white silvery

---

[10] He held this post until his death on 24 October 1996.   [11] Harry Stanbury died in 2006, aged 90.

Fig. 4.16
James Gordon Ingram-Marriott, first invited to demonstrate the use of fins for cave diving on 12 Oct 1946 on Operation Flippers Magnum.
Photo from Bob Davies album in Wells & Mendip Museum

ripples dancing in our light beams against the deep black water in the shadow, and it took a practised eye to see that even so it was not quite as clear as it was last October.

After final adjustments to his equipment, the first diver [Jack Sheppard] went down and we were treated to the grand spectacle – for cave explorers at least – of seeing him submerge and, as he crawled down the slope in the Fifth Chamber, become a flattened and distorted figure silhouetted by his own lights, his outline wobbling with the ripples he left behind him. He looked like some elongated jelly fish, disappearing into the green. He seemed to move horizontally, whereas we know he was going down a very steep slope. He grew smaller and his lights fainter until at last he disappeared from sight under the far wall of the chamber and only the glow of his lights still showed pale green below the undercut wall. Then that too disappeared and we settled down uneasily to wait.

His job was to run out a tracer wire to the Gateway to the Seventh Chamber and to leave his line-laying apparatus there for me to take forward.

At last we saw his light appear, not the blaze of light in which he went down, but just a pin-point for he was using his reserve hand-torch. He was soon at the surface and it was my turn to go down. Slowly I sank into the water, into the now familiar green under-water world; the bubbles gurgled from my rubber helmet and then stopped as the dress emptied. Down the slope I sank, the breathing apparatus on my back scraping on a low portcullis of rock as I passed over some fallen blocks, to the First Deep, which is 15 feet below surface level.

The water pressed me tight and my ears felt the unaccustomed pressure as I passed under the low archway from the Fifth to the Sixth Chamber, then it eased again as I climbed up the slope to the Gateway to the Seventh. At the Gateway, I stood by the rock Scylla, and looked down the Seventh where a huge fallen block, Charybdis, blocked the route; then down the deep green hole to the left where the river current flows. I debated a moment which route to take, for this was our Operation Janus and we had to look both ways. I chose the deep green hole and jumped, taking the line gear with me.

In this way I floated down the first three steps of the Giant's Staircase, to reach a low passage leading a little to the right, then down again where the meter read 18 feet depth of water. At each turning of the way I anchored the tracer wire to a spike or block of rock; the way on seemed endless, huge mud walls then sheer rock, low roof then a great height of water above me, I could not be sure whether there was air above.

I suppose I could have sent up some bubbles to disturb the surface if there was one, or to cling like silvery stars to the roof if there were not, but I was beginning to realise the awful loneliness of this place, so far from safety, no possible escape if anything went wrong. At last at a point 210 feet from the start, but seeming like a mile, I halted. I had turned a corner and ahead looming through the mirky green water was a mighty barricade of tumbled pinnacles and blocks; on the right a steep mud slope leading up into the gloom which my light could not penetrate.

I tied off the line, switched off the main lights on the navigating gear, left it there for the next man to take or bring back and turned back to the base, using my reserve light for the return journey.

I had covered perhaps only 30 or 40 feet when without warning my light went out and I was plunged into utter darkness. I fumbled with the connections but without avail. I was completely without light, deep down in a rugged and dangerous underground watercourse. I summed up the possiblities, and decided to continue back to base, partly urged by fear of the up-river, partly by an urge to make this grim journey in the darkest of all dark, amid boulders and treacherous mud slopes, trusting to the slender wire to guide me back to safety. Inch by inch, foot by foot I followed it back, tracing each turn round the belay points, never letting go the one side before being absolutely sure I had hold of the other, for without that wire I was lost, to miss it by an inch would be to lose it altogether. I wondered how many belays there were before the Giant's Staircase, and whether I would be able to climb that Staircase without a light.

It was a trying journey. I comforted myself that I had been liberal in belaying the wire and there was not really any chance of it pulling across a corner and leading me into some impossible place as once before a line had done. I went on and on recognising nothing until suddenly my hand touched a smooth ring. That was no natural belay, it was the weight we had used to belay at the Gateway to the Seventh. I had passed up the staircase without knowing it and knew now that I was comparatively safe. My heart beat rapidly with excitement, but I could not lose caution; [I] slowly slid down the slope to the First Deep where I passed the last belay and saw the first glimmer of distant light. Relieved, but not entirely out of trouble, I groped my way through the tumbled blocks until I broke surface, startling the base party who had no warning of my coming except the twitching of the wire and were still anxiously peering down to catch the first glimpse of my light.

We recovered the navigating gear and packed up for the night, disappointed for we had hoped to reach the promised land, and now felt that for ever it would be beyond our reach. On further thought it struck us that a frequent cause of mud accumulation like those I had seen is a fall of a free-running river into a spacious submerged channel, like the detritus at a river mouth. Maybe our promised land is at the top of that mud slope after all. Hopes again run high, and we are preparing for the next attack, when among other things our lighting system will be better safeguarded.

Later investigations showed that Graham Balcombe had certainly entered the Eighth Chamber on the push of 22nd February 1947, and there was a distinct probability that he had been on the verge of entering a ninth chamber. He could not be sure, however, as no surfaces overhead had been broken whilst he had been pinned to the bottom. Getting bearings on line directions and surveying beyond the Seventh Chamber kept everyone busy until Balcombe's claims to have reached the Eighth were definitely confirmed by J2 on 15th November and J3 on 6th December. His sketch-plans made before and after J3 show the gains that had been made during 1947 and the nagging uncertainties still to be resolved.

Plans were made for checking actual surfaces at the start of the new year. These 'Surfacing Ops' coded 'S' would stake out possible above-water 'landings' or standing ledges as the need for an advanced base was under consideration. If needs be, they would break surface by either jumping or using fins. Note the latter. Inflatable dinghies would help, too, and balloons would be released to measure the heights of chambers. All the divers took fair shares in this support work, although Harry Stanbury and Bill Weaver did most in the event. Graham Balcombe and Don Coase were the forward men sorting out the Janus route in the main. S1 was completed successfully in the Sixth Chamber on 7th February 1948 and S2 saw an inspection of the Seventh early in March. After a cancellation owing to poor visibility, Balcombe and Coase reconnoitred the way to the supposed Ninth Chamber on J4, then carefully familiarised themselves with the whole route from base on J5 on 10th April. All was ready for the big forward push of 24th April. With luck, J6 would also be 'Avanti' or A1. It was!

Optimism was running high before the push. It would be the last chance for regular meetings at Wookey Hole until after the closed Summer season. Every loose end had to be tied up, and everyone involved would be kept busy. Graham Balcombe's notice of operations was brief but the programme packed. Its code 'J6S4A1F1' covered everything; F1 being 'Footrule', the record of line directions and distances on the push. The notice concludes:

Fig. 4.17
A 1947 survey of the chambers and sumps with provisional pencilled outlines of the Eighth and Ninth chambers added later by Balcombe and Coase. From Cave Diving Group records

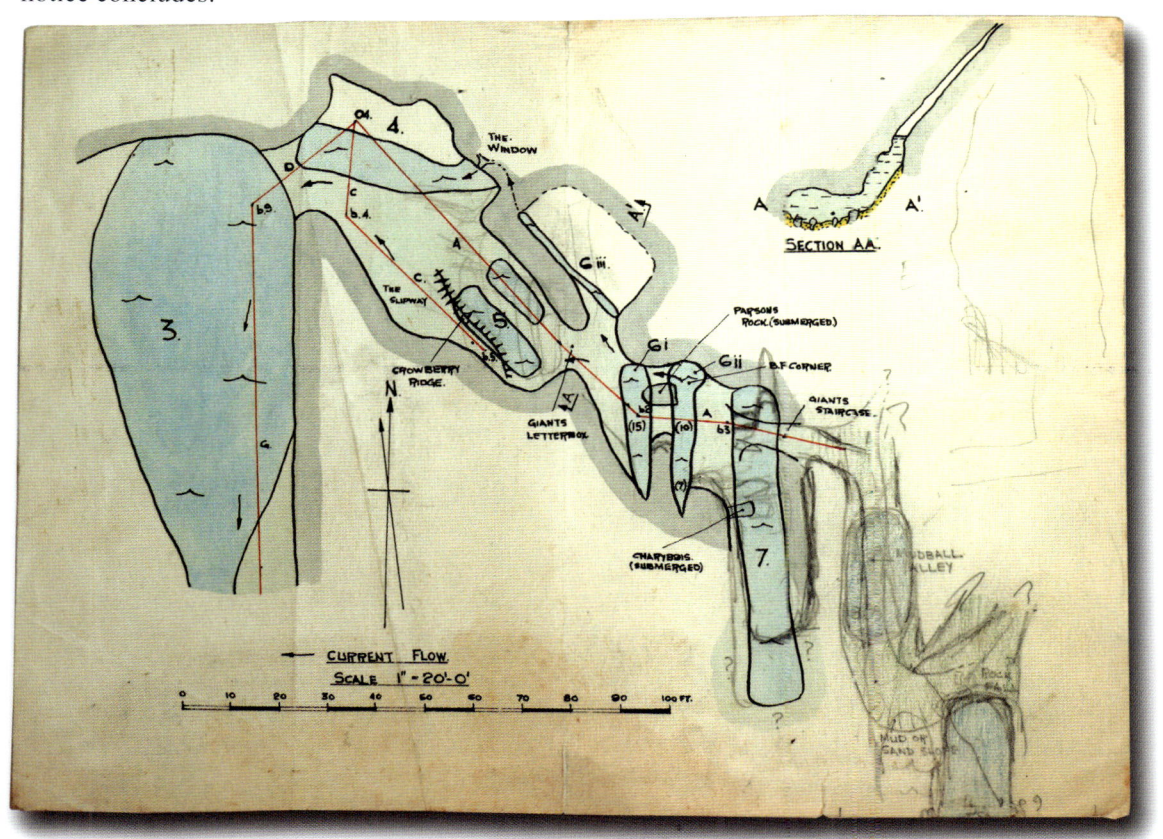

Except in case of emergency, attendants and other non-diving members will not leave the hards. No smoking, except in trapped chambers.

So, tough luck if you wanted a smoke and did not dive!

Once again, Graham Balcombe and Don Coase led the push. They walked to a record depth of 26 feet beneath the Eighth Chamber, then started the underwater climb to the promised land. Let Balcombe's own notes capture the moment:

Rising up at the end of the Eighth there was the mud bank to our front and right and it too led to a surface. The smooth, rounded and polished mud looked treacherous but excitement rather than caution ruled us. I signalled to Coase, 'Up!' and he promptly agreed. I fumbled with the line while tying on and at last had a belay serviceable, but doing no-one credit, and off we went, Don first ... I soon found the mud not too horrible to climb direct and made towards the larger part of the surface and a good landing. Don was heading for some nasty rocks but as we rose so did our excitement. It was a matter of moments now. Recklessly, I grabbed Don by his breathing bag and yanked him off the wall and we stood up with our heads above water.

In front was a rampart of muddy pinnacles rising ten to fifteen feet fom the water's edge; what was beyond we could only guess but the roof soared up to a height and away into the darkness.

After an awkward climb over the pile of boulders in their diving gear, they reached a small platform like a beach:

Before us at slightly lower level stretched a long expanse of sand, dipping gently to the east. Its surface was pitted by many craters made by the drip from the roof, most about 4 inches deep. Beyond was the far boundary wall, rising slowly at first then steepening and shooting up to the roof, leaving one or two deeper ways on only to peter out at quite short distances. Left and right, the walls ran sheer up to the roof and up there were two huge wedged blocks, or part of the roof, waiting somewhat sinisterly I thought. The sandbank at the top of the sloping floor on the right finished at a narrow raised shelf against the wall. At the lower end it finished at a beautiful green well ...

*Fig. 4.18
Following the discovery of the Ninth Chamber, Don Coase sends triumphant greetings to fellow cavers for Christmas 1948. Photo from Bob Davies album in Wells & Mendip Museum*

Visitors to Wookey Hole can now see and admire this vaulted chamber for themselves, of course, and I hope that they wonder about its discovery. When most cave explorers set eyes on beautiful caverns never seen before, there is invariably a speechless moment among all concerned as each collects his own thoughts. So it will be here save to invite the reader to visit the Ninth Chamber of the Great Cave, to stand, stare and try to imagine what Don Coase and Graham Balcombe thought as they climbed out of the water into this place. And what if, like Balcombe, it had taken you thirteen years to get there!

Needless to say, there was a specially arranged return trip for the support divers to see the new chamber and to get photographs. This was approved by the 'Wing Co' and took place on 19th June 1948. Graham Balcombe's time-table for this may be best summed up by his promise that:

There will be a long interval while divers are in Nine.

The controller and attendants were invited to retire if they wished. The divers had a lot to 'take in'.

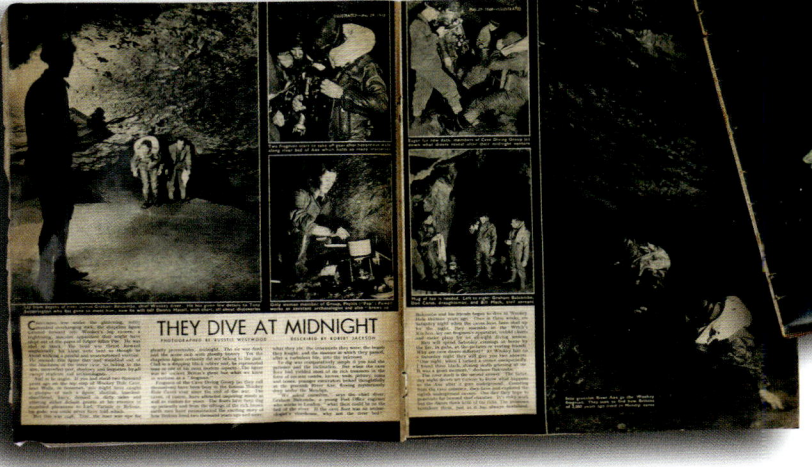

Fig 4.19
Nationwide publicity for the new discoveries appears in the popular weekly magazine Illustrated on 29 May 1948, in an article by Robert Jackson entitled 'They Dive at Midnight' with photographs taken by Russell Westwood. From CDG records collected by Ted Mason

And then there was the pay-off, for the 'Wing Co' thought that the whole nation should get a glimpse of the action going on in his cave. So, Pathé Pictures came to make the first commercial film of cave diving there. Everyone duly performed, called it 'Operation Nitrate' and some may even have seen the clip released in London on 20th September 1948 as it was part of 'Pathé News'. With that unique urgency conveyed by the commentaries on such newsreels, the country got to know more about the divers' discoveries at Wookey Hole. In those days, there were always long queues for 'the flicks' and cinemas were packed.

The next step beyond the Ninth Chamber would be a big one and needed much more back-up: a forward base was essential, a telephone line for communications was vital, and waterproof bags would be required to carry all sorts of supplies to an advanced base in Nine itself. 'Operation Rearguard', or 'RG' succeeded in erecting a platform above the water in the Sixth Chamber with a ladder down to the river bed. RG4 saw its completion on 15th January 1949 after a hard winter of ferrying parts through. Some divers used the platform for a smoke, others to have a nap or, according to the hard men, it was the 'funk hole'! The telephone

Fig. 4.20
The unmistakable profile of Dennis 'Dan' Hasell (wearing shorts), CDG Controller on the night of the publicity event illustrated in Fig. 4.19. Graham Balcombe (left) and Tony Setterington stoop into the Third Chamber.
Both 'Dan' and 'Sett' (top of opposite page) were BEC members and stalwarts of CDG operations at Wookey Hole throughout the post-War period

was then connected to the platform and through to the Ninth Chamber in a series of six dives called 'Operation Linlay', or simply 'L', and finished by 9th April. Most of these dives were combined with sherpa-ing supplies through to Nine as part of 'Operation Stockpile'. The 'SP-Ops' attracted some ingenious designs for waterproof carriers, such as the one by George Lucy, yet another BEC diver. Lucy's 'compensated oxygen waterproof frogsack' was called 'Copwac', whilst the Balcombe-Coase version took a more graphic acronymn, 'Cowsack'. Another Balcombe-Coase model, the 'trapped air compensated waterproof apparatus carrier', was called 'Tacwack' or, more irreverently, 'tick-tack' or 'tick-tock'. And whatever the Admiralty could do for codenaming equipment, the CDG would go one better with names for suits such as Sefus-Sesal and Shawads-Sesal. It was the era of the cipher and acrostics, after all, for such had been the language of the War.

It was during time off from such chores that Graham Balcombe went around a short loop to the downstream end of the Ninth Chamber on 15th January 1949 to establish the existence of submerged ox-bows or 'loopways'. This became Loop One, or LP1. Following this up on 19th March, Don Coase then found a longer loop that linked the first entry point of the divers to the Ninth, called Nine-One, with the larger upstream bay and beach in Nine-Two. This useful bypass to the awkward boulder scramble over the Pinnacles became LP2 or, later, Coase's Loop. Nine-Two is now the farthest part of the show cave and tourists can look down from a cat-walk onto the beach-head from which all explorations since 1949 have started. The tunnel from the Third Chamber to the Ninth Chamber was driven in 1974-75 and bridges give splendid views of the river that the divers followed to the Sixth, Seventh, Eighth and Ninth chambers. It also takes in a line of natural high-level links between each chamber that were found much later by John Parker. His story must not be scooped here, however, save to note that the upper route well above the river that Herbert Balch had predicted in 1930 is actually there! Colin Priddle was the first caver to get from the Third Chamber to the Ninth entirely by free-diving, climbing and roof-traversing on the 3rd November 1973.

Fig. 4.21
Tony 'Sett' Setterington of the BEC who assisted on most diving operations at Wookey Hole in the late 1940s.
Photo by Luke Devenish

Coase's Loop, in fact, re-enters the Ninth Chamber from its upstream end and includes an offset chamber with a low roof and no landing which was numbered the Tenth. After he had found this on 19th March, he returned with Graham Balcombe, and both set off further upstream on 'Avanti Two' or A2. The passage began to slope downwards against the upwelling current, and after a short distance approached the critical 30 foot depth limit for oxygen. Suddenly they came to the lip of a much deeper and larger place, entirely submerged but deserving to be called the Eleventh Chamber. It was as far as Graham Balcombe was destined to go in Wookey Hole Caves, so let the description be his:

A2 (19.3.49) This op. followed on LP2; with little delay the divers got off the mark and after a false start caused by mud which rose up while the leading diver (Coase) tried to surface in Ten by ascending a knife-edged arete, they went off upstream. After the blackness of the slope from Nine-Two (as black as the Bear Pit, and for the same reason) and the browner slope through Ten we suddenly burst into clean water and golden sand underfoot. The transformation was magic, the scenery became strikingly mountainous rather than cave-like, with greeness of the water giving an illusion of grass or forest and the scattered light from our Aflos giving the effect of daylight. We passed a huge block with an eyehole in it, and went onwards slightly downhill still on lovely sand until Coase halted abruptly on what looked like the edge of a precipice with the green water going down far below. It was like looking down into Ebbor Gorge from the top of the flanking slope, even the size of the place seemed right on that first glimpse. Although we even now are not able to make true allowance for the illusions of a new under-water scene, we know that some adjustment always follows. We were in the limestone. Clearly we had run through the side wall of the ancient gorge of Wookey and must have been not far from the line of the present cliff face heading the valley. In front of us was an archway at our level, 25 ft. deep, way down below it was a dark hole to the SE where presumably the Axe comes in. We returned to the mundane mudslopes of Nine-Two. This Eleventh Chamber is too deep for our present techniques with oxygen equipment. We withdrew to make new plans.

This account is vintage Balcombe from his *Letter to Members* of the Cave Diving Group, No. 13, of 31st March 1949. The advance had taken them through the thick mask of conglomerate that flanks Mendip's slopes; a warm, red rock that adds unusual colour to the Great Cave. They were now into the massive grey limestone that forms most of Mendip, and whose sloping beds can be seen on the cliffs of both Cheddar and Ebbor gorges. From now on, the subterranean River Axe would plunge down much deeper, then up again to more caverns beyond. During floods, vast quantities of fine mud are flushed through these deep sumps and the mud banks shift a lot. For instance, the vertical cliff described so graphically by Balcombe was covered up by a thick mud slope when Bob Davies subsequently visited the Eleventh Chamber. He did not recognise it from the description given and went down to a depth of 50 feet, unaware that he had reached the limestone. But Graham Balcombe had seen the rock, and at last it was 'ITTL' (In To The Limestone!).

Fig. 4.22 Preparations outside Crooks' Rest for Operation Innominate, 9 April 1949. Left to right: 'Half-pint' John Dwyer, 'Digger' Harris, Ray Nunwick, Graham Balcombe, Sybil Bowden-Lyle, Tony 'Sett' Setterington and the rear of Luke Devenish's ex-Army Jeep. Photo from Luke Devenish

The first week of April 1949 saw preparations for the next and uncertain push. 'Operation Innominate', therefore, would be 'an inspection of Eleven, especially the cliff face for the best drop'; divers would 'swim out singly over the deep water', but 'must not descend below the 30-foot level.' The three divers would be Graham Balcombe, Don Coase and Robert E. Davies. Bob Davies, often referred to as 'RED', was an experienced Peak Cavern man from Derbyshire though a relative newcomer to Wookey Hole. It was noted that P-Party equipment which could be converted to an oxygen-nitrogen mixture would have to be used for the eventual descent of Eleven below 30 feet. Bottom walking divers in support at the beach-head in Nine would be Tom Grosvenor and George Lucy. Gordon Ingram-Marriott was invited to join them again, for although he was not a caver, he had recently visited Peak Cavern and now knew what to expect at Wookey Hole. The wide publicity after the discovery of Nine had renewed the interest of both the Navy and former frogmen such as Marriott who probably needed some excitement from his mundane demonstration dives at Butlins. The 'Wing Co' approved, of course. Moreover, Marriott knew all about 'finning' and 'mixture breathing' and the CDG was anxious to learn more at first hand. Marriott would go with the advance party so that the more experienced CDG members could help him to adjust and observe how to tackle cave diving using swim-fins. But Marriott would not go beyond the Tenth Chamber and would wait with the support divers in Nine.

On the actual day, however, things did not go right from the start. After a bit of to-ing and fro-ing to the platform in the Sixth Chamber, Balcombe withdrew because he could not cope with his unaccustomed fins to an adequate degree of safety and Grosvenor cried off with face mask problems. Marriott had a light failure whilst on the platform, but this was put right by Coase. During this, Coase himself had created 'plenty of splash' and had been 'a very comical sight' to his recorder, Sybil Bowden-Lyle. He also lost a weight whilst sorting out Marriott's lights. Eventually, when they had got their kit and revised plans together, Coase led the way from base followed by Marriott, Davies and Lucy. By now it had become a 'look and see' trip. Whilst Marriott and Lucy surfaced in Nine-One to take the overland route to Nine-Two, Coase showed Davies around his Loop. After making some kit adjustments in Nine-Two, they both dived on to the Eleventh, but found no way on above the 30 foot

Fig. 4.23
The principal divers make last-minute plans in the Third Chamber, 9 April 1949. From the left: James Gordon Ingram-Marriott, Graham Balcombe, Don Coase and Bob Davies.
Photo from Luke Devenish

level and so returned for a rest and chat on the beach-head in the Ninth Chamber. Coase had been discomforted by frontal sinus pains during his dive and his log sheet records slight nose bleeding which got worse on going to Eleven. So, he took the short way back, accompanied by Lucy. Meanwhile, Marriott elected to go around the longer loop with Davies so that he could see Ten before returning. Both divers had mastered the extremely difficult technique of communicating underwater by spitting out their gag behind the face mask, then placing the plate against the head of a colleague and talking. Speech was thus conducted through the bone of the skull and clearly received. Conversations were limited, of course, and the gag had to be relocated by using one's tongue; an awkward manoeuvre that needed skill and practice, to say nothing of confidence in each other's ability. Bob Davies and Don Coase frequently talked like this underwater, for example. The divers had incorporated a flutter valve which enabled water that leaked into the face mask at any time, especially during 'conversations', to be expelled.

Fig. 4.24
Operation Innominate begins in earnest, 9 April 1949. Ingram-Marriott (left) and Davies (right) wear fins ready to dive upstream from the Third Chamber. Dan Hasell, the Controller, is assisted by Nell Davies and Sybil Bowden-Lyle. Photo from Luke Devenish

After the usual equipment checks and breathing drills, Marriott set off with Davies close behind. On reaching the Tenth Chamber, they had a brief underwater chat, then headed for home. By now, the visibility was very bad, but Davies was reassured by the tugs on the line ahead that Marriott was making good progress. When they stopped he concluded that his companion had made it to the next belay and, closely following the guide line himself, he arrived back in Three. But Marriott wasn't there!

Despite being low on oxygen reserves Davies dived back immediately to look for Marriott, hoping to find him on the platform, or somewhere safe. Balcombe and Coase, neither fully fit, kitted-up in record time and followed. But it must have felt like an age in the circumstances. Davies and Coase found Gordon Marriott's body just a few feet upstream of the platform, face down, pointing upstream and his kit spent of oxygen. He was hauled quickly to the platform, but attempts to resuscitate him failed. Marriott had drowned and Wookey Hole Caves had claimed the first cave diving tragedy. His missing reserve cylinder was discovered by Bob Davies on 7th May 1949, a fortnight after the inquest.

Fig. 4.25
Marriott steps into the River Axe in the Third Chamber following Coase, whilst Davies waits his turn to dive. The relaxed shore party is unaware of the impending tragedy.
An enlargement of this photograph was made for the subsequent inquest on Marriott's death, as proof that the diver had a reserve oxygen cylinder and gauge suspended from the back of his waist belt.
Photo from Luke Devenish

Fig. 4.26
Davies finds Marriott's missing reserve oxygen cylinder on 7 May 1949, and shows that the gauge records 117 atmospheres of oxygen left.
Photo from Bob Davies

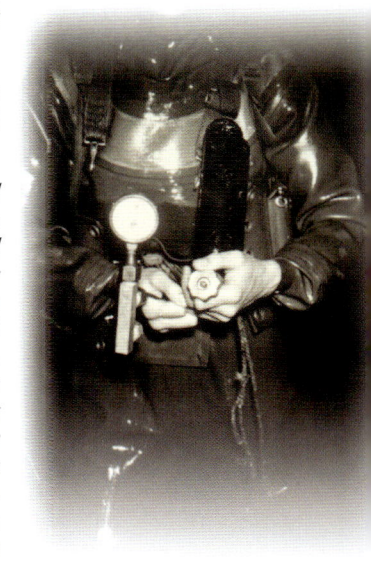

It must have been difficult for all involved to imagine that anything other than equipment failure had caused such a highly trained and experienced Royal Marines diver to lose the line and drown. The inquest thus records:

Death was due to anoxic anoxaemia, accidentally sustained during diving operations when his oxygen supply became exhausted due to a fault in the test pressure gauge.

But Bob Davies's subsequent discovery of the ample reserve of gas in the missing cylinder leads to a different interpretation. To the vexed question of, 'Why did Marriott die?' Davies replies:

Well, first of all it was the flow-meter that was at fault but it certainly played no role in the situation. I swam right behind Marriott to Wookey Nine and he was not blowing off oxygen. At the time of the inquest the missing oxygen cylinder had not been found. He had started with a spare one tied to his waist and no cylinder was there when Don Coase and I found him. Thus at the time of the inquest no one knew whether he had used it or not. Now, I found that cylinder on 7.5.49 and have a Kodachrome slide showing it to contain 117 atmospheres of oxygen. This changes the whole complexion of the case. Marriott knew he had very little left when he did his last breathing drills. Yet he still spent many minutes looking round in Wookey Ten before we started back. As stated earlier his gas supply ran out where later calculations showed it would. The cylinder was under the Eighth Chamber surface. His body was under the Seventh Chamber surface. Dropping the cylinder would make him float. His light was still working hours later. We should have found him swimming on one of the many surfaces if he had lost the wire and if it really was an accident. I personally don't believe that his death was due to 'loss of line'. At best it was gross negligence in not changing the cylinder; at worst it was deliberate. I do not wish to believe the latter because both Don Coase and I went back for him with very little reserve gas and seriously risked our lives for him. Yet his dresser, Nell Davies, said afterwards that he absolutely couldn't care less about anything that night. We may never know the whole truth. Certainly his last words to me were 'I can't see a thing in this fucking mud' as he moved ahead of me towards the Third Chamber and the safety he never reached alive.

Frogman (He wanted new thrill) dies in Cave No 7

AS DIVERS SWIM BY

```
            MENDIP HILLS        BODY FOUND NEAR
                                LADDER TO SAFETY
Entrance                                PLATFORM
to Caves  1   2        5   6   7    8   9   10  11
               3  4
Dam  RIVER AXE
     Runs under   River-bed   BASE FOR      MARRIOTT STARTED
     the Hills                OPERATIONS    BACK FROM HERE
```

From Daily Mail Reporter

WOOKEY HOLE CAVES, Somerset, Sunday.

TWENTY-SEVEN-YEAR-OLD Gordon Marriott, ex-Marine frogman, had spent over 500 hours of his life under water. He helped to clear the Normandy beaches for D-Day. Recently he volunteered for a job merely for the thrill of diving in his frogsuit again.

He joined the diving team exploring the newly discovered 11th cavern in Wookey Hole Caves and yesterday was his first major operation with them.

He went to the ninth cave while two other divers carried on to No 11.

When the order to turn back was given Marriott apparently lost his hold on the guide-wire at the side, and in the seventh cavern the oxygen supply in his frogsuit began to fail.

After a fight for life in the swift-flowing, murky waters of the subterranean River Axe, he died within ten feet of safety.

He was found in 12ft. of water almost at the foot of a ladder which would have taken him to a safety platform above water level.

Marriott acted as supporter to two divers who made the hazardous journey through a succession of caves, diving and swimming from one cavern to the next.

Swam on

When the order to return to base, in the sixth cavern, was given by field telephone, the advance party presumed Marriott was in front of them.

Mud and silt stirred by their inward journey made it impossible for torches to penetrate the water more than a few inches.

The journey had to be made back by touch, keeping a hand continuously on the guide wire, and the two divers passed through the seventh cavern while Marriott, unknown to them, was floundering below.

As Marriott groped for the wire his oxygen supply was failing.

Then, it is believed, he began to crawl by instinct towards the ladder which led to safety.

When Marriott failed to surface at the base in Cavern No. 6 the two other divers turned back and a third followed with oxygen.

Fifteen minutes later their groping fingers found Marriott—dead.

He leaves a widow and a baby son, who live in Exeter.

Fig. 4.27
Marriott's tragic death is graphically reported in the Daily Mail, 11 April 1949.

So the conclusion is that you had better be an experienced and committed caver first before taking to cave diving. Nor was it something to do when preoccupied by great personal stress. And once you have learnt such lessons the hard way, it is best to forget the worst and look ahead. For Graham Balcombe, however, this tragedy had a profound effect. In 1933 he had felt much the same about climbing, and so turned to caving. He returned to Swildon's Hole and made another attempt on Sumps Two and Three with Bob Davies on 27th May 1954. On this dive, Davies led the push and discovered Bob's Bell, but the way on downstream was blocked by impassable mud banks which stirred up and ruined the visibility. The old enemy of being unable to see where to go thwarted them. On another day, the route ahead could just as likely have been open and clearly visible. Graham Balcombe might well have accompanied Bob Davies into the handsome stream passage of Swildon's Four that day. But such are the fortunes of cave exploration. Graham Balcombe had completed his last pushing operation at the very place where cave diving on self-contained respirators began in October 1936. It was time to train the up-and-coming youngsters and put pen to paper for *British Caving*. Their stories could not have been written if Graham Balcombe had not got things going.

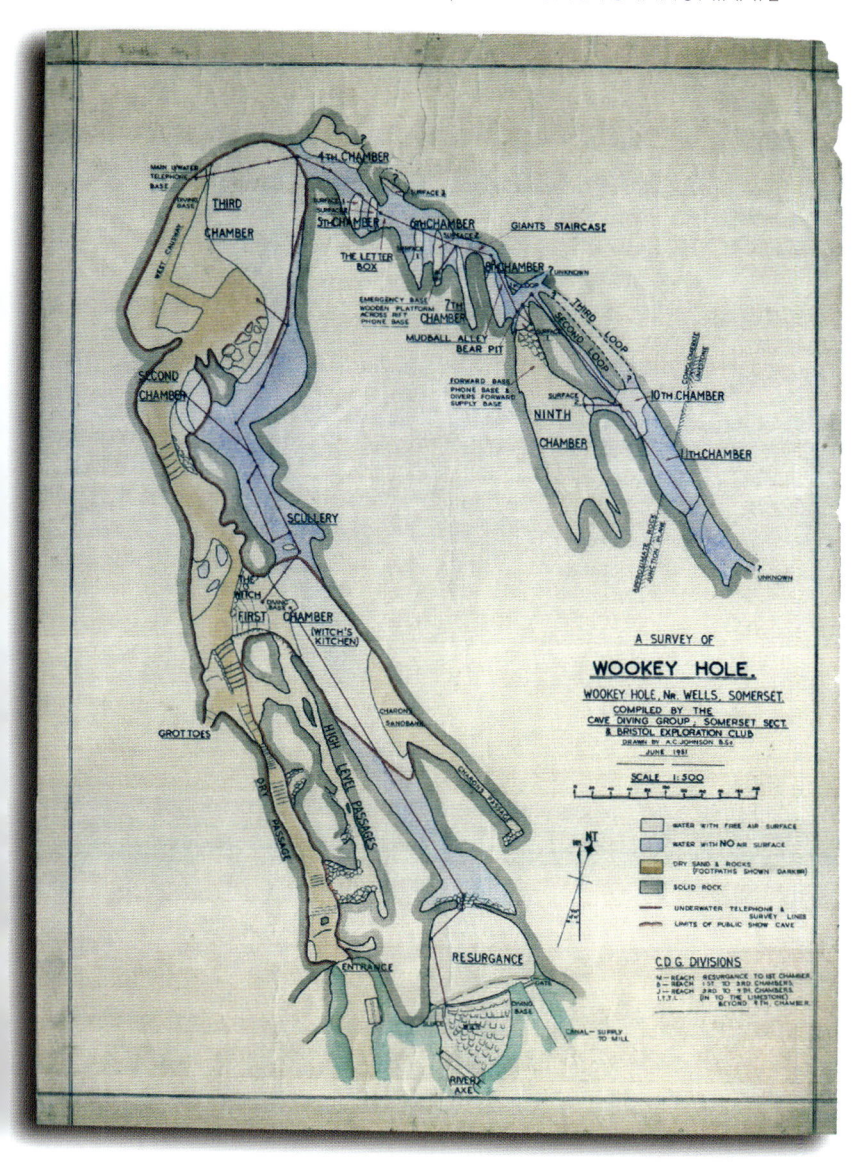

Fig. 4.28
A survey of the known cave, compiled by the CDG Somerset Section and BEC, is published in June 1951.
Drawn by A.C. Johnson

A few restless ones carped a bit at the time and, maybe, carried the 'dictator' jibe beyond its happier origins, for Graham Balcombe stoutly refused to take account of opening hours when time-tabling Wookey 'Ops'. And this of an old hand of the 'Northern Tavern and Hell Club'! He grew tired of running the CDG in 1950 and, in his last set of accounts, he said of Wookey Hole Caves:

> My prospect of re-starting the ITTL-ops. is receding; I have been unable to throw off the effects of the extreme strain of the last op. season and unless another diver is able to take over the lead, the question of liquidation will soon arise. (At the moment of writing, my face is smothered with eczema again – my 'Wookey Legacy' – and my right eye completely hors-de-combat!)

A few recollections given to me by Graham Balcombe when he was eighty one shortly after reading the above will give some inkling of the realities of cave diving that his successors probably face too:

I was ever a rather apprehensive diver, the apprehensions outweighed by the consuming desire to push ahead. Possibly all divers are much the same, but I never saw signs of it in the Coases or the Davies's ... My sights were on upstream Wookey; the lower sections were logical necessities in the course of exploration. Don Coase from his first dip never looked back; he took the moral and physical lead in the B-Reach and chased me hard in the M-Reach. In the surfaces exploration of Six and Seven, especially Seven, I flagged distinctly but went into the lead again as we pushed further upstream with Don close on my heels until we reached Nine-One. The size of the place and the problems of the way on rather appalled me. I needed to pause and take stock. Not so Don, he was merely invigorated and chased off round his brilliant circuit of the loop to Nine-Two. Then he later showed me round his loop and continued to lead to the brink of Eleven.

Eleven was another shock for me; I had tried, or was trying, to learn to use fins but had failed ignominiously in cave conditions; I was a bottom-walking oxygen diver and had little wish to change. As time goes on, the various incidents that chequer the way, not only my own near misses but those of my companions too, erode the sharp edge of enthusiasm, and the balance swings slowly but inevitably. On reaching Nine, my star was already setting; Don's had not yet reached its meridian. With the prospects of deep Eleven and beyond, I was quite content to let others take over. Then came the final crash with the loss of Marriott. It fell to me to bring his body out and that journey dispelled for ever the happy memories of the scenes from Six to Four, once so brilliantly lit by our original explorations in 1935. A dull pall had been laid over those lovely boulders and the sands of Five. After that, my little diving was a duty rather than a pleasure.

So, if this chapter is a long one, then it deserves to be, for it covers six years of extraordinary work in the story of the Wookey Hole divers. It is a shame that the seventh year was the unlucky break for the one man who had led it all up to then. This is a tribute to everyone who played a part in what was undoubtedly Graham Balcombe's era. Let him sign off with comments made after his very own 'Operation Swansong' to the Ninth Chamber in Wookey Hole on Saturday 29th June 1957:

On Sunday it took more or less all day to pack up, hand in my kit, and natter, feed and natter again. For this was my last dive; my fiftieth birthday had passed a little while ago and I could not go on any longer at the prodigious pace of the younger men; I was a handicap to them. When I look around and note the magnificent achievements of cave divers and, in particular, of the Cave Diving Group, I shall be ever proud to have been in from the beginning with such splendid companions.

*Fig. 4.29
Mavis and Graham Balcombe at home in retirement, at Goffs Oak, Waltham Cross, Herts.
Photo by Martyn Farr, 1975*

Chapter Five
Sacrifices to the Witch?

by Ted Mason

For ten or twelve years I had been excavating caves in search of ancient remains and nothing was further from my thoughts than cave diving. I knew, of course, that divers were exploring the subterranean River Axe within the Great Cave of Wookey Hole in search of new chambers and I don't suppose anything was further from their thoughts than underwater archaeology. I had, in fact, done a little diving myself with Arthur Hill under the instruction of Graham Balcombe and Jack Sheppard at Ffynnon Ddu in the Swansea Valley over Easter 1946 when they attempted to get into the then unknown Ogof Ffynnon Ddu system. This was not realised until Peter Harvey and Bill Weaver dug their way in during the August Bank Holiday weekend that Summer. However, the diving meet at Easter saw the birth of the new Cave Diving Group. But I had no ambition to pursue diving like the others.

Fig. 5.1
The iconic silhouette of The Witch of Wookey Hole stalagmite (left) in the First Chamber, also called the Witch's Kitchen.
Human remains were first found on the river bed beyond the boat (right), Whitsun 1946.
Photo by Peter Baker from Wookey Hole Caves

Fate was to play a hand for in 1946 while exploring Wookey Hole Graham Balcombe stumbled across human remains and pottery. And so I was asked by the Cave Diving Group whether I would take control of the archaeological problems, to which I consented and was given the imposing title of 'Archaeological Controller'. Obviously, my little knowledge of cave diving would have to be improved so I attended the group training sessions at the Mayor's Paddock Baths in Bristol. After encountering problems over chlorine deposits accumulating in our diving kit there, we transferred to the Minery pool at Priddy where the pre-War team, led by Balcombe, had started it all.

I was put in the water at the end of a rope and told to make my way underwater to the opposite bank of the Minery pond. Not only was the water murky, but long reeds grew from the bottom and I had to fight my way through an entanglement of sub-aquatic foliage. More alarmingly, people had been using it as a rubbish dump for years; bedsteads and crates were concealed among the reeds, not the kind of material that would attract an archaeologist let alone an inexperienced diver! At one point my foot was snared in the pool's plant life and, falling backwards over a box, I landed in a pram which had settled right-way-up on the bed of the pool. There I was firmly seated in the pram, my head sticking up at one end and my feet at the other. Now getting out of a pram when in full diving gear is not at all easy. I tried rocking it so as to fall out, but the weeds had grown into the wheels and held it firmly to the ground. I tried raising my body out of it but was jammed solidly between the sides. If only I could get the wheels loose, at least I would have been able to return to shore, even if the obstacle was still attached to my seat and would no doubt have caused surprise and amusement to those waiting for me to surface. Probably any casual passer by would have thought it part of my equipment!

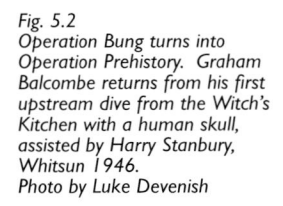

Fig. 5.2
Operation Bung turns into Operation Prehistory. Graham Balcombe returns from his first upstream dive from the Witch's Kitchen with a human skull, assisted by Harry Stanbury, Whitsun 1946.
Photo by Luke Devenish

As a matter of fact, perhaps it is not a bad idea to have such an attachment for those intent on long dives for they would be able to rest in comfort when necessary. My position was hardly comfortable at the time, however, as the exertion of rocking the pram led me to consume more gas and the cylinder would soon run out. I made one last stupendous effort. Thankfully, the front wheels became free, the pram tilted and I slid over the handle bar, not at all gracefully but at last able to continue my journey to the opposite bank of the pool. When I surfaced, I was pleased to see that the shore party was there to meet me, only to be told that they had not moved! They had, in fact, only paid out about 20 feet of rope for I had simply completed a circle underwater. I learnt a lot about the dangers of becoming disorientated in cave dives from this experience.

My next exercise was at Wookey Hole. I was pleased when the trainer told me to follow him and do everything that he did. We both jumped into the water and I turned towards him to see what he did next. Through my visor, I was surprised to see a pair of boots the wrong way up. He was standing on his head which appeared to be a difficult feat for my very first exercise! I gave the emergency signal to be hauled out and both of us were helped out onto the bank. It transpired that the dressers had forgotten to give my companion the lead soles for his boots. When they removed his visor, the flow of language was quite terrific and I don't think that he repeated himself once!

Then, later, I had my first solo dive. I was no sooner in the water and submerged when I had difficulty in breathing. Once again, I gave the required five tugs to be pulled up. Dan Hasell was the operation Controller. 'What's the matter?' he asked. 'Can't breathe' I replied. So, Dan asked the helpers to strip down my equipment to see what was wrong. They took everything to pieces, spread it all out on the bank and then reassembled it all checking every fitting. It must have taken about half an hour before I had it on again. Dan slapped me on the back saying: 'It's alright now. Is the breathing O.K?' 'Yes, perfect!' 'All right, in you go'. And I had a good trip.

A year or so later, when Dorrien and I were sorting relics on the same bank, a new diver entered the water on his first solo trip. After about a minute, I heard the controller, Dan Hasell again, say to the others: 'I thought he would have been out by now'. And sure enough, he had hardly said it when the diver signalled to be pulled out complaining that he couldn't breathe. I glanced up and noticed that his breathing bag was inflating and deflating quite normally. There was obviously nothing wrong with the circuit. Then, to my surprise, Dan said: 'All right lads. Strip down his equipment and see what's wrong', and they went through the same routine as they had done in my case. I realised for the first time that this was exactly what had happened to me, and that it was just a case of panic on one's first dive alone. The breathing 'fault' was purely psychological and I thought how well the Controller had handled the matter for, by his actions and that of the team, he had completely restored the diver's confidence.

Fig. 5.3
Further finds are recorded and packed up for examination by Dorrien Mason. Ted Mason (standing) and Tony Setterington (on the boat), Whitsun 1946.
Photo by Luke Devenish

Dan Hasell has spent hours in Wookey Hole Caves controlling numerous diving operations. He reckons that you can't control divers, in fact, because they are there to get on with the diving in their own different ways. Well, who knows how many have had their kit checked out by Dan! That's just one reason why he became President of the Cave Diving Group.

The archaeological operations at Wookey Hole were a challenge as no other such organised underwater investigations had ever been carried out before in a cave; nor probably since in terms of the amount of relics found in the Great Cave. There had been under-sea operations on ancient sunken vessels, but these did not involve actual excavation. This merely meant picking up material already exposed on the sea bed. The difficulty was how to adapt systematic archaeological methods used on the surface where dating and the relationships among finds depend upon stratigraphy and each soil layer is removed in turn to obtain relative dates, getting older as the dig deepens. On the bed of the River Axe in Wookey Hole Caves, however, changing currents kept shifting the sediment and, after floods for example, whole mud banks would appear in different places. The river bed often altered from one week to the next. So we had to keep looking over the same ground and, by carefully locating every find on plans drawn up by Jack Duck, hope that the changing picture would reveal their source and the story behind them. Each find was carefully recorded and identified for its likely date.

Herbert Balch had found relics from the Iron Age to medieval times in the Great Cave, but the main material consisted of Romano-British remains excavated in the dry passages. It seemed unlikely to me that finds of any other period than the Romano-British would turn up in the river bed except perhaps some medieval to modern material from more recent visitors. Balch's work was well documented and authenticated, so there was every reason to expect that anything we found would, therefore, be dateable not only to the period but also to the century by comparison with known items.

Fig. 5.4 A mud-filled human skull is carefully brought ashore by Ted Mason, 10 April 1948. Photo by Luke Devenish

Positive or absolute dating is only one aspect of an archaeological investigation, however. Another important aim is to build up the background to objects found. How they came there and why. It is the story behind the evidence that we want and so the objects themselves are only a means to this end. In particular, we hope to find out what the people were like who may have inhabited or used the cave, the conditions in which they lived and why. Details of their skeletal remains would tell us about their stature, life expectancy, race and health, but it is also the context of such material that is needed to determine their possible way of life. How, for example, did they bury their dead?

Some skeletal remains had been washed out of the cave years ago following a flood, but from what part of the system was not known. Balch had suggested some associations with the legendary Witch of Wookey Hole Caves and, whilst he was typically matter-of-fact about such things, it was customary in his day for others to give archaeological interpretations that were either romantic or gruesome. Even he felt it possible that the bones had belonged to victims of a sacrifice; maybe they had been tied to a stake below the grotesque stalagmite supposed to be the Witch in the First Chamber, or even decapitated and thrown into the river beneath her. A not unreasonable theory given the long history of mysteries about the Great Cave, but one I wished to test against the evidence that still remained on the bed of the subterranean Axe. Cave diving all the way upstream would be necessary to prove things one way or another.

Fig. 5.5
Ted and Dorrien Mason raise doubts about human sacrifices having taken placed in the cave, 10 April 1948.
Photo by Luke Devenish

Socket holes for a stake had supposedly been seen above the water's edge below the Witch; but I found no such evidence, although tell-tale marks could have been obliterated during a cave 'clean-up'. More reliable evidence was the location of the relics found by Graham Balcombe on his dives over Whitsun 1946. He had discovered the remains of three humans at the bottom of an underwater slope several yards upstream of the First Chamber in a place known as the Witch's Scullery. It seemed unlikely that any eddies could transport such remains so far against the prevailing current. So, unless the river in Romano-British times defied gravity and flowed uphill, the sacrifice theory was now doubtful. This is what brought me to the Great Cave.

I also came to reject the decapitation possibility on examining the skulls that were found by Balcombe, and later by other divers. If the heads had been cut off, one would expect to have seen evidence for this on the topmost vertebrae of any severed spinal column. For example, the atlas vertebra in the hollow under the cranium would be protected, but those immediately below like the axis cervical vertebra would not. Also, the base of the cranium or occipital would probably show scrapes and even cut marks. Although there was an absence of the smaller vertebrae washed away by floods or buried without trace, none of the eighteen skulls examined had signs of decapitation. So what was the likely reason for their profusion in the cave?

Fig. 5.6
Human skulls and bones mount up, giving rise to theories that a human cemetery had existed upstream.
Photos by Luke Devenish

Olive Hodgkinson subsequently donated several of these skulls to the Metropolitan Museum, New York, USA

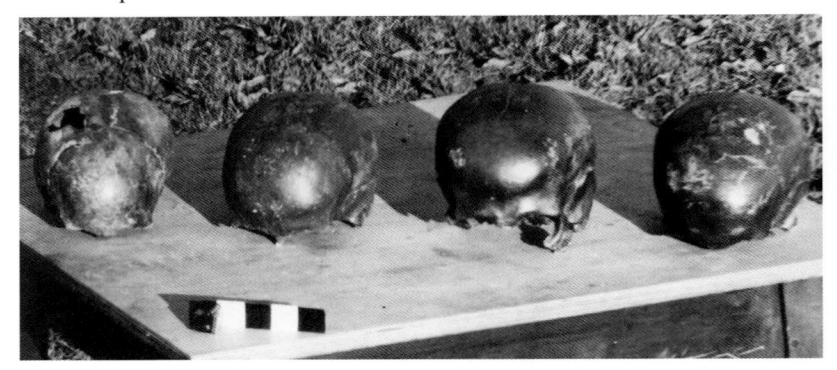

As our work developed, the remains discovered led me to believe that there had been a specific burial ground somewhere in the cave during Romano-British times. It must have been accessible and upstream of the First Chamber. As the Second Chamber known as 'The Hall' could also be ruled out on these grounds, this left either the Third Chamber or partly submerged Fourth Chamber as sites for my cemetery. All subsequent work narrowed the spot to the Fourth which could only be reached by diving unless water levels were lowered by lifting the sluices outside.

Now, I have no doubt that the 'witch's skeleton' found by Balch in the Vestibule near the entrance, and later displayed in Wells Museum, is Early Iron Age or Romano-British because of the pottery 'milking pail' found with it. Nor do I dispute the existence of the so-called Witch of Wookey Hole for legends usually have some foundation and a dry cave would make an ideal shelter for an old recluse. Perhaps she was demented and bore the blame for everything that went wrong in the village. Our diving and discoveries found nothing to disprove this old legend just the unlikelihood of sacrifices. For some reason, the innermost chambers had been used as a communal burial ground by Romano-British peoples.

Before carrying out the search of the river bed, a scale plan was produced and mounted on linen so that every item found could be marked in. Normally, on a surface site the ground and plan is carefully laid out with a measured grid. This was impossible in our case, but lines had been laid with tallies at yard intervals along the river bed by bottom walking divers when exploring the cave. These lines were carefully surveyed and plotted onto the map so that offsets to each object recovered could be measured to locate their exact position. We found material to be in much more scattered positions than is the case with a sunken ship.

The very first skull found presented other difficulties not encountered at surface excavations. It was the skull of a young person in which the sutures joining separate parts of the cranium were partly open. It was laid on the ground ready for measurements to be recorded, teeth present or lost to be counted and any other general information. When dry it would be numbered SK1 (skull number one) and its position plotted on the map where found by the diver. So far so good; but, when I was about to pick up the cranium, it was no longer intact. Because the outside water pressure had been removed, the wet mud that filled the inside of the skull pushed the specimen apart. And when dry it is almost impossible to fit back the pieces because they will not join up as before. Some method had to be found to avoid re-occurrences of this problem.

Fig. 5.7
A detailed underwater survey locating the relics is made from the Second Chamber to the Resurgence, 1948-9.
Photo by Luke Devenish

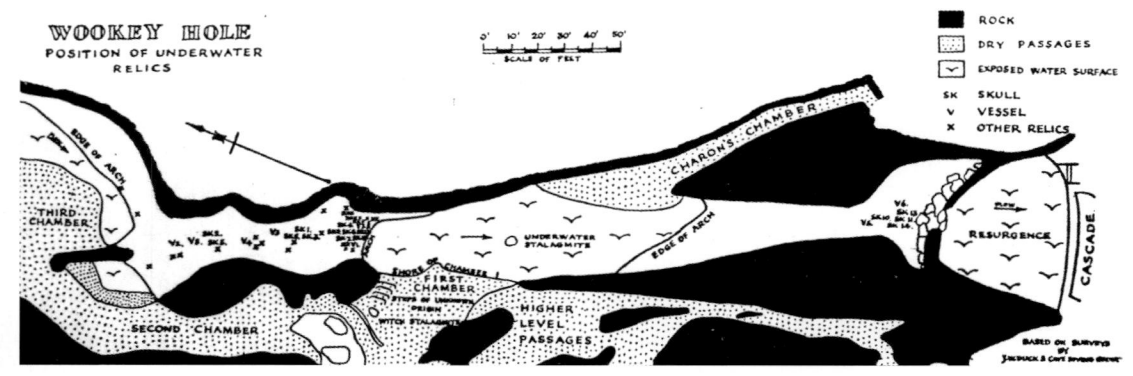

Fig. 5.8 Plan and account of the underwater survey is published in the Proceedings of the Somerset Archaeological and Natural History Society, Vol. XCVI (1951), pp.238-43 (See Fig. 5.12)

I made a wooden open frame with a leather strap handle, large enough to take the biggest skull likely to be found. Inside was a linen bag made to the required size by my wife and nailed to the frame with brass studs. Any skull could now be carried by the diver without buffeting. A couple of boxes were added outside for lead weights, enough to produce 'neutral buoyancy' so that the frame could be propelled about 3 feet above the river bed. The woodwork was painted black and white so that it could be more easily seen in a mud cloud. As soon as it arrived onshore, the mud was washed out of the inside of the skull whilst it was held firmly within the frame, and we had no more problems regarding drying out.

Trial trenches in the Third Chamber across two mud banks that were the most likely burial sites produced no evidence. It was obvious that the sediment where any burials may have taken place had long since been washed out by floods. Rising water after heavy rains on Mendip also causes the famous 'noises' of Wookey Hole as trapped air gets displaced from one partly submerged chamber to the next and the 'gloop' is amplified. The Fourth Chamber is one such place and, as Balch had done earlier, it was agreed that the sluices outside would be opened so that we could look at the mud banks inside. Accompanied by Olive Hodgkinson and Luke Devenish, I helped steer the boat under the now open arch and began to excavate the older silt of an exposed bank. Within half-an-inch of the top we found some small fragments of a skull, neck bones and one upper arm in juxtaposition with two beads resting on the neck. Although the rest of the skeleton had been washed away by floods, I now felt sure that the Fourth Chamber had been the burial ground and source of the other finds downstream. So, it was through an old cemetery that cave divers set out to explore the unknown reaches of the subterranean River Axe. This has since been substantiated by Professor E.K. Tratman's findings already cited in the previous chapter.

Fig. 5.9
Ted Mason's human 'skull carrier', designed to prevent disintegration on surfacing and drying out, 19/20 June 1948.
Photo by Luke Devenish

The eighteen skulls we recovered by diving downstream of the cemetery were probably our most notable finds. All but two of them exhibited the characteristics of Romano-British peoples, similar to those of the earlier Iron Age marsh village near Glastonbury. Two were different, however, so we used to call our eighteen people the sixteen 'locals' and two 'lodgers'. Other finds were various pots and two lead ewers. Many of the bones and vessels are displayed with my skull carrier in the Caves Museum[12].

[12] The skull carrier is now in Wells & Mendip Museum.

More recent material was also found, of course. In the Seventeenth and Eighteenth centuries, for instance, it was common for small parties to dine in the Third Chamber with wine. Apparently, some must have thrown their empties into the river for we found two bottles. One was completely empty but the other was corked and, being half-filled, was resting at an inclined angle on the submerged bank. Alas, it contained no vintage wine, merely dirty river water that had seeped in.

Having discovered all we could on the river bed and proved that the Fourth Chamber was a burial ground, and maybe the Third too, we decided to probe into the sediment beneath. But how could we stop any dig filling almost as quickly as you dug? So, I hit on the idea, unsuccessful I'm afraid, of scouring the surface of the river bed by means of pressure jets. Why not try the pump used for washing down the show cave? So, two divers went into the water in the First Chamber where, with the cave fully illuminated and the underwater lights on, we could watch the effects of the experiment. They held the nozzle of the pipe in position against the river bed and the pump was switched on.

The effects were certainly quite dramatic; in fact, one might say artistic! The two divers poised for action and holding the hose between them suddenly lifted off their feet. With the natural tendency to hang on to anything, instead of letting go of the hose they gripped it more tightly and, jet propelled, they made a complete circle in a kind of underwater flight. But that was not all. The circle was only a prelude of delights to come. The two man subaqua ballet performed spirals, figures-of-eight and graceful curvilinear dives in all directions. With the hose constantly weaving and the attractive green tint of the water, their performance was quite spectacular and a welcome change. Someone standing next to me muttered: 'Ballet Russe should have seen this!' I was so engrossed and moved that I have forgotten who the performers were. But I do remember the thought crossing my mind that, if only Olive had seen it, she would have realised its potential immediately. Imagine cave divers staging subaqua ballets to audiences seated around the Witch's Kitchen. A much nicer attraction than stories about sacrifices!

The pump operator had been so fascinated, too, that he could not take his eyes off the fun and forgot to turn off the pump. At last the divers managed to let go and, as they struggled to the shore, very dizzy no doubt, the hose slowly sank writhing like a snake and spraying fountains of white water everywhere. It was a finale that clinched its failure as an archaeologist's aid. And the divers didn't think much of the ballet idea either!

Before abandoning the aptly named 'Operation Sandblast', however, we toyed with various ideas. What if we reduced the pressure to enable divers to keep their feet? Well it helped a little but did not overcome the great bugbear of underwater archaeology that we faced: the mud cloud. You had only to kneel down for a few minutes searching for relics, and the cloud would swirl like curling cotton wool around you; first the knees, then all over and gradually thicker so that the powerful light at your side turned from brilliant white through stages to orange to deep red and finally invisible as if going out. We thought of covering the jet with a glass box and even using underwater fans to disperse the clouds, but they were obviously impractical. As archaeologists we could not use

Fig. 5.10
Professor Edgar K. Tratman at work in the Fourth Chamber. 'Trat' led the subsequent archaeological excavation by the University of Bristol Spelaeological Society during 1973-76, proving that the chamber had been used as a cemetery.
Photo from Wookey Hole Museum

the 'mud racing' tactics of the exploring divers. We consoled ourselves that the material embedded in the mud was unlikely to add much that was new to the story of Wookey Holes Caves. Yet, who knows. Maybe if like Jacques Cousteau we had thought of sucking-up the mud through filtered pipes, there would be more to tell. 'Operation Vacuum Clean' perhaps?

All the cave diving 'ops' at Wookey Hole Caves attracted quite a lot of publicity, for the newspapers, film and television companies never tired of new stories to cover, whether about discoveries or the human remains found. Olive Hodgkinson masterminded a lot of these, of course. There was a Pathé Pictorial Newsreel filming session, though few of us saw it screened, and another by Twentieth Century in a series of support pictures with the dreadful title: 'Spotlight on Danger Men'! On another occasion, we refused to follow a prepared script by a young television entertainer who wished to introduce hyaenas and Old Stone Age finds into the Great Cave, things that had never been found there. I was always relieved when Dan Hasell managed to tone things down, usually by ruling out certain diving scenes as unsafe and suggesting his own instead. It always worked!

I think that Wookey Hole Caves was the first underground location from which a live television broadcast was made in this country when the BBC gave a factual report on our underwater archaeological work. Bob Davies joined us for the filming and a 'talking rope' was used to broadcast from underwater. On this occasion, everything had to be carefully set up because there could be no editing of what 'went out'. The script was narrated by Cliff Michelmore in the First Chamber. As he introduced the report, the cameraman would pan to the Witch's stalagmite as he came to the words:

Fig. 5.11
Dorrien Mason at work in the Witch's Kitchen, 1949.
Photo by Luke Devenish

'Here we have the Witch of Wookey', then down to the shore where my wife Dorrien was sorting relics, around the chamber in silence and then back for him to pick up the next part of the story. But one of the cameramen suggested playing a trick and, as it was not an insult to the narrator, we agreed. When he came to the key phrase, the camera actually zoomed in on Dorrien! Fortunately, the flabbergasted narrator had time to compose himself before the camera returned to him.

So, jokes about the Witch have become more commonplace than stories about sacrifices to her. All the old generation of divers used to spit on her for luck as they passed her to start a diving 'op'! Olive herself would sometimes put on a grotesque mask and wig to scare visitors she was escorting around the cave. She knew exactly how to strike a balance between the lure of the Great Cave and its challenge to cave divers. Moreover, Olive Hodgkinson was at pains not to sacrifice her cave for she appreciated the appeal of the unspoilt and unknown as we did. After Wing Commander Gerard Hodgkinson died in October 1960, aged 77, Olive continued to encourage and entertain cave divers although, as we shall see later, owing to changes in diving techniques at the time, there was virtually no cave diving at Wookey Hole from 1962 to 1966. In 1971 she published her own version of *The Story of Wookey Hole*, and wrote:

In 1928 [actually 1927] my husband began the process of opening the Caves to the general public on a commercial basis. I hesitate to use the word commercial because it conjures up the wrong impression. Visitors to Wookey Hole comment frequently on the fact that the whole place has remained unspoiled in spite of the vast numbers who come here yearly, and since my husband's death I have striven to comply with his ideas that although Wookey Hole is one of the great attractions, if not of the whole world, certainly of Great Britain, the whole place should remain as unspoiled as possible.

Olive left Wookey Hole shortly afterwards in 1973, so ending the hundred year dynasty of Hodgkinsons in the village. Sadly, her retirement to Jersey was brief but typically memorable. When Jim Hanwell visited her there, the showpiece in the lovely garden was a swimming pool designed with underwater lights to resemble the view from the Third Chamber in Wookey Hole Caves towards the Fourth and that part of the Great Cave that was the preserve of the cave divers. They spent the evening looking and reminiscing[13].

[13] Olive Hodgkinson died in 1976.

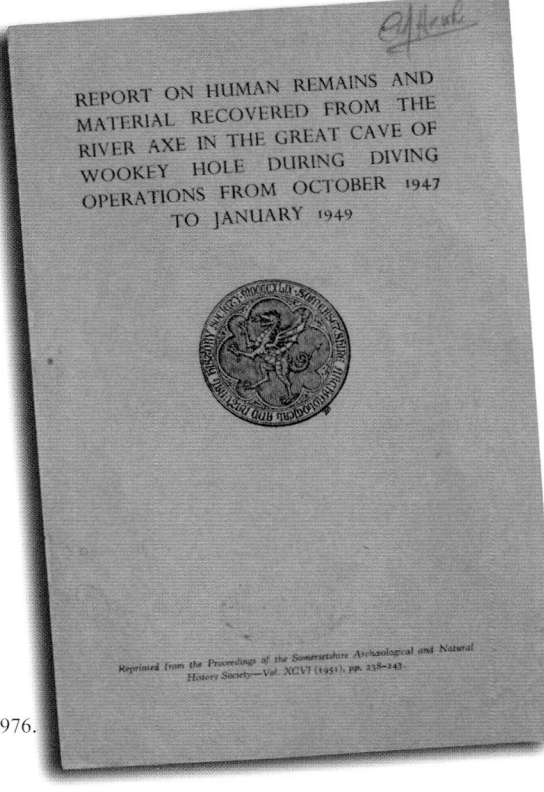

Fig. 5.12
Ted and Dorrien Mason's Report as 'Joint Archaeological Controllers' of the Somerset Section, Cave Diving Group excavations in Wookey Hole, 1947-49.
From Christopher Hawkes

Fig. 5.13
Bob Davies brings out 'the skull which disintegrated'.
Photo by Luke Devenish

Chapter Six
Photographing the Ninth Chamber
by Bob Davies

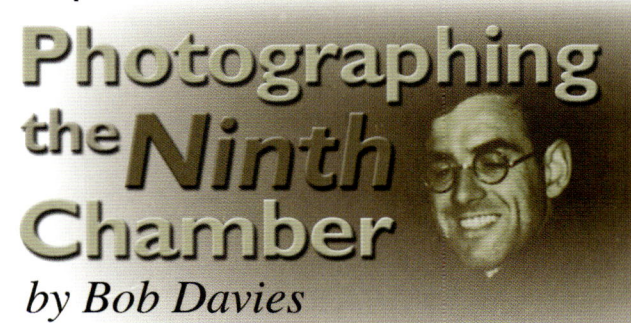

The Cave Diving Group *Letter to Members* No. 11 records the discovery of the Ninth Chamber by F.G. Balcombe and D.A. Coase on 24th April, 1948. This chamber is 330 feet from the telephone base in the Third Chamber and plans were made at once to photograph it. However, five years passed before this was accomplished on 28th June 1953, on the first visit to the chamber since G.I. Marriott and I left it on 9th April 1949, but only I got back alive to the Third Chamber.

On 31st January, 1953, Balcombe and I had started to renew the guide and telephone wire to the Ninth Chamber in preparation for photography, but owing to shortage of time and an incident involving a supporting diver the line reel was taken only part of the way and left on the floor below the Seventh Chamber.

Fg. 6.1
Robert E. Davies, or 'RED', had a two-seater Austin Swallow car nick-named 'Scarlet', in which he made regular weekend trips to Wookey Hole from Sheffield following his first dive there on 1 Jan 1948.
Photo from Bob Davies album in Wells & Mendip Museum

The successful operation was on 27th/28th June, 1953, and was carried out by the divers: F.G. Balcombe, J.S. Buxton, D.A. Coase, R.E. Davies (in charge) and J.A. Thompson, assisted on shore by members of the Bristol Exploration Club and the Sheffield University Mountaineering Club. After some time spent assembling and testing the oxygen breathing apparatus for low and high pressure leaks, Balcombe and Coase set off along the bottom at 23.35 hours to belay the new wire and get it to the Ninth Chamber. At 00.35 hours they reported by phone that they had arrived. They had had a difficult job climbing the mud slope to Nine and reported the wire they had laid was safe to pull on. The sound-powered phone at the Base in Nine had thus remained functional in its waterproof box although the signals from it were very faint and it had no bell; so a battery-powered field telephone was taken through by the next party who also carried the cameras, flash bulbs, tools and emergency supplies.

Small cameras were put in Kilner-type, large sealed jam jars, wrapped in rubber bags and put in the waterproof 'frog-sacks' which were large rucksack-size sealable bags. They thus remained at constant pressure. My larger camera, along with a tin of silica gel, was sealed with a candle flame into a bag of thermopliable plastic (Pliofilm), and put in a tied rubber bag inside a pressure-compensated ammunition tin, in a frog-sack. This camera had the external pressure of 26 feet of water acting on it, but the silica gel prevented the condensation of water inside which might have affected the mechanism.

Cameras were taken in the frog-sack which had a 6 cubic foot oxygen cylinder connected to it and was fitted with a flutter-valve near the opening. As the sack was taken deep and the walls collapsed, it could be blown up again with oxygen and this helped to keep it waterproof and, by increasing its buoyancy, made it easier to handle. Tools, tripods and less delicate equipment were carried in an unmodified frog-sack which collapsed and became much heavier the deeper it was taken.

The next party dived at 01.02 hours. Buxton, the least experienced, went first with the unmodified frog-sack, so he would have a clear view. Jack Thompson followed operating the compensated frog-sack and I came last, available to help should the need arise. We moved steadily away from Base until we reached the Gateway, just past the Seventh Chamber, where Thompson signalled that Buxton could not find the way ahead as the mud had risen. Neither of these two divers had been further than this before so I felt my way along the wire, found the way down to Mudball Alley, slid down under the low roof and guided them through. Then along we crawled and up the long slope to the Ninth Chamber where we turned to air breathing at 01.30.

Fig. 6.2
Bob Davies, a key diver in the Derbyshire Section of the CDG, wearing his two-piece 'Sefus' suit, 1948.
Photo from Bob Davies album in Wells & Mendip Museum

Fig. 6.3
'RED's' two-piece suit and weight belt, 1948.
Photo from Bob Davies album in Wells & Mendip Museum

The new telephone was connected and tested so Base no longer needed to listen to their phone continuously. The next four hours were spent in exposing the negatives of the photographs. They were all taken with either large or small size flash bulbs and, in general, each flash was used by all cameras to ensure success. There were four cameras taken, but unfortunately the best one, belonging to Coase, developed a mechanical fault and could not be used. After the photography, a low-grade survey was made of the Chamber and at 05.30 we reported to Base that we were about to leave and expected to return at about 06.00 to 06.10. We were all rather tired by now and had some glucose and Benzedrine before starting. During preliminary testing Thompson's Aflo had failed and he had got to Nine with a hand torch. My Aflo had a six volt battery, several light bulbs and, a compass, watch, writing pad, pencils, depth gauges, bell, electric horn and line reel. Thompson's hand torch became temperamental and finally defunct during our stay in Nine, so the plan for return was: Thompson and I to go together using my light, followed by Buxton and Coase with the normal and compensated frog-sacks with Balcombe as rearguard.

Fig. 6.4
RED's' closed circuit respirator (with pliers attached) and face-mask, 1949.
Photo in Bob Davies album in Wells & Mendip Museum

Thompson and I slithered down the mud slope for 26 feet into the Bear Pit with no trouble except for pain in the ears owing to sluggish Eustachian tubes. However, at the bottom of the Staircase from Eight to Seven, Thompson found that the wire went into a narrow crack and could not be followed. Luckily, Balcombe had warned me of this danger so after a bit of gentle persuasion we managed to free it and found the way up into Seven. Once through I stood up in the gully where Marriott's body had been found and looked up and back. There seemed to be a clear open route to the top of the Eighth Chamber over a low arch of rock under which we had just emerged. This fitted with my memory of finding Marriott's lost oxygen cylinder below the Eighth Chamber by swimming high above the floor through the Seventh Chamber and then straight down to it without following the low, narrow, awkward route taken by the wire.

The rest of the trip back was very pleasant with relatively clear water and the growing relief from strain which comes at the end of any long operation underwater. We got out at 06.15 with the good news that we had taken the photographs, and waited for the other three divers to come. Well, they didn't, so at 07.07 I went in to find out what was wrong and bumped into them coming through the muddy water.

As soon as all the divers were out of the water Sybil Bowden-Lyle went off at once as arranged to send the usual telegram to my wife that we were all safe, but it turned out to be premature as both frog-sacks were lost!

In the Bear Pit, Coase had had trouble with a leaking face-piece whilst holding the wire, his Aflo and the compensated frog-sack. This left no free hand, and finally he had to let go the sack and blow a lot of bubbles to clear the mask. Buxton saw this, thought there was a serious emergency, dropped his sack and went to help. They all got back to air in Nine, regrouped and came back to Base through thick, muddy water with no sign of either sack. This was a blow. The cameras were worth well over £100 but the negatives were almost priceless. However, it was no use searching in that muck so we left the cave for breakfast, the first food for 16 hours, whilst the water cleared itself.

The less valuable unmodified frog-sack clearly was on the bottom somewhere, but the compensated one could either have sunk down or floated up. It certainly wasn't in Nine, but it might have been in the air surface of Eight above 24 feet of water. This was awkward, because whilst only a swimmer could reach it, he probably couldn't sink it. The following plan was therefore evolved. Coase and I, as the more experienced cave diving swimmers, would swim by the high route to Eight, look for one or possibly two, sacks on the floor, and clip them on to the wire with rope loops and karabiners. Every effort was to be made not to stir up the mud until the sacks had been sighted as it is a major problem for bottom-walking divers. If only one sack was seen Coase was to deal with it whilst I went up to the Eighth Chamber surface with 50 feet of wire on a reel, tied this to the bag and swam back by the top route to the Seventh Chamber. Here Balcombe and Thompson were to wait for us and act as beacons, since both Coase and I might be high above the wire in muddy water and thus need some target to aim for. These two divers were heavily weighted so they could pull down the floating bag and retrieve it. Coase and I were to act independently and on the way out we were to signal to the beacon divers what we had found so they could follow the wire and collect the sacks. Any diver in trouble who couldn't get to Base was to get to Nine and use the phone. Base was therefore to man the phone continuously. We all renewed our oxygen and soda-lime supplies and hoped we had thought of all contingencies.

Fig. 6.5
Pristine mud formations on the floor of the Ninth Chamber, 1953.
Photo by Bob Davies

Fig. 6.6
George Lucy mans the telephone line from Base in the Third Chamber to the Ninth Chamber, a task he did often during 1949-53.
Photo by Luke Devenish

Coase and I did a practice swim in the Third Chamber to check trim and buoyancy, which can be tricky in dry suits, and the four of us set off at 11.09 in fairly clear water. We had an uneventful trip through Four, Five, Six and Seven, but got a shock when, on aiming towards the top route I went straight into a black wall of rock! No way on was visible so I did a nose-dive for the bottom route with Coase at my flippers and got through a very tight place a long way to the right of the wire. This led to Mudball Alley all right and 20 feet ahead, well off to the right of the wire was one, and only one, frog-sack, not the one with the cameras in it. I went for this, looked around for Coase, but saw only the mud I had stirred. Whilst I was dragging the frog-sack to the left, feeling about in the mud until I met the wire, Coase had passed me, reached the Bear Pit, seen nothing and returned. I touched the wire as he was passing back along it, but could not make him understand that I had found one of the sacks.

It turned out later that the fastening of his main cylinder had come loose and it was flapping about so he was returning quickly. At Seven he gave the thumbs down to indicate that he had found nothing and started back, but dropped his Aflo. His signal and his lack of leg movement while retrieving his Aflo from the crack alarmed Balcombe who followed him back to Base to make sure he was all right. On returning to his post he met Thompson coming out because he had been so confused with the movements in the mud that he thought that both Coase and I had come out. However, they soon settled that misunderstanding and went back to shine their lights as a guide for me.

In the meantime I had floated to the surface of Eight and was surprised to find no trace of the missing sack. After swimming right round and taking a spot compass bearing for Nine (east) I went down to 15 feet and along horizontally until I hit a mud slope which to my great pleasure led to the Nine-One surface. Here I had a rest and looked around, but not a trace of the frog-sack was to be seen. This was very tricky as it meant that the container was probably jammed somewhere under the water which was now getting rapidly opaque.

There was nothing for it but to spiral round the walls of the Bear Pit and hope. Before very long I saw it between the Eighth and Ninth hidden in a pocket in the roof at a depth of 12 feet and invisible except from immediately below. It was still the right way up. Very thankful and pleased I lay face up against the roof, pulled it down and towards Nine, then shot up out of control in a cloud of bubbles back to Nine once more. It had expanded a lot when floating up from 24 feet down in the Bear Pit and was several pounds buoyant. It dragged me up scraping the walls and when I surfaced I found that I had lost the line reel and dropped my Aflo. The line reel was lost, but as my Aflo was clipped onto my kit with a short line I soon pulled this back into reach (When the line reel was eventually recovered on 11th December, 1955, it was covered with a filamentous saprophyte new to science; see *Nature* Volume 178, 28th July 1956, pages 215-216 article by F.E. Round and A.J. Willis). It was easy to clip the sack to the wire in Nine and I lifted it out of the water and had a rest for two minutes.

Whilst this caper was going on Balcombe and Thompson got worried because I had not glided down from above them and decided to come and see what I was doing. It must have been disturbing to them when they got to Mudball Alley and saw no light at all above them, just cloudy water and rock. They then started back to their post again and were at the narrowest part when I caught up with them. My fingers sliding along the wire in the (genuinely) inky blackness touched another hand which at once seized mine and gave two taps, meaning go back to Base. I returned the signal and we crawled to the Gateway again where there was more room and visibility was at least one foot. I signalled two bags found, (two fingers up, and thumbs up) and tried to indicate that one was in the Ninth Chamber itself. This needed five, then four fingers, but as five fingers means a dire emergency, this only led to confusion. The original signal was then repeated, I pointed the way for them to go, waved goodbye, swam back to Base, checked that Coase was back, turned to air at 11.46 and told them the news.

At 11.56 Thompson appeared with a bad headache and dazed from carbon dioxide excess. Two minutes later Balcombe arrived with the first frog-sack. They had thought my signals meant that the two sacks were tied together and when they found one they returned. Thompson brought it back,

Fig. 6.7
Looking south to the far end of the Ninth Chamber from the telephone terminal on the mudbank, 28 June 1953. John Buxton stands below the steep rock ribs (subsequently climbed by John Parker and Brian Woodward in November 1970 – see Chapter 13).
Photo by Bob Davies

but as it was very heavy in deep water and as he moved too quickly, he developed carbon dioxide excess, so dropped the sack for Balcombe to bring, and came out on his own. The important frog-sack was still in the Ninth Chamber and since the rest of the team was a dazed diver, a diver with unreliable kit and an inexperienced diver, there was nothing for it but for Balcombe and me to return to Nine again to collect it.

I removed my fins, added more weights to the belt and once more we fitted full oxygen cylinders and replenished the soda-lime canisters to absorb carbon dioxide. We then tested the kit for leaks and entered the extremely dirty water en route for Nine. By now it was 13.01 hours, but this was a perfect trip without incident. Balcombe collected the sack and operated the pressure compensator whilst I followed behind and helped it through the narrow places. We reached Base again at 13.25 and found that both sacks were still dry inside.

Although we had been assembling and testing kit, photographing and diving for 20 hours this was not the end as there was still the six hour, 200-mile drive back to Sheffield. In fact, this expedition involved me in 42 hours without sleep, 400 miles driving, 16 hours in the cave, 5 hours in the Ninth Chamber and 1 hour 45 minutes underwater!

Still, it certainly was a fine operation, full of excitement, with the outcome uncertain until the very end. I hope the unique photographs we were able to obtain will give some idea of the Ninth Chamber of Wookey Hole Caves and will be of added interest because of their dramatic history. Their very existence depended on a piece of team work unparalleled in the history of cave diving.

Fig. 6.8
Looking downstream from the Ninth Chamber towards Nine-One, 28 June 1953. The improved forward telephone base is at the foot of the boulders, and the way on upstream follows the guide wires to the bottom right.
This was the first photograph taken of the large and vaulted Ninth Chamber (which the general public have been able to visit since 1975 – see Chapter 14).
Photo by Bob Davies

Chapter Seven

In the Waters Under the Earth
The Thirteenth Chamber of Wookey Hole

by Bob Davies

It really amazes me that I am still alive. There have been so many times when I could easily have died: nearby fire and explosive bombs and hostile machine gunners in World War II; being scarred and nearly impaled by a bamboo pole that broke when I was pole vaulting; being blown off a rock when leading on old-fashioned hemp climbing line; being hit and badly burned by lightning on the top of the Grand Tetons; escaping very close rock falls on the Matterhorn and ice avalanches on Mount Robson; having troubles parachute jumping; being trapped under an overturned raft under Velvet Falls, Idaho; rowing Lava Falls in the Grand Canyon of the Colorado in a tiny raft; going backwards over Raine Falls, Oregon, etc... but I'll always remember Wookey Thirteen.

That was in 1955 at a time when the Cave Diving Group had routinely explored caves by bottom-walking but had faced a new problem in Wookey Eleven. It seemed clear that the bottom of Wookey Eleven was too deep for pure oxygen and too far from the dry land in Wookey Three for the types of aqualungs then available.

Fig. 7.1
Amazed to be still alive! Bob Davies relaxes at Crooks' Rest after being trapped overnight in the Thirteenth Chamber on 10/11 December 1955. His Aflo, Aqualung and flippers are on the floor.
Photo from Bob Davies

Fig. 7.2
John Buxton (left) and Bob Davies (right) respectively use 'old and new' respirators on dives in Cheddar Caves earlier in the year, 21 May 1955. Oliver Wells (far left) holds the logbook, and Audrey Buxton (centre) assists.
Photo by Luke Devenish

The best plan, we thought, was to take an aqualung to Wookey Nine on oxygen and swim underwater from there on air. But the best laid plans...

The following is the account of what happened overnight during the weekend of 10th/11th December 1955 and is based on the report I dictated between 6 and 7 a.m. on the Sunday morning.

At 21.17 hours I left the Third Chamber wearing a P-Party oxygen breathing apparatus full of O_2 and towing an Aqualung (twin 40 cubic feet cylinders) to which was tied my swim-fins. The diving order was: R.E. Davies, O.C. Wells and J.S. Buxton (with frog-sack containing spare oxygen and soda-lime etc.), and we had a straightforward trip. I then tested my Aqualung to make sure that it had not been damaged during the dive and checked the amount of lead that I needed to sink by making a trial dive in Nine-One.

Fig. 7.3 Bob Davies sets out from the Third Chamber wearing a P-Party breathing apparatus and towing his Aqualung at 21.17 hours on 10 Dec 1955. Photo by Luke Devenish

My P-Party set was left on a flake near the Nine-One shore and we carried all the diving apparatus over the rocks and down to the mud shore of Nine-Two. Here we noticed that all the mud formations, that had caused us so much trouble to photograph, had been completely destroyed by a recent flood. I had found a remnant of the photographic operation at the bottom of Nine-One, the small line reel which we had thought we might need to recover the lost cameras. It was spongy with a growth of 'algae' which were later identified at Bristol University as mentioned in the previous chapter.

The plan of this operation was to find out if there was a way ahead at the bottom of the deep Eleventh Chamber and to see if the passages turned upwards within a short distance. It was important to know this, because if they did it would be possible to continue exploration with oxygen apparatus which would be a great advantage. Now it is difficult and dangerous to lay a wire when swimming in a complicated cave because the wire might get tangled round your feet, yet it was thought necessary to swim to explore the Eleventh because it has very steep sides and might be difficult to climb out of if the normal bottom-walking technique were used. It was therefore agreed that Buxton should lay the wire to the rim of the Eleventh and that I should tie my wire to his at this point for the quick look-see which was all that a solo diver could be expected to do under these conditions (since the CDG had no other apparatus than my Aqualung suitable for deep diving).

Since only F.G. Balcombe, D.A. Coase and myself had ever been to Eleven, it was clear that I would have to lead the way and go on alone. We agreed that after we became separated they would wait for me for 45 minutes before returning to Nine then Three to raise the alarm. These were our last words before entering the water at 23.45 hours. My Aqualung contained air at 120 atmospheres pressure in each cylinder and I started off on only one of these so that I would know when I had used up half of the available supply.

I swam slowly down to the Tenth Chamber and guided the other two divers towards Eleven past a large boulder which partially blocks the passage which then descends steeply. The distance to Eleven is much shorter than we previously thought and I soon found myself on flat ground before a flake which crossed the passage from side to side. I was surprised to see from the gauge that I was at a depth of 47 feet. This meant that I must already be at the bottom of Eleven since the rim was known to be at 24-25 feet depth. I took a compass bearing upstream and found that the passage continued ahead in the direction 10 degrees North of Magnetic East. What had been a very steep rock wall was now a much gentler mud slope, no doubt a result of the last flood. The beacon divers were way above me on the rim so I flipped back up to them through the rising mud to make contact. On reaching them I found it impossible to pull the end of my wire through the stop hole I had drilled in the side of the bobbin. The insulation must have got frayed during the Labouiche Expedition I had been on to France so I cut the wire with my pliers. By this stage visibility was down to a few inches and I noticed that the balance of my Aflo was wrong. I felt about and realised that it had opened up so the bolts must have been badly seated and had slid back. I had designed my Aflo with this possibility in mind so the battery remained wedged in place and the light still operated; however, it was a nuisance. Numerous attempts to re-close the Aflo failed and the extra energy expended caused me to breathe fast and deeply so I became buoyant. I had had one arm round Buxton's wire all this time but now found that the free end of the Aflo had got itself wedged somewhere in amongst his cylinders. By the time I had extricated it I had also lost contact for, in a few seconds I realised that I was up against the roof, with the depth gauge reading 10 feet and visibility 6 inches. I descended but did not find the divers, their wire or the way back. In fact they had taken my ascent to mean that I had started back for Nine-Two and had returned there themselves almost at once. This sort of misunderstanding is very easy to make when there is a confused mix up of divers in thick mud; it was a pity, however, that it happened just then.

I floated up again but visibility was so bad that I could not read any of the instruments and was reduced to spiralling round the walls by touch alone. It is now clear that I descended right down the mud slope of Eleven to the bottom and went up again on the other side of the flake in the muddy water. During all this time I was still rather despairingly trying to close the Aflo bolts but they would not slide home, so I had to hold the thing together by hand. At this stage breathing became difficult

Fig. 7.4
Dan Hasell, the Controller on the 'Night of the Thirteenth', awaits a long overdue Bob Davies.
Photo by Luke Devenish

which meant that the first air cylinder was very nearly empty, and I had used half the supply. I opened the valve of the second cylinder, equalised the pressure in the two cylinders, and then closed the valve again. There was nothing else I could usefully do except continue to spiral round and try to find the divers, the wire, the exit, or clear water; but none of these appeared. Soon the first cylinder became empty again, which meant that I had only a quarter of the air supply left, barely enough to get back to Nine-Two and the situation was clearly critical. Luckily I cruised into a clearer patch of water and followed it to find, as it only could, that it led upstream. I'll never forget how beautifully clear and green-blue it looked.

I decided that something must be done soon, since all attempts to return had failed, and by now there was almost no chance of getting back to Nine through the mud even if there was a wire I could follow. The only possible way out was to go upstream. This was a very forlorn hope because the cave was now in the limestone, as we had left the conglomerate between Ten and Eleven, and had started to go deep after the fashion of many French *fontaines vauclusiennes* nearly all of which had proved impassable. After 25 feet or so I was down at 50 feet depth and saw a cross rift above me that seemed to go high, though I was too deep to see if it had a surface. Anyway there was by now no other alternative so up I went in a beautiful cloud of air bubbles and waited more or less impatiently to see whether I would hit a rock roof or not.

It was a relief to bounce through an air surface and to see that the chamber had enough air in to last at least for a few hours. I wedged myself across the rift, turned the face-mask valve to air, closed both main cylinder valves and gasped for several minutes. There was a 3 foot diameter passage leading off out of sight starting 6 feet above the water, but rock climbing in swim-fins and wearing 80 lbs or so of equipment was out of the question. It was convenient that Aqualungs have quick release straps so I wriggled out of it, chimneyed my way to the ledge carrying my Aflo, then pulled the Aqualung after me by its longest strap. The time was 00.24 hours on 11th December by my waterproof watch so I recorded this on the Aflopad and spent the next half-hour in darkness shivering with cold and, no doubt, relief. During this time there were two series of glugs from the water which I now know were caused by Buxton and Wells entering and leaving Nine-Two on their last trip to look for me. It showed that they were alive and taking action.

Fig. 7.5
Pensive Wessex Cave Club members await news at the club's Hillgrove Hut.
Left to right: Denis Warburton, Dave Willis, Derek Ford, and Luke Devenish (front).
Photo from Luke Devenish

After pondering on the eternal verities, the nature of luck, and the relation between chance and lack of information, it was clearly time to survey the situation. The very first thing was to find out for how long the air in the chamber would last me. The passage I was in went backwards towards Nine and after 30 feet came to a small aven with a pool in the bottom. This I named the Twelfth Chamber and I am glad that I did not go up the rift below it because it looks impossible to climb from the water to the passage. I would have been like a frog in a half-filled vertical drainpipe. The other end of the passage went round the corner to the right above the surface of the water I had swum up. This I named the Thirteenth Chamber. The visible part of this chamber contains enough air to last a single diver for about three days so there was time for me to think about what I should do.

One of the brass side-strips of the Aflo had got bent right over and it was this which had prevented me from closing it before. This was soon remedied. The Aqualung cylinders were found to contain 35 atmospheres pressure each, which was a little more than I expected and meant that the first pressure equalization had not been quite complete. I then worked out the following calculations: I had used 120 minus 35, i.e. 85 atmospheres pressure, to get from Nine-Two to Thirteen. Thus a return trip would have to be done at 2.4 times the speed that I came. I had come in 39 minutes and therefore would have to return in $39 \times 35/85 = 16$ minutes. But some of the time spent on the journey here had been spent almost stationary when I would be using only small amounts of air, so I reduced the likely time that the air would last at the depth of the passages to 12 minutes swimming time. I estimated the distance to Nine-Two to be 200 to 250 feet. Now from Three to Nine-One is about 350 feet and takes 14 minutes. Thus a return to Nine-Two whilst reeling out a wire might take $250/350 \times 14 = 10$ minutes. The only other information available was that when the first cylinder had become empty I had used the equivalent of 60 atmospheres pressure in two cylinders and thus would have to get back in $35/60 = 0.58$ of the time that that had taken.

These calculations told me that I might hope to get back with less than a 20% safety factor if I could see where I was going and did not get lost again. This seemed too dangerously near the limit but there were other considerations.

The other members of the CDG are trained only in the use of oxygen apparatus and the rescue route went down to 50 feet which is well below the 33 foot safety limit for such apparatus. They could not know how deep they might have to go, nor how far to find me, so the nervous strain of such an attempt would be very great and the risk of oxygen convulsions incalculable by them. I would not have wanted a fellow diver to die in the attempt to find me. There was a probability that the Controller (Dan Hasell) would call out Naval divers with oxygen/nitrogen mixture apparatus. It would not have been reasonable to allow men without cave diving experience to try to find me though I'm sure they would have tried anyway. Another chance was that Buxton and Balcombe, who would no doubt be called out, could learn to use the special Naval apparatus (not yet available to civilians). I did not see how they could do all this, find me, then bring another apparatus for me to practice with and then use to get back to Nine, in less than two or three days. By then I would probably be very short of air and, since I was shivering about one third of the time in my sodden underwear, may well be suffering severely from exposure, hunger and lack of sleep. Besides there was no certainty that I would be found in time. All in all there seemed to be strong reasons for trying to get back with the resources available to me. If successful, this would stop a lot of people, including my wife, from having a very worrying few days; if unsuccessful, I would not be worse off provided I got back to Thirteen.

By now it was 02.50 hours and I recorded on the Aflopad some details of the route, the pressure in the cylinders and my intention of trying to return. This was in case I did not get back to Nine or to Thirteen and would help in reconstructing what had happened. I also wrote: 'I will lay a line until one of the two cylinders is empty and will then go on or back according to where I am at the time.' It would have been quicker without a wire, and there would not be the chance of getting tangled, but there would have been no possibility of return once I had started. An extra hazard was that I would be going as fast as possible and a rapidly revolving line reel can over-run and jam the wire.

I therefore put my underwater horn (Aflohonk) in its standby position at the front of the handle so that I could now slip off the line reel and leave it behind if I had to. This would be quicker than cutting the wire, which would be a pity because it would mean losing contact with Thirteen.

The optimum time for the attempt was fixed at about three hours after arrival. The water should be clear enough by then, and there would not have been enough time for any preliminary rescue attempt so I would not have to waste time getting past other divers or getting slowed down by the loss of visibility they would cause. I spent this time in darkness, to conserve my lights, and rechecked my calculations several times as I wished to convince myself that I really had done them correctly. The 'inexorable nature of time' to quote Balcombe was symbolised for me by a drip which splashed eerily and regularly into Twelve, and quite soon it was necessary to do something. The line was lashed very firmly to a boulder, and I lowered the Aqualung into the water to wash the mud off the face-mask window. My spectacles were fixed to the Aflo, they were covered in mud and there would not be time to evaporate condensed mist onto the window, and the compass was set to 10 degrees South of magnetic West. The time had come.

I dressed myself lying down, opened the right-side cylinder, and slithered into the water. The nearly empty cylinders were now rather buoyant and it was possible to sink only by turning right upside down and swimming down head first. The pressure equalization in my ears was a bit slow on the rapid descent but at last they clicked nicely and I got to the bottom. The compass bearing proved correct and I recognised the slope to Eleven, but found no wires. There were two alternative routes and, as this was downstream diving the mud I had scraped off the sides of the rift was already running ahead. I chose the right route, which was wrong; it became sealed, so I flipped round for the left, which was right.

At this stage I saw 2 feet of wire left by Coase in 1949 which was otherwise buried in the mud, and my first cylinder gave out. Although less than half way I decided to continue and pulled on the wire to increase speed but it snapped. Visibility was now 4 feet and in another 15 feet the line reel jammed solidly. There was no question of returning to 13 now in any case, so I removed it and left it there. The passage led on to the beginning of the Second Loop (which goes from Ten to Nine-One). I recognised it, turned smartly left and started to ascend from 20 feet down, at which stage breathing became difficult, thus heralding the impending exhaustion of the air supply. I went up obliquely as fast as was reasonable, dropping the Aflo en route and hit an air-surface at speed in complete darkness. I hoped it was Nine-Two. After a pause to recover breath (believe me you tend to gasp when you are frightened), I removed the Aqualung, waded ashore, fished out my emergency torch from my weight belt and confirmed that I had reached Nine.

After 5 minutes the mud cleared sufficiently to see the light of my Aflo about 12 feet down. Several coils of wire had grown beside the belay point (they had been left by Buxton). I removed the weight belt, swam out onto Nine-Two and fished for the Aflo with the wire for a while. This was successful and I got it ashore. The watch read 04.07 hours. I had recorded my departure from Thirteen as 03.50 hours and I estimated the journey took 9 to 10 minutes. The Aqualung pressure gauge read 3 atmospheres which, whilst not much, was enough; I had plenty of oxygen in the P-Party set over the rocks in Nine-One.

I considered returning cleanly with all my gear, but decided to return as soon as possible in case anyone was getting alarmed or had already decided to eat my sandwiches.

The climb over to Nine-One was easy without kit and there I found a lighted candle near the water. This meant that the other two divers had also got safely to Nine and had left for Three.

The P-Party set is heavy and cumbersome but somehow I managed to get it on without help, all but one strap that I couldn't reach, turned to say farewell, hit a rock, knocked open the high-pressure bypass, and fell over.

Having lost what I estimated to be 20 atmospheres I closed it and decided that not much more could happen, and that if it did, it didn't matter, I was past caring, so returned uneventfully to Three

in water visibility of 3 feet between 04.49 and 05.02 hours.

Later that day Buxton, Wells and I returned to the cave, found the line reel and brought out the frog-sack, fins and Aqualung from the Ninth Chamber. The markers on the wire I had laid showed that the distance from Thirteen to Nine-Two is 202 feet.

Fig. 7.6
Oliver Lloyd (left) and Derek Ford (right) take the late-shift on the overnight vigil for Bob Davies in the Third Chamber.
Bob eventually returns at 5 am on 11 Dec 1955 to greet them with the wry remark that 'The Devil is a Gentlemen'.
Photo by Luke Devenish

Fig. 7.7
Bob Davies' Aflo used on the 'Night of the Thirteenth' (now in Wells & Mendip Museum).
Photo from Bob Davies album in Wells & Mendip Museum

Amongst the lessons to be learnt from this escapade are:-
1. Fix the bolts on Aflo properly.

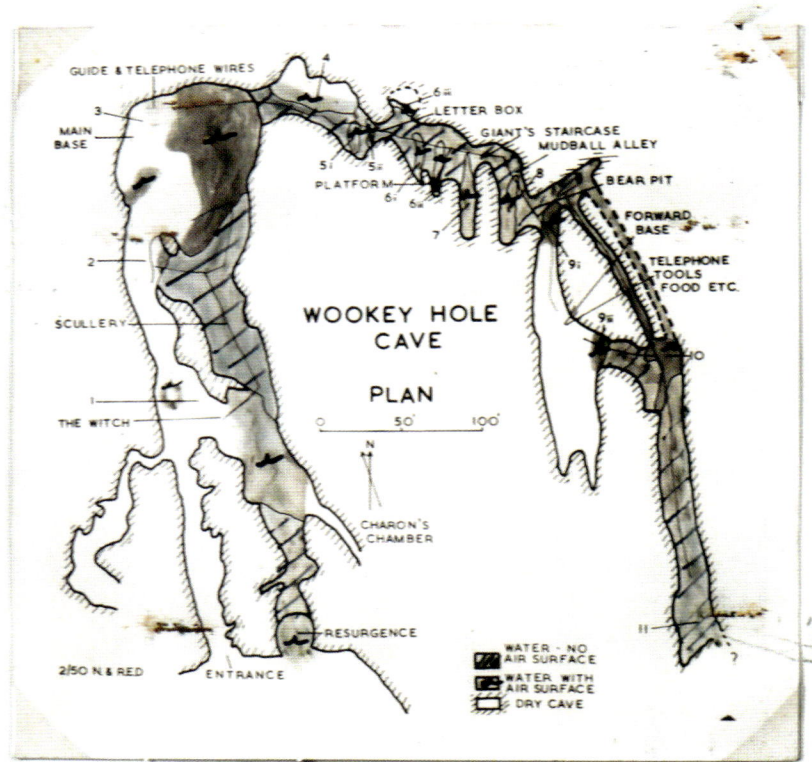

Fig. 7.8
Bob Davies adds a pencil sketch of the Twelfth and Thirteenth chambers to the survey of the known cave at the end of 1955.
Photo in Bob Davies album in Wells & Mendip Museum

2. See that the line reel works smoothly and that the wire can be withdrawn.
3. Swimming in unknown cave territory is very hazardous because of the danger of floating away, thus bottom-walking divers should always hold hard onto swimmers when they require it.
4. If a diver is lost, a wire with a light on the end (torch or Aflo) should be left as soon as possible as near to the place where he disappeared. He may want such a guide badly.
5. Spit on the Witch (I forgot this for the first time!)

[The above account was typed-up a few days later on 19th December 1955 whilst everything that happened was fresh in the memory, but it was never published at the time. Over thirty-two years after the event, Bob Davies writes as follows to explain his choice of the title for the chapter.]

The phrase '(In) the waters under the Earth' comes, of course, from Exodus 20:4. When I was lying in the Thirteenth Chamber waiting for the mud to settle to give me a chance to return I thought how lucky I was that it was there and thought how deeply religious types might think that it had been made for them there in the waters under the Earth. Of course it's only those who escape who talk about how God saved them. I have no such beliefs. I even toyed with calling it the cave of the Eumenides (Erynes or Furies) from Aeschylus's play in the Oresteian Trilogy since the Furies went underground after Athene voted to save Orestes. 'I promise you here in this upright land, a home and bright thrones in a holy cavern, where you shall receive for ever homage from our citizens' and they finally went 'on to the deep of earth to the immemorial cavern...' But it all seemed very pretentious so I never told anyone about it until you asked, so many years later.

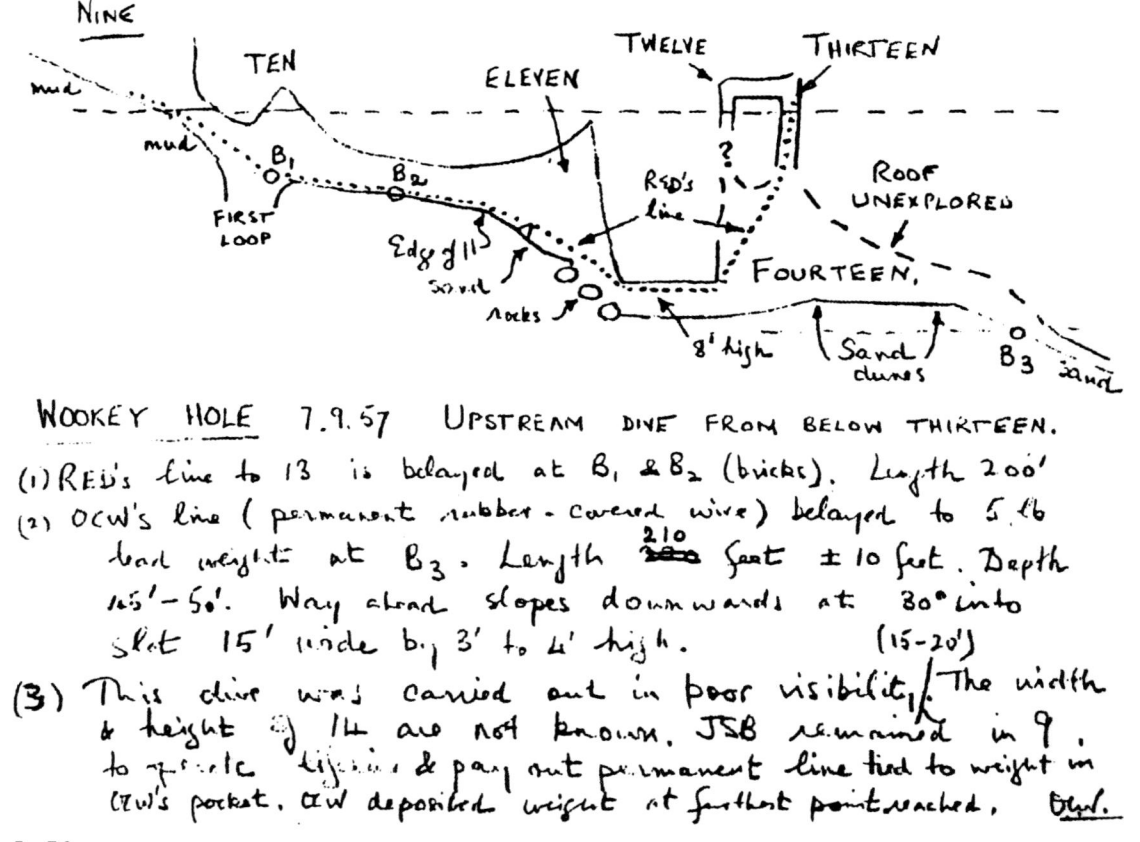

Fig. 7.9
Oliver Wells illustrates Bob Davies' epic upstream dive to the Twelfth and Thirteenth chambers on an extended section, drawn in 1957 – see Chapter 8

Wookey Hole Caves Ltd.
NEAR WELLS, SOMERSET

DIRECTORS:
G. W. HODGKINSON, O.B.E., M.C.
OLIVE HODGKINSON
C. G. S. HODGKINSON
J. R. HODGKINSON

TELEPHONES: WELLS 2243-4
TELEGRAMS 2243 WELLS
STATIONS: WOOKEY (W.R.) 1 MILE
WELLS (W.R.) 2 MILES

YOUR REF:
OUR REF:

CAVE DIVING STORY
10t December 1955.

As given to B.B.C. News Room at 7.45 a.m. on 11th.

Professor R.E. Davies, XXXXXXX, Mr. John S. Buxton and Mrs. O.C. Wells, members of the Cave Diving Group continued the explorations in W.H. Caves. They revisted the very deep 11th chamber and Professor Davies discovered the 12th and 13th chambers with continuing passages upstream.

This exploration was unique in that the distance travelled underwater was too far for Aqualungs (compressed air aparatus) and too deep for oxygen equipment because of the danger of oxygen poisoning at great depths. Oxygen equipment was used from the limit of the public part of the cave to the 9th chamber and from then on compressed air equipment was used to pass under the flooded passages and up to the 12th chamber.

This exploration involved travelling 1200 feet under water in mud with visibility nil for much of the time, down to depths of 50 feet. Due to the opaque mud that was stirred up it was necessary for Professor Davies to remain alone in the 13th chamber for 3½ hours in complete darkness until the water was clear enough for him to return to the rest of the diving team who were preparing rescure operations since he had been away from base for three times as long as he could sruvive with his underwater breathing aparatus.

Fig. 7.10
Wookey Hole Caves Ltd., issue a press release to the BBC announcing the safe return of Bob Davies, early on 11 December 1955

Fig. 7.11 A typical shore scene of cave divers in the mid 1950s. Photo by Bill Darby of West Advertising Limited, Bath

Chapter Eight

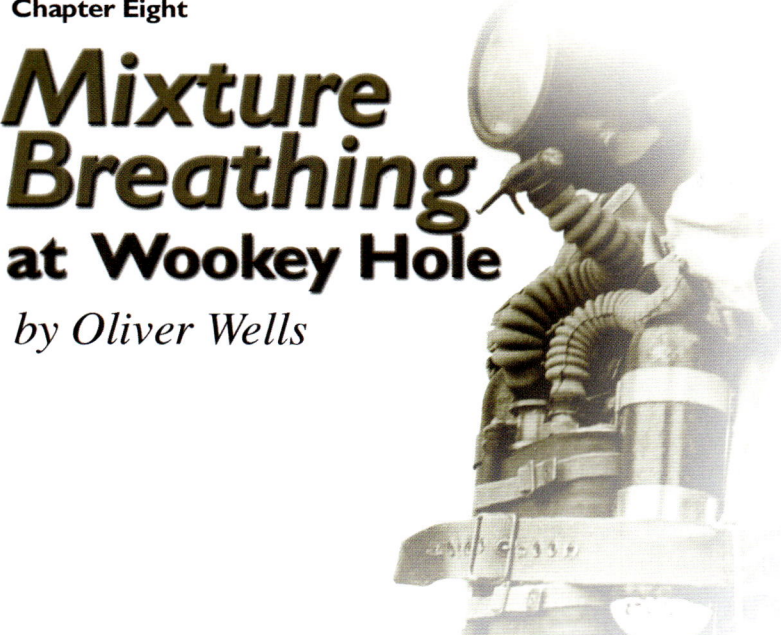

Mixture Breathing at Wookey Hole

by Oliver Wells

*Fig. 8.1
The kit used by Oliver Wells on the 'Night of the Thirteenth', as described in the previous chapter, 10 Dec 1955.
Photo from Bob Davies album in Wells & Mendip Museum*

In this chapter I shall explain why, in 1956, we chose to explore beyond the totally submerged Eleventh Chamber of Wookey Hole using mixed-gas rebreathers and with boots on our feet, rather than swimming with flippers and aqualung as had been tried by Bob Davies in 1955 and is the method used nowadays. The word 'we' here refers to John Buxton and myself, and the time when these events took place was shortly after Bob had discovered Thirteen under somewhat traumatic circumstances, after which he went to America. Had Bob remained in England, then he would no doubt have continued with aqualung and fins. But as it was, I was left with John Buxton to decide what to do.

The dives which are described in this chapter were made into the upstream pool in the Ninth Chamber, which is known as Nine-Two. The Eleventh Chamber is about 80 feet upstream from this point. Visitors to the cave can now reach the dry area in Nine through the tunnels which were driven later. In 1955, the downstream pool in the Ninth Chamber, called Nine-One, was reached by a dive of about 18 minutes duration from Three, after which you could either climb over the rocks to reach Nine-Two, or you could follow the underwater route round Coase's Loop. We generally carried an Aflo, which incorporated the diver's light, line reel, depth gauge, compass, watch, note pad, and various other items such as the electrical signalling device which I built in 1958.

My first visit to Wookey Hole was in 1940 at the age of 9. The cave guide told us about the chambers that had been found by diving. He also told us that the most powerful electric lights had been unable to reach the roof. I had been reading about the use of searchlights in air-defence, and I was left with the impression that these chambers must be very big indeed.

I started caving on Mendip in 1951, and in due course became increasingly intrigued by the stories of dives in Swildon's Hole and Wookey Hole by Graham Balcombe, Jack Sheppard and their friends. Cave diving on Mendip was at a fairly low level at that time and information was essentially limited to comments made by Don Coase and Luke Devenish on the few occasions when I was able to draw them out on the subject. Both of these gentlemen had the distinction of having 'pushed' an unexplored terminal sump, Casteret-wise, without either a respirator or protection from the cold water; Don in the Stoke Lane Slocker sump on Mendip in 1947, and Luke in the Grotte de Sainte-Héléne in France in 1949. Other Mendip cavers who held this distinction included Don Thomson and Willie Stanton for penetrating the sump in Ludwell Cave, and Oliver Lloyd

Fig. 8.2
Oliver Wells goes caving and cave diving whilst at Cambridge University in the early 1950s. Photo from Oliver Wells

for free-diving a side-sump in Stoke Lane Slocker. All benefited from a familiarity with sumps having free-dived Sump 1 in Swildon's Hole on many occasions.

In the days before exposure suits and wet suits the diving of Sump 1 was a great psychological barrier; once accomplished, you had joined the ranks of the few. Don Coase had also shared the honours with Graham Balcombe in leading exploratory dives in Wookey Hole Caves from the Resurgence to Eleven between 1946 and 1949. Unfortunately Don had to retire from cave diving for health reasons shortly after the photographic trip to Nine in 1953. After we had discovered Fourteen (as described in this chapter) he told us that he was going to have an operation and then take up diving again: 'My doctor thinks that he can repair my aneurysm or at least make it safe'. Unfortunately he did not survive the operation. Graham Balcombe wrote in 1981: 'Coase was a first-class diver and a great companion always, and a steadfast support when I was stressed'. Bob Davies wrote: 'Don Coase was a major force in the post-World War II diving explorations associated with the Cave Diving Group in England. He was bold, reliable and always ready to push ahead. His early death during surgery was a great loss'. He had been at the centre of most Bristol Exploration Club achievements too, notably the initial exploration of St. Cuthbert's Swallet.

In the early 1950s, the centre of active cave diving in England had moved to Derbyshire. Thus, when Wookey Nine was photographed in 1953 only Don Coase lived near Mendip, Graham Balcombe came from London, while Bob Davies, Jack Thompson and John Buxton were all members of the Derbyshire Section of the CDG. Trevor Ford was active as Secretary in Derbyshire for many years; a geologist and authority on the Peak District.

Without Don Coase cave diving on Mendip remained at a low level until Bob Davies asked Oliver Lloyd to organise support parties for a pushing dive on Sump 2 in Swildon's Hole. Oliver did this in magnificent style on the weekend of 26th and 27th June 1954. Bob Davies then started on a series of dives in Somerset supported by John Buxton. Oliver Lloyd organised the local sherpa teams; indeed, he was the first to borrow the term 'sherpa' from mountaineers to describe the portering role of cavers on big diving operations under Mendip for, at the time, Swildon's was the deepest cave in the country. So it was that several younger Mendip cavers became interested in cave diving, myself included. This was only the beginning of many years in which Oliver Lloyd contributed significantly to cave diving, as will be told in subsequent chapters.

It turned out that Graham Balcombe lived about ten miles from my parents' house in London, and I paid several visits to him and his wife Mavis, each of whom gave me their impressions on the subject of diving in caves. Mavis was very enthusiastic when I expressed appreciation of Graham's successes in the past, was overjoyed when I took items of diving equipment out of their house, but was not quite so cheerful when I encouraged him to come on dives.

It was at this time that Graham Balcombe gave me the 'bicycle respirator' which I refurbished in 1960 and used on a dive that was photographed underwater by Jim Stark in a swimming pool near Pittsburgh, Pennsylvania. I gave this historic apparatus to the Wells

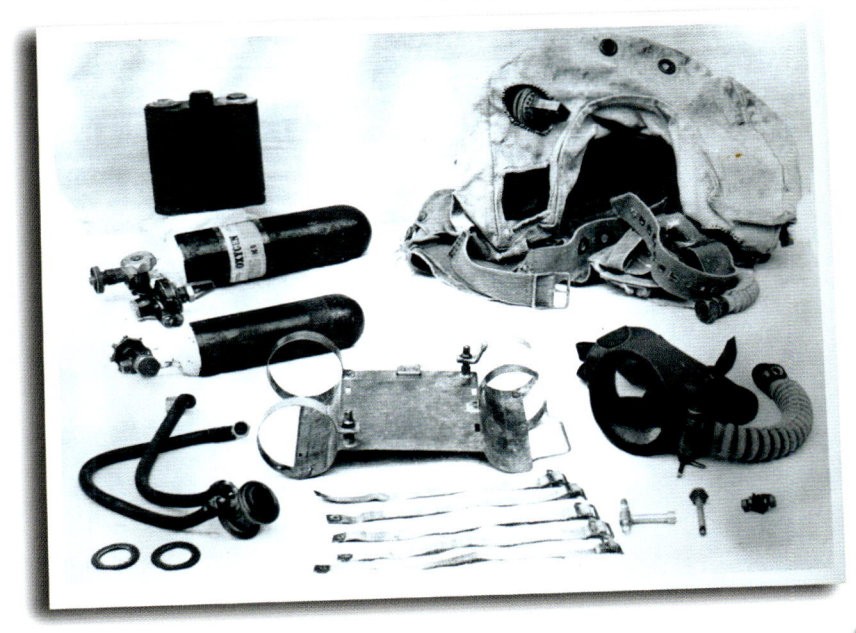

Fig. 8.3
(a) The kit is dismantled and displayed by Oliver Wells, then...
(b) Reassembled ready for use (as in Fig. 8.1)

Museum in 1985 and it was appropriately taken back to Sump 1 in Swildon's so that Sid Perou could re-enact the pioneering cave dives for his BBC television series.

Graham told me about his dives at Keld Head in Yorkshire in 1944 and 1945, to which he travelled for several miles with Mavis on a bicycle made for two. He told me about his dreadful moment when he suddenly found that he could not breath either in or out, and so felt over the breathing tubes to see if either of them had been obstructed in some way. Then he found that he could breathe again. It turned out that he had fitted a shutoff valve onto the mouthpiece, and this had accidentally been knocked shut. Equally accidentally, he had knocked it open again. During the emergency, he had forgotten about the existence of the valve completely. Careful training under controlled conditions had understandably become paramount, each experienced cave diver taking turns to hand on his knowledge and skills to new trainees.

My 'safe water training' was organised by Jack Thompson in Sheffield. The CDG had a training schedule in those days which called for five hours of underwater experience with a diver tied onto the end of a rope, together with several more or less disagreeable events such as losing consciousness in a closed-circuit respirator from carbon dioxide excess and from oxygen lack; both of these tests are dangerous and require medical supervision. The War Surplus oxygen respirators that were in the greatest supply were Siebe, Gorman Amphibian Mk. II (SGAMTU). I reorganised one of these with the main cylinder, its reducing valve, the emergency cylinder and the soda-lime canister on a brass plate on the chest. The breathing bag stayed around the neck.

There were two sorts of diving dress in use in the CDG. The Sladen Suit was a bulky dry suit with a strange-looking rubber helmet with a little window on the front which opened on hinges. You could not disconnect the respirator without the use of a spanner. This had been developed for use on Human Torpedoes with a long-duration respirator, and the watertight nature of the helmet was then an advantage. Graham Balcombe and Jack Sheppard had used such outfits to discover Eight. By 1954, it had become the standard practice to replace the helmet with a thin rubber hood which was used with an easily removable full face mask. I was given a Sladen Suit, and my first task was to glue on the rubber hood. Meanwhile, the established members of the CDG were issued with two-piece frogman suits, which were less bulky and more suitable for swimming. It is much easier to swim with a wet suit, but the CDG did not have them in those days.

We were also called upon to jump into the water from a height of a few feet, which I did from a diving board into the river at Stratford-on-Avon under the watchful eyes of John and Audrey Buxton. Finally came the first cave dive, which was a walk with Jack Thompson into the Swine Hole at Peak Cavern. Graham Balcombe and John Buxton took me for a dive to the Seventh Chamber of Wookey Hole, after which I achieved the grand status of 'Junior Employee', and was permitted to go with Bob Davies and John Buxton on the 'Night of the Thirteenth'. Next day, Bob Davies resigned his position as Chairman of the CDG in view of his impending departure to America, and I was elected to replace him.

Concerning the Eleventh Chamber, our thinking was heavily affected (perhaps too much so) by the traumatic events of 10th and 11th December 1955, when Bob Davies vanished in a cloud of bubbles while swimming in the muddy water in Eleven. He was given up for lost, and then caused astonishment to everyone when he reappeared in the early hours of the following morning looking somewhat embarrassed, perhaps, but unmistakably still alive. We had all been invited into their house by Wing-Commander and Mrs. Hodgkinson, and their domestic staff had reappeared, even though it was many hours after their normal tour of duty had ended. I recall that when Audrey Buxton came up to me shortly after 5 a.m. and said 'Bob is back in Three', I replied: 'That is impossible'.

Bob's dramatic reappearance has been described by Derek Ford as follows:

> *I was one of the two who stayed (in) Three to look for signs of Bob ... After a couple of hours we saw Bob's woolly cap come out of Sump 3 and fished it out. A while later the guide wire twitched and he emerged and beached slowly and steadily. He was too tired or overwhelmed to stand so we hoisted him ashore. But he very quickly recovered and took command.*

My main impression from that dive is how slowly the mind comes to an unwelcome conclusion under unexpected conditions. John Buxton realised that a problem had arisen much sooner than I did, but could not tell me about it while we were underwater. After we returned to Nine-Two, John took the first step of a correct analysis when he said: 'Bob must have seen something down there that was worth exploring or he would not have tried to tie on his line in order to continue with the dive.' At the time it never occurred to either of us that it might be a good idea to leave a line with a light on the end of it in Eleven. After discussing the situation, we replaced our main cylinders and soda-lime canisters, and started to climb back over the rocks on our way back to Nine-One, and from there to Three.

Fig. 8.4
The decision in the mid-1950s to go deep, either by bottom-walking or swimming with fins, is subsequently illustrated graphically by Oliver Wells

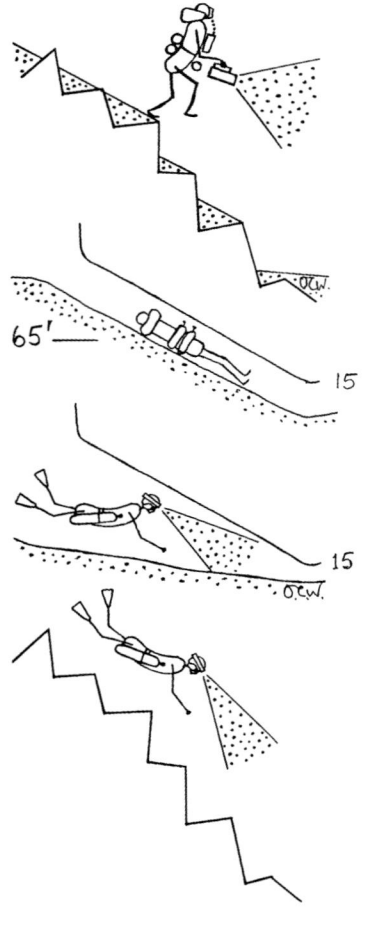

Afterwards, both of us were determined to continue with the exploration. The first question was: 'Should we walk or swim?'. Present-day divers are expert swimmers, but this was not always so. In 1955, it was only two years since the publication in England of *The Silent World* by Jacques-Yves Cousteau and Frédéric Dumas in which they described their explorations in the Mediterranean using aqualung and fins. Naval 'frogmen' were proficient at swimming with closed-circuit oxygen apparatus, using Universal Breathing Apparatus (UBA).

Within the CDG, Don Coase and Bob Davies had become expert at swimming in caves, and had demonstrated the reduction in mud-stirring and the higher speeds that this provides. On the other hand, Graham Balcombe never had liked fins, and said so:

It is easier to carry loads if you walk along the passage floor, and you are less likely to get lost in muddy water.

I had been somewhat surprised when he had first said this to me because, like most people, I had been greatly impressed by the pictures taken by Cousteau. Later, however, I tended to agree with Balcombe on this point of view for on the three occasions between 1956 and 1959 that I discovered dry land at the far end of a submerged passage, it was always under conditions when walking was satisfactory, and I would not have obtained any advantage from the use of fins.

Naval divers in those days were also divided into those who liked swimming and those who did not. I can remember being told by an officer at a recruiting stand in Chatham:

Flippers are splendid for the glamour boys, but for real work underwater there is NOTHING LIKE CARRYING EXTRA WEIGHT AND GETTING SOMETHING DONE! You do understand that, young man, don't you?

You can obtain some understanding of this point of view from the issue of *National Geographic Magazine* dated April 1964. In the early pages there is a wonderful article by Cousteau, describing an underwater habitat, and supported by beautiful pictures of divers, fishes, and the good life below. In the later pages there is an article on building the tunnel under Chesapeake Bay, which seemed more appropriate to cave diving in my experience:

Under a diver's guiding hand, the linking pin on the upper rim of a newly dropped section of tunnel glides towards its hole on the lower rim of a bedded section. In inky darkness, the diver performs his dangerous job by touch. Later the joint will be protected by steel plates and sealed in concrete.

This diver is shown as wearing a helmet and boots, with clouds of bubbles and a hose from up above. This emphasises once again the 'Two Cultures' situation which existed for so many years on this question.

Before leaving this subject, perhaps I might mention that Bob's adventure on the 'Night of the Thirteenth' was the third occasion on which divers had been lost in muddy water at Wookey Hole. The first of these took place on 9th April 1949, when Don Coase led with Bob Davies in support to see if there was a way upstream from Eleven without going below 30 feet. None was found, but the trip was eventful for other reasons. A line had been laid as far as the near side of Eleven on an earlier dive, and according to Don Coase, the plan was for him to swim round the chamber at the 30 foot level, while Bob Davies was to remain at the end of the line. According to Don, the plan was then for Bob to swim at the higher level to see if there was a higher way on. Now it is easy to get things wrong underwater especially when more than one diver is involved and, in the event, this trip ended up with TWO divers swimming in circles, hunting for the line. The second occasion was the Marriott tragedy. The third occasion was when Bob Davies discovered Thirteen.

So all of this did not leave us with any very wild enthusiasm for the idea of fin-swimming in caves unless there was an inescapable reason for it.

The next question was: 'What sort of respirator should we use?' In 1955, the choice for diving below 30 feet was the twin-hose aqualung or mixture-breathing apparatus. Today, mixture-breathing respirators of the sort described here have been replaced by sensor-controlled rebreathers of greatly improved design. This is referred to in Chapter Twenty One.

Fig. 8.5
John Buxton (wearing fins) and Oliver Wells (in boots) demonstrate the closed circuit oxygen respirators then used by cave divers, variously based on the Siebe, Gorman Amphibian Mk.II, and safe down to 30 feet with moderate exertion. Deeper dives required the use of P-Party sets with an oxygen/nitrogen mixture delivered by a special reducing valve.
Two gas mixture cylinders, each containing 250 litres, were slung horizontally on the lower back within easy reach.
Photo from Oliver Wells

The main point to be considered was the duration of the dive. It made a great impression on us that at no time were either John Buxton or myself in any real danger from running out of oxygen on the 'Night of the Thirteenth', while Bob Davies with his aqualung was in a very unhappy situation at the same time. This led us to over-react on the question of duration, and we ended up by believing that the 55 minutes that Bob's aqualung lasted was not long enough for a serious dive of this kind. The present arrangement in which the diver carries more than one respirator and changes the mouthpiece underwater was never even suggested, so far as I can remember.

As an alternative to the aqualung, the CDG possessed two mixture breathing respirators that could last up to two hours at depths down to 80 feet or more, limited mainly by the size of the soda-lime canister and by the need for decompression stops during the ascent. The larger soda-lime canisters were good for 3 hours regardless of depth. These were the P-Party sets which had been developed for harbour clearance during the Second World War. Graham Balcombe had bought two of these in 1946 and had used them with Jack Sheppard to discover Eight. They had only been used by the CDG with pure oxygen up until that time. Graham Balcombe suggested to us that we should use the P-Party sets. When I asked Bob Davies about it, he replied that they would be 'excellent'.

The technology of oxygen apparatus and of mixture-breathing apparatus is as follows. With a closed-circuit oxygen respirator, the diver breathes from a rubber bag or 'counterlung' with a soda-lime canister in the circuit to absorb the carbon dioxide. The soda-lime must be kept dry. This is the reason why the mouthpiece is entirely surrounded by air in the Admiralty Pattern face mask. This helps to keep the water out of the apparatus, but the rescue procedure in which divers exchange mouthpieces underwater ('buddy breathing') is not possible.

Fig. 8.6
Oliver Wells and Phil Davies test their respirators for leaks before diving at Wookey Hole Cave in 1953. Apart from gas leaks, even the smallest quantity of water could be harmful if it entered the soda-lime canister.
Photo from Phil Davies

Oxygen must be supplied for two reasons. First and foremost the oxygen that is consumed by the diver must be replaced. Second, the volume of gas in the system must be maintained as the diver goes deeper. This is accomplished by a reducing valve which gives a steady flow of typically 0.8 litres per minute, to which additions can be made by opening a bypass valve by hand. You must 'crack the bypass' in this way either to increase your buoyancy when swimming, or if the bag goes flat when you try to breathe in. Thus, if you are consuming oxygen at a rate of 1.4 litres per minute, the balance of 0.6 litres per minute must be let in by hand. You are in great danger if any air remains in the system, because you can burn up the oxygen and leave yourself with essentially only the nitrogen, which leads to unconsciousness without warning; a condition called 'anoxia'.

With mixture breathing sets, the danger of anoxia is avoided by blowing an oxygen/nitrogen mixture into a rebreathing respirator fast enough to provide an acceptable partial pressure of oxygen from the surface to the maximum working depth. Sir Robert Davis briefly described this in his article 'Divers and Diving Apparatus' for the 1929 edition of *The Encyclopedia Britannica*. A diagram shows the standard diving dress modified for self-contained diving. In the text, it says: 'The apparatus can be used at depths down to 150 feet'.

The theory and practice of mixture breathing respirators as they existed in the 1950s is described by Stanley Miles in his book: *Underwater Medicine*, (1962; pages 271-273) and by Mark Terrell in *The Principles of Diving*, (1967; pages 98-117). As an example of this technique, Terrell discussed a dive to 165 feet. The oxygen/nitrogen ratio for such a depth is 34.4% to 65.6%. This must be blown into the breathing circuit at a constant mass rate equivalent to 16.67 litres per minute at the surface. This is 0.6 cubic feet per minute as measured at atmospheric pressure. Since an aqualung diver usually needs about a cubic foot of air per minute as measured at the depth of the diver, it follows that the rate of gas consumption of the mixture-breathing diver at a depth of 165 feet is about half of what an aqualung diver consumes at the surface.

The P-Party sets were only suitable for use by walking divers, but this did not worry us too greatly since we had decided not to swim. Even in the Navy, where the P-Party set had been replaced by the UBA, swimming with mixtures below 30 feet was experimental as late as 1958. Thus, in the *RN Diving Magazine* for July 1958, we find:

> *Perhaps the most interesting have been the development of deep-swimming to 120 feet using a new type of reducer with a 12 litre flow, and investigations into the problems of controlled ascent and decompression stops from that depth introduced by the expansion of the gas in the counter-lung and suit inflation.*

There are, of course, several advantages of the aqualung which we did not appreciate at that time. Thus, you can hold the cylinder at arms length in a tight submerged passageway (as was done by Mike Boon through Sump 6 in Swildon's Hole later in 1961) or even exchange the whole thing for another one underwater (as was done by Oliver Statham and Geoff Yeadon on their through-dive at Keld Head in Yorkshire in 1979, and nowadays is commonplace in the deeper cave dives in Florida and elsewhere). Bob Davies said: 'It is the simplicity of the aqualung that appeals to me'. Jack Thompson also used an aqualung in caves. Luke Devenish, who dived commercially as well as in caves, preferred to breathe air. The serious nature of mixture-breathing was expressed by Empleton et al in their book *The New Science of Skin and SCUBA Diving* (1962, page 105) as follows:

> *Only a very few semiclosed-circuit scuba are in use by sport divers. Most of these are used by former commercial or military divers who have had several years of training and experience in the use of the equipment. The safe use of such equipment requires knowledge, training, and experience under the supervision of a competent instructor...The many disadvantages and limiting factors encountered in the safe use of closed- and semiclosed-circuit make their use for sport diving both impractical and hazardous.*

In the event, we decided to explore the River Axe upstream from Eleven using the two P-Party sets, with an extra oxygen cylinder for shallow diving and for emergencies at depth. In our preparations for mixture-breathing at Wookey Hole, we were greatly helped by Sir Robert H. Davis, who by then had been appointed as the Chairman at Siebe, Gorman and Co. Graham Balcombe wrote to him on 19th March 1956, and received the reply:

The reducing valves fitted to the P-Party apparatus made during the War were in fact 'constant mass' reducers, not 'constant volume'... I would suggest that if you are considering the use of mixtures in deep water, you should send your reducers back to us for overhaul and re-setting, since their efficient functioning is vital for the safety of the diver.

This was duly done, and on 19th September 1956, Sir Robert wrote to me:

Your reducers have been adjusted to pass 3.3 litres per minute, and are going to you by parcel post today.

This was for a 70/30 oxygen/nitrogen mixture which is safe down to 65 feet. Sir Robert also gave us various spare parts for the P-Party sets, and two 2 cubic feet oxygen cylinders which I used in my lightweight oxygen respirator for discovering Swildon's Six in 1958. In addition to the above, members of the CDG were very kindly received by Naval divers at HMS Vernon in Portsmouth on many occasions.

*Fig. 8.7
Oliver Lloyd, Oliver Wells, John and Audrey Buxton make plans to go deep down the steep slope beyond the Eleventh Chamber in Wookey Hole, at the Wessex Cave Club's Hillgrove Hut, 1956. Photo by Luke Devenish*

The first step in the cave was to discover whether the downwards slope in the Eleventh Chamber really was as steep as it had appeared. If it was gentle enough to walk down, our job would be easier. On 30th June 1956, John Buxton and I returned to the cave breathing oxygen. He lifelined me from Nine-Two as I walked to Eleven and looked over the edge. I let some water into my mask, looked down the slope, and estimated from the angle of my local water surface that the slope was no steeper than 45 degrees. At my feet I could see the line that Bob Davies had laid on his return from Thirteen, and it disappeared downwards at an angle that was only slightly less steep than the slope itself. The water was its usual slightly greenish colour, but on that day was also slightly milky. I could not see the hole at the bottom. All that I could see down the slope was a small triangular boulder pointing upwards out of the sand almost at the limit of my visibility. By looking to the left and right I could just about distinguish the vertical rock walls. Unfortunately, the light on my Aflo used a silver-plated motorcycle reflector which gave a rather wide beam – unlike the beautifully sharp beam on Bob Davies' Aflo.

Immediately on my left there was a rock buttress six feet high overhanging the slope by a couple of feet. John Buxton and I had spent many anxious minutes beside this buttress on the night we lost Bob

Davies, and he had investigated a blind alley on its far side on his return from Thirteen. I also verified that there are no problems when returning in a straight line from this place to the shore in Nine-Two. This occasion was also memorable because our supporters set fire to some jackdaws' nests in the passages above the entrance to the cave, so that the first three chambers were filled with smoke (Oliver Lloyd describes this in a later chapter).

In August and September, 1956, I took one of the P-Party sets on a caving trip to Norway, and spent more than 2 hours at depths down to 65 feet in milky water. On one of these dives, I was lowered on a rope straight down into 45 feet of water and, on the way up, became buoyant when the air in my dress expanded. The solution in such a case is to empty the breathing bag by blowing past the gag, and I can remember being surprised by the speed with which I carried this out. Needless to say, you must keep your hand on the bypass valve at a time like this in case you start to go down again. John Buxton practiced with the other P-Party set on various occasions during bone-hunting dives in the first four chambers. In his log sheets for 18th November 1956, we find:

Mixture not the best thing for bone hunting. Bubbles disturb mud on roof.

Fig. 8.8
John Buxton (foreground) and Olver Wells prepare to dive upstream from the First Chamber on 4 February 1956.
Photo by Bill Darby of West Advertising Ltd., Bath

Our first attempt at exploration using mixture-breathing apparatus at Wookey Hole on 6th October 1956 was not exactly the world's greatest success. I went through to Nine with John Buxton and Dave Morris. Our first mistake was to take a large bundle of bricks to put at the top of the Eleven slope. It was not needed. In the event, I dropped it into the upstream entrance to Coase's Loop between Nine-Two and Ten. John Buxton rose to the occasion in the comments recorded in his log sheets:

OCW went too far down 9:2 ... found a vertical hole ... says he couldn't get through, but large belay which he pushed in front could and did and he lost it.

We also became entangled with decaying lines of vintage 1949. So we wrote in the report:

An ambitious plan to explore Eleven and Twelve foundered when the leading diver lost the belay he was carrying ... and the two divers returned to dry land amid a fantastic tangle of lines, both old and new.

On the return trip through Eight through muddy water and, therefore, in total darkness, I passed my Aflo from one hand to the other without noticing that I had done so under the line, and found myself securely anchored by the safety lanyard to the next belay block a few moments later.

On 12th January 1957, John Buxton and Luke Devenish took six bricks through to Nine for use as belays.

After these various episodes, we decided that further practice with the P-Party sets was in order, which we did in the Ladybower Reservoir in Derbyshire. We walked in turn from the shore down a gentle slope to a depth of about 50 or 60 feet, following a hemp rope to a boulder which we dropped from the surface. Audrey Buxton life-lined us using thin ex-Army telephone line from the reel on one of our Aflos. From my log sheet:

Visibility 5 ft ... Floor very muddy below 30 ft ... Below 40 ft pitch dark ... Reached 50 ft ... was pulled out by JSB rather too fast.

In John Buxton's log sheet we find:

At depth 55 ft walked backwards over a steep cliff.

However, it was good practice for the dive in Hurtle Pot on the following day and this was our first properly conducted cave dive using mixtures. It was 20th April 1957. Here we walked under conditions of limited visibility along a horizontal passage at a depth of 45 feet for several tens of yards; a route that Jack Thompson had found earlier when swimming with an aqualung.

On the following weekend, I was lowered on the end of a rope to a depth of 60 feet in the Main Rising at Speedwell Cavern. My feet did not touch the ground, so I rotated slowly as the rope untwisted. The visibility was about two feet, so I had no idea about the size of the space that I was in. Every half minute or so, a steeply sloping muddy ledge appeared in front of me, and slowly moved out of my line of vision to reappear later. We had arranged to signal by sharp tugs on the rope, but owing to friction at the top, these signals were not received. After a while I became bored with the view, and climbed up the rope hand-over-hand.

On 29th June 1957, Graham Balcombe, John Buxton and Phil Davies visited Wookey Hole Caves and went to Nine. As Trevor Ford wrote in the CDG *Review*:

A sad event, Balcombe's final cave dive before handing in kit, the last active dive in an underwater career which began in Sump 2 in Swildon's Hole in 1934. Out of the water, however, he continues to instruct the Group in matters of safety and his advice was taken in the planning of several of the dives described below...

The Fourteenth Chamber in Wookey Hole was discovered shortly afterwards in an uneventful manner on the night of 9th September 1957. Once again, John Buxton stayed in Nine and operated the lifeline. According to the log sheet, I was away from him altogether for eleven minutes while he paid out 220 feet of line and then hauled it in again. I can remember walking fairly fast on this dive, but even so, 40 feet per minute seems rather high! He also paid out a thick rubberised cable which had been tied to an extra lead weight which I put into one of my weight pockets. I left this at the furthest point reached.

Fig. 8.9
Aflo containers were filled with corks to improve their buoyancy, 1957.
Photo by Oliver Wells from Bob Davies album in Wells & Mendip Museum

Underwater, I walked down the slope in Eleven past the small triangular boulder that I had seen earlier, and then down a steep staircase of sand-covered boulders. The horizontal passage at the bottom measured six feet high by fifteen feet wide. I walked along it for a few yards until the roof lifted abruptly and by looking up I could see Davies's wire disappearing upwards into the haze. The depth here was 45 feet. The floor started to rise again, but only for the briefest distance, and soon I found myself looking downwards again across a gently descending sandy floor.

I called this place the Fourteenth Chamber[14]. The sand was heavily ripple-marked, and I admired the way in which it was scoured out beside a boulder. After only a few yards, the passage shrank once again and measured only ten feet wide by perhaps a yard high, sloping more steeply downhill. From the ripple marks on the sandy floor of this slot and from the sculpture of the sand banks generally, I concluded that the river entered the known cave at that point. The depth was 50 feet. I deposited the lead weight, gave two pulls on the line, and returned through muddy water in a straight line to John Buxton in Nine-Two, without any obstacles or problems of any kind. This had been an all-night dive, and we were finally in bed at 6.30 a.m. That evening we returned to Nine, but I was too tired to continue.

This point was passed on 14th March 1958, by John Buxton and myself diving together. We walked down the slope in Eleven and across Fourteen. I then entered the slot, which went down at 45 degrees, going to our depth limit of 65 feet on a 70/30 oxygen/nitrogen mixture. Both of my depth gauges gave this same 65 feet figure. Every time I moved, puffs of muddy water swept slowly past me up towards John. Looking downwards, I could see that there was a wide open space below me, and that the roof seemed to level out at a depth of perhaps 70 feet. So we called this place Fifteen. We were concerned both by the depth and by the need to prevent the line from pulling across into the narrow side of the slot.

After the discovery of Fifteen, we were naturally disturbed to find ourselves at our maximum depth limit once again. We ordered some more gas mixture with a lower oxygen/nitrogen ratio, and bought some larger cylinders for the P-Party sets to maintain the duration of the dive. We discussed various ideas for belaying the line at the entrance of Fifteen, most of which would sound extraordinary to present-day divers who swim straight through at that point. Graham Balcombe encouraged us by saying that great things were waiting to be found.

Meanwhile, we turned our attention to Swildon's Hole, and here I had the good fortune to be the leading diver during the discoveries of Swildon's Five on 13th September 1958 with John Buxton, and Swildon's Six on 8th November 1958 with Phil Davies. For the discovery of Six, I used a lightweight oxygen respirator, John Buxton's cap lamp instead of my Aflo, and an underwater signalling device with a sound-powered telephone for use from dry land. Later that month I was

[14] The numbering of Chambers in Wookey Hole is often somewhat arbitrary, often representing limits of exploration rather than air surfaces.

married, and while I did not expect this to affect my cave diving activities in any way, we did start to make plans for a two-year visit to the New World. When we left for America in July 1959, we hardly expected that we would stay there until the present day.

These dives in Swildon's Hole represented only a small part of the explorations that were being carried out by several very enthusiastic parties of cavers at that time, particularly after the discovery of the extensive Paradise Regained Series on 5th-6th March 1955. A few weeks later, Dennis Kemp explored this new series to impenetrable squeezes at the end on 24th April, but heard the noise of a stream beyond. Enticed by this, he and his team laid seige to the constrictions and, over several weekends, even camped underground whilst blasting and digging open the tortuous Blue Pencil Passage. They triumphantly regained the main streamway in Swildon's Four on 15th June 1957, thus completing the exploration of the high-level route begun by Edgar Tratman in Tratman's Temple above the streamway in Swildon's One on 12th November 1921.

The discovery of Swildon's Four opened the way for divers, and members of Dennis Kemp's dedicated team took part in the diving operations that followed. Norman Brooks, for example, free-dived into the upstream sump to verify that there was room for divers to push back towards the known cave, but the downstream Sump 4 and the prospect of new passages beyond was the greater attraction, of course. Len Dawes organised these expeditions and was the first to free-dive the awkward Sump 4 hot on the heels of the divers in September 1958. Len is still an active caver and climber and has long since moved from London to a new life in Derbyshire. Frank Darbon, who accompanied Len on most of his Swildon's pushes and was a London cab driver at the time, also found a more rewarding life in British Columbia, Canada, later. Almost every autumn since, however, Frank has taken his holidays back on Mendip with friends from the fifties who still live locally. Likewise, Dennis Kemp returned often with news of his many travels around the world walking and climbing high mountains in remote places[15].

While most of the above expeditions were satisfactory from the safety point of view, there was a tragedy on 17th January 1959, when the weather unexpectedly broke during an expedition to revisit Swildon's Six, and John Frank Wallington lost his life from exposure. A rapid thaw of snow on frozen ground was accompanied by unexpected storms and floods on the surface. This tragedy emphasises once again the difficulty of planning lengthy operations underground with large support teams drawn together from many parts of the country for a particular event. Such expeditions acquired a momentum difficult to resist even if the weather looked bad. The smaller party typical of later expeditions is more flexible and able to give itself greater safety margins. The maxim 'safety in numbers' does not work well underground.

Although I have written this chapter from a very personal point of view, it is apparent from the few examples given above for Swildon's alone that cave exploration on Mendip was at a high level generally, owing to the widespread enthusiasm of a large number of cavers and caving clubs there. Members of the CDG were active in other caves on many more occasions that I have described here, and the CDG *Reviews*

Fig. 8.10
Underwater signalling devices in the 1950s:
(a) The Aflohonk, as used by Bob Davies in Wookey Hole Cave (1951), and
(b) The sound-powered BEEP 'diverphone' used by Oliver Wells in the discovery of Swildon's Six at Priddy (1958).
Photos by Oliver Wells in Bob Davies album in Wells & Mendip Museum

[15] Dennis Kemp died in 1990 whilst climbing in Australia, aged 67.

for August 1955 until December 1959 contain reports from 33 different diving sites, for example. Especially, we owed a great deal to Graham Balcombe, Jack Sheppard, Don Coase, Bob Davies, and others who had set up the CDG and obtained the equipment that was in use at the time. A special mention must be made of Dan Hasell and Oliver Lloyd as controllers and organisers of so many diving operations; both happily remained active in CDG affairs for many years.

My ambition to progress beyond the deeply submerged Fifteenth Chamber in Wookey Hole Caves is still very much alive and was realised over thirty years later as described in *The Last Adventure* edited by Alan Thomas and in Cave Diving Group publications. Back in the fifties, my own particular high point then was to become the third cave diver, after Graham Balcombe and Bob Davies, to hold simultaneously both the downstream and the upstream penetration records for the Swildon's-Wookey Hole cave system. After I left, John Buxton reached the lowest point in Fifteen using a 60/40 oxygen/nitrogen mixture on 17th December 1960, accompanied by Mike Thompson (P-Party set) and Charlie George (UBA). In Swildon's Hole, Len Dawes and his successors continued to organise the sump-diving operations that eventually led to the discovery of Swildon's Seven through to Twelve by the sixties. So as expected, my record did not stand for very long at either end of the system – but this is not my story to tell.

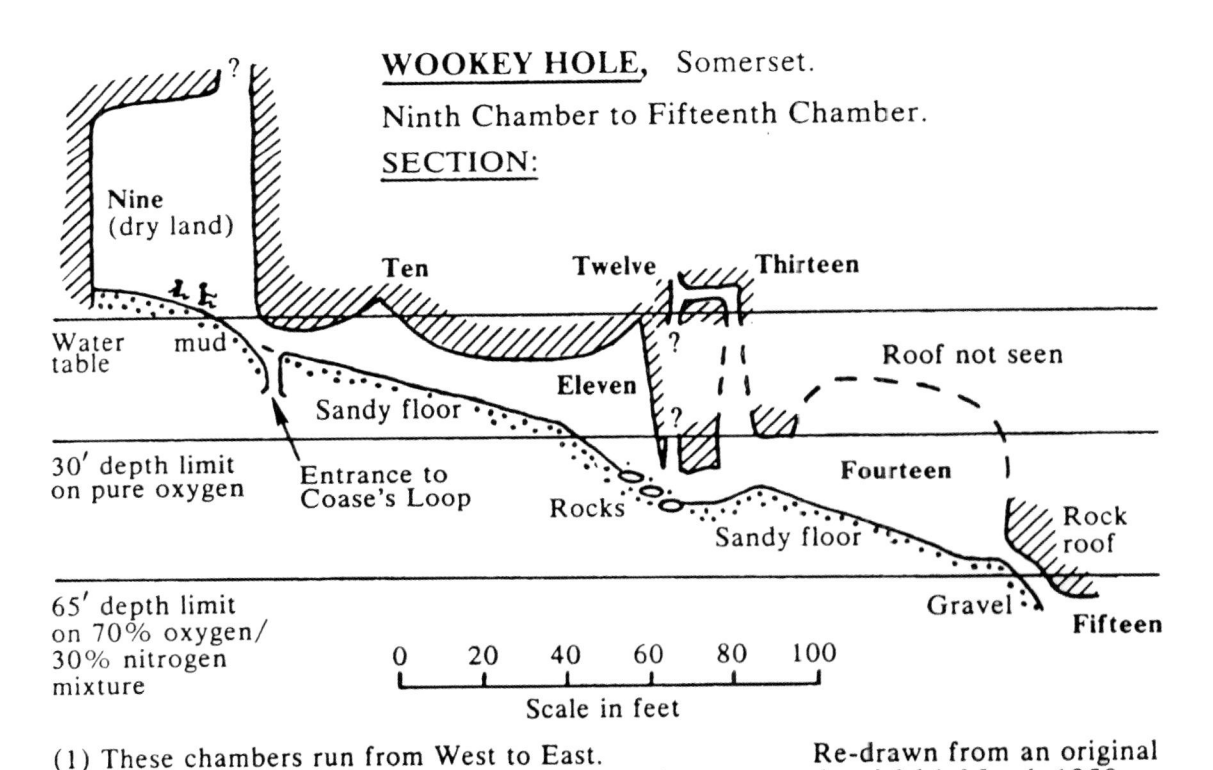

Fig. 8.11 Mixture breathing respirators enable Oliver Wells and John Buxton to break a new depth record at the submerged Fifteenth Chamber in Wookey Hole Cave, March 1958. Section drawn by Oliver Wells – see also Fig. 7.10 in Chapter 7

Fig. 8.12
Charles George (left) and Brian de Graaf in South Wales take on the task of training new cave divers and developing the use of finning over bottom-walking, 1961.
Photo by David Hunt

Chapter Nine
Twenty-Five Years
and the Turning Point
based on Mike Thompson

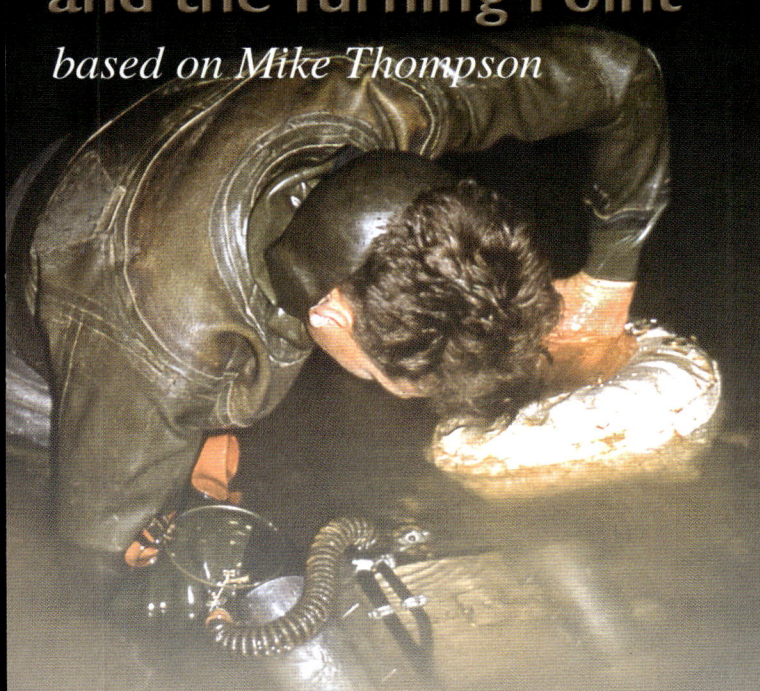

Fig. 9.1
A 25-year old Mike Thompson turns to cave diving at Wookey Hole Cave in October 1958, and fully qualifies by October 1959. Seen here testing his gear for leaks in March 1959. Photo by Jim Hanwell

With the arrival of 1960, we reach the halfway stage in this story commemorating fifty years of cave diving in Wookey Hole and caves under Mendip. The subterranean River Axe had been explored underwater against the upwelling current to its deepest point yet at -70 feet. But in those days, dives to reach record depths in Wookey Hole were less important than the information that was gleaned about the way on. More significantly, and very tantalizingly in retrospect, the lowest part of the submerged Fifteenth Chamber reached seemed to 'bottom out'. The way ahead was upwards beyond this deep. Further progress would have to wait for others, however, for times, too, were changing.

The new decade heralded a 'turning point' to yet another phase of development in cave diving. The second twenty-five years would be very different from the first; particularly in equipment, but also in approach. John Buxton led the last group of divers using closed-circuit sets whilst it fell to Mike Thompson more than anyone else to ensure that the Cave Diving Group itself altered to meet the changes afoot. Their dive together to the bottom of Fifteen came at the end for John and at the beginning for Mike. Thus progress was sustained by building on the past. At the time, keen observers such as Christopher Hawkes noted that cave diving had taken on a 'new look'.

John Buxton had started diving in Derbyshire after being lured by streamway wallows in Peak Cavern with fellow students from Nottingham University. His degree in Horticulture subsequently led him and his wife Audrey to a busy life skilfully breeding bees and specialising in making different varieties of honey. Would that there was room here for their fascinating stories about the behaviour of bees, especially how they navigate to and from the hives when at work. Both come back to Mendip for special occasions and John's recollections of his last deep dive in 1960 are still vivid as recalled towards the end of this chapter.

Mike Thompson, then articled to solicitors in Surrey, had been won over by Mendip and its caves during his National Service training at R.A.F. Locking, near Weston-Super-Mare. Regular weekend visits from London with Dennis Kemp and Len Dawes put a go-ahead Westminster Speleological Group on the map. Many WSG cavers also joined Mendip-based clubs and Mike, for example, became a member of the close-knit and highly active Shepton Mallet Caving Club of the day. Their old hut adjacent The Beeches overlooked the Priddy Mineries, St. Cuthbert's Swallet and, of course, the Bristol Exploration Club's own hut, the incomparable wooden Belfry. The Wessex Cave Club's headquarters was along the Bristol Road at Hillgrove Farm and much to-ing and fro-ing among the three huts took place when trips were being planned. Anyone who compiles the history of Mendip caving, during this period especially, must use the three hut log books as primary sources; most notably the log of the Shepton Mallet Caving Club. After each trip, those involved wrote up their discoveries and commented upon the prospects for further work. The excitement that accompanied each new discovery just lingers in the memory but lives afresh in letters sent to friends unable to share the moment, yet who treasure the record. Individuals often kept their own diaries; another priceless source of information that is rarely tapped. Mike Thompson's diary affords such a first hand glimpse of his dives in Wookey Hole Caves in the extracts that follow.

Apart from John Buxton's guiding hand from afar and help from Brian de Graaf in South Wales, Jack Waddon of the BEC, Luke Devenish and Phil Davies in the Wessex Cave Club were the most experienced and active local men once Oliver Wells had left for America. Jack and Phil teamed-up for many dives around the caving regions where the chances for shallower dives using closed-circuit oxygen respirators were clearly more optimistic. Why take on a deep dive in Wookey Hole when so much remained to be found through shorter, shallower sumps elsewhere? On Mendip, for example, the second sump in Stoke Lane Slocker had yet to be investigated, the first sump in St. Cuthbert's Swallet was under siege and the Swildon's Hole streamway remained as attractive as ever.

At the time, then, Wookey Hole was low down the list of priorities. The downstream sumps in Swildon's became the main lure, and five years elapsed before a new generation of divers returned to the upstream challenge in Wookey Hole. They were important years too, not least because the early sixties brought a freer phase in our social history. Austerity was largely a thing of the past, and new opportunities for caving opened up for those who were prepared. Contributing to this was the ending of National Service; the noticeable increase in car ownership rather than the motor-cycle; less hitch-hiking with full kit; the availablity of lightweight exposure suits surplus to military needs, called Goon Suits by cavers; the boon of second-hand NiFe cells and lamps following coal mine closures and from Fire Brigades; the variety of man-made fibre ropes, and, not least, the Vibram sole on cheap working boots. All played their part as reviewed in an essay 'Eighty Years of British Caving' by Jim Hanwell for the *Alpine Journal* in 1975.

From Mike Thompson's diaries we can savour a taste of the times on Mendip with respect to diving and Wookey Hole Caves. He starts thus, with a party comprising Oliver Wells and Phil Davies from Mendip, John Bevan and Brian de Graaf from South Wales, and a fellow solicitor friend, Ron Penhale:

Sunday 26th October 1958. Wookey Hole.

During the Wessex dinner Oliver Wells invited Ron and myself to come and watch some diving in Wookey Hole. With a great deal of groaning we got out of bed at 8.30, arriving at Wookey Hole at 9.15. I did not know that the CDG had a hut in the village, so we walked up to the cave. The door was open but there was no one in evidence until the guide appeared. He put on the lights and went as far as the Third Chamber. The cave which I had not previously visited is very decorative, particularly the Third Chamber. No one appeared so we were directed to the hut. We waited for half an hour and then carried some gear back to the cave, taking it to the Third. Wells and Bevan checked their gear and then disappeared under the far wall on a training dive as far as the Fifth Chamber. All very interesting but if one is not participating, inclined to be a little dull. Ron asked Oliver about the chances of becoming a diver, but got no satisfactory answer. They are so undermanned that they cannot train new divers. At about 11.30 we left and returned to Beech Barrow to find Luke festering. At this point I succeeded in falling the 10 ft from the loft thereby 'improving' my ankle. Hence to the Hunters.

First impressions were obviously mixed and it was clear that getting to dive would not be easy given that so few qualified members of the CDG were available. Many training dives did, in fact, take place. But the qualified trainers concerned then also had a conscious and sensibly practical policy of not making it too easy for would-be divers to train. Expensive equipment was in short supply and maintaining it to a high standard was time consuming. Oliver Wells has recently summed up the situation that prevailed:

Trainer divers more or less automatically took a not-too-positive attitude with would-be divers to see if they were keen enough to make the grade. All had to overcome difficulties ... not a bad thing in view of the nature of diving in a cave.

Nowadays, the availability of equipment and better training facilities permits a different approach, of course.

Kenneth Dawe, also a Shepton Mallet Caving Club member at the time, was living in Bath and had more opportunities to get some basic training in locally. Both Ken and Mike returned to Wookey Hole early the following year with John and Audrey Buxton, Brian and Valerie de Graaf, Ted and Dorrien Mason, Luke Devenish and Dan Hasell who is referred to affectionately as 'Pop' in the following extract. The main purpose of the meeting was the visit of someone writing an article for the *National Geographic Magazine* of America. So, it was a much more light-hearted occasion than their more workman-like training sessions; just right to encourage anyone who might have had second thoughts:

Sunday 4th January 1959. Wookey Hole.

I accompanied Ken down to Wookey for his second training dive ... We walked over Rookham and then down Ebbor Gorge, arriving at the CDG hut at 10 o'clock. After some mucking about, a chap arrived to accompany the divers. I didn't catch his name but he was writing an account of the diving at Wookey Hole. Everybody seemed a great deal friendlier than my previous visit. Ted Mason then arrived with wife and children (part author of a regional book on The Mendips and translator of Lavaur's Caves and Cave Diving). They brought various bones and skulls which were to be used for faking the photographs. We all marched up to the cave and Ken changed into his diving suit and breathing set. With 'Pop' on the end of the life line he began travelling round the exposed part of the river in Wookey One. While he was floundering about I approached Pop as to whether I could have a dive and he said yes. I subsequently spoke to John and was advised to talk to Oliver at the next Swildon's op. While all this was going on the divers were putting on a show for the proposed article. Ken was down for two periods lasting about an hour. When he had finished we walked back to Priddy arriving at about 5 o'clock.

Mike Thompson's chance to dive came on a special training meet over the weekend of the 14th and 15th March 1959 when the same party returned with the addition of Jack and Dorothy Waddon, Geoff and Sherry Roberts, Sybil Bowden-Lyle, Prof. E.K. Tratman and Dr. Oliver Lloyd. He used an oxygen rebreather.

Saturday and Sunday 14th & 15th March 1959. Wookey Hole.

I went with Ken to Wookey on a training meet of the CDG. I had been agitating for a dive and this time I got one; in fact, I think I am now in. We got to Wookey at about 12.30. After the usual amount of mucking about went on we went to the cave where I had a go with a breathing apparatus on dry land. After Ken and Geoff had had a go and we'd had a bit of refreshment I eventually got into the water. After having had some trouble with the breathing apparatus I spent 35 minutes submerged. On Sunday I had another go and spent an uneventful 25 minutes underwater, coming out when I found some difficulty in breathing. I thought that the cylinder was empty and told John so. He tried the by-pass and said that there was about 40 atmospheres left. I did another breathing drill which in fact exhausted the cylinder – one up to me. The next dive will be on 4th April and I have been asked to go.

Sybil Bowden-Lyle who had kept Mike's log on both occasions noted: 'He pranced away underwater like a natural..., reminiscent of Don Coase'. Praise indeed. The April operation was another turn out for the *National Geographic Magazine* article. It was also Oliver and Pamela Wells' last visit to the cave before leaving for a new life in the United States:

Fig. 9.2
Mike Thompson dons the two-piece 'Sefus' suit he inherited from Bob Davies, assisted by Brenda Willis, 1959.
Photo by Jim Hanwell

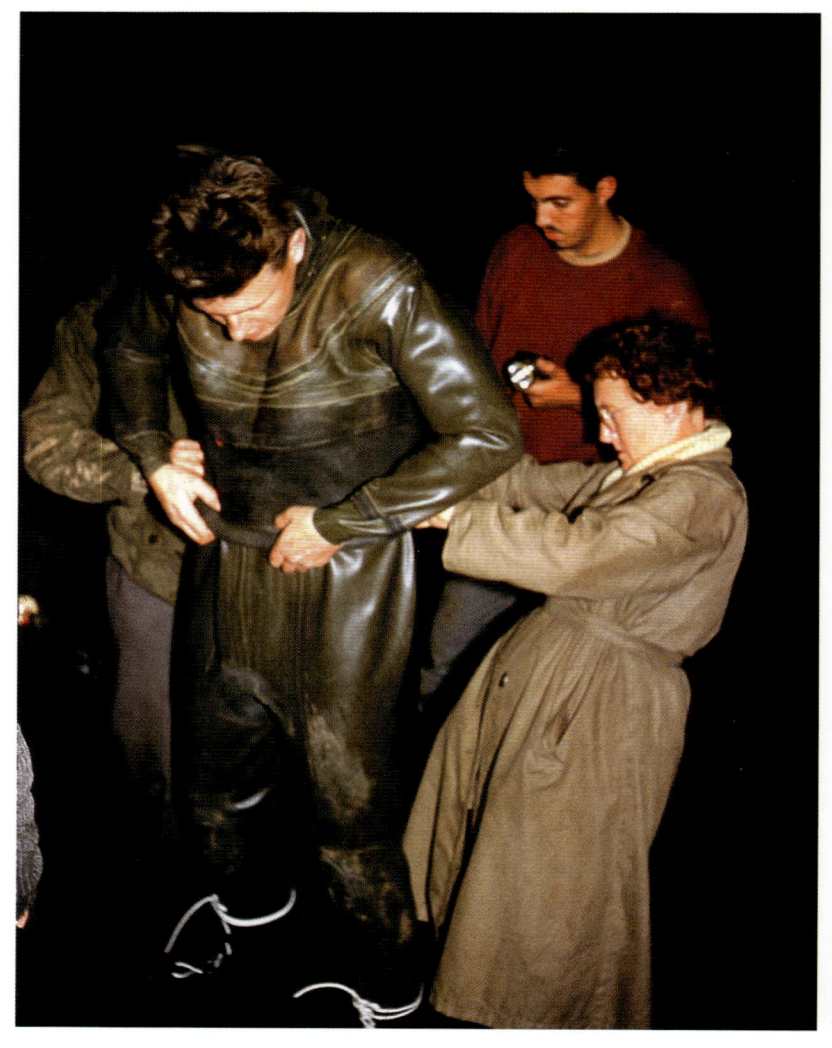

Saturday 4th and Sunday 5th April 1959. Wookey Hole.

I walked down from Priddy and arrived at Crooks' Rest at 12.30 just as Ken arrived. There was no diving until 5 o'clock and even then it was unlikely that we should get a dive. We spent the afternoon with Mike Holland and Oliver and Pam Wells. When we got back we helped to carry gear up and generally made ourselves useful. The following morning we returned and then much to my delight John presented me with a two-piece 'Sefus' suit. It formerly belonged to R.E. Davies and then to Oliver. It is the suit which accompanied Davies into Wookey 13. I used Luke's B.A. which is specially built for large people. I spent 27 minutes submerged thereby bringing my total to 1½ hours.

Further training dives with Ken Dawe under Jack Waddon and Brian de Graaf's guidance took place in the cave late in the evening of 23rd May. Fred Davies and Roy Taylor, also from the Shepton Mallet Caving Club, supported. Brian de Graaf was an architect and vintage motor-cycle enthusiast from Breconshire. The observations on experimenting with the use of fins bear out what Oliver Wells and John Buxton found when using closed-circuit respirators which had long and narrow soda-lime canisters rather than short but wide ones:

Saturday 23rd May 7pm to 1 am. Wookey Hole.

While John, Brian and Jack went to Nine to do some odd jobs, Ken and I continued our training in Wookey One. I was using my own kit as yet unmodified. We used up half empty cylinders and so obtained experience of cylinders emptying and having no emergency supply. We both tried fins and as a result suffered from CO_2 excess: remedy, sit on the bottom and think about it. Afterwards I had a swim around without a B.A. and Fred and I launched the boat and made a landing on the far mud bank. Fred kept the logs. The following day we scrounged more parts. I now have enough to start my modifications. In all I got a further hour towards the five.

In those days, would-be cave divers had to complete five hours' safe water training before being allowed into a sump. The summer of 1959 gave ample opportunity to exceed this requirement in the Priddy Mineries, a stone's throw from the Shepton Mallet Caving Club hut. Thus, Mike Thompson's sixth visit to Wookey Hole in the autumn was memorable. Dan Hasell controlled the operation whilst Jack Waddon, Brian de Graaf and Luke Devenish observed the new trainees, Ken Dawe and Mike Thompson:

Saturday 10th October. Wookey Hole.

A great occasion, our first genuine cave dive. The date for this operation had been changed, then Wally's wedding was cancelled and so both Ken and I were on Mendip. Brian was also there so we were able to arrange the dive for Saturday night. We were down at Crooks' Rest at 6.30 and into the cave by about 8 o'clock. My set was the only one available, so I had first dive with Coase's Aflo. On a line I first went into Four following the guide wire and on into Six. The water level was so low that one could walk into Four. On the first trip we went to the head of the First Deep. The water was beautifully clear, a pale sea green. Standing on the bank, we could look down into the First Deep, we then returned having been under for perhaps 5 minutes. After a sit and rest I took the line off and Brian and I set out again. We got to the head of the First Deep and started to descend into a large submerged chamber. We crossed this chamber, the bottom of Six and Seven and then came to the head of the Second Deep where the water was 25 feet deep (the First Deep is 18 feet). This was as far as we were going on this trip. I could see across the deep; the deep itself being a sort of ravine cutting across the floor. After a minute or two we started to come back and very shortly reached the First Deep. Coming back, the climb is steeper, but a few pushes and one glides upwards. We had been submerged for 12 minutes. Another short rest and we then went into One and dived though to Three. The water was very murky and I couldn't see much. I succeeded in cutting my hand rather badly and came out in a hurry. Ken had his dive afterwards.

Fig. 9.3
The first photograph of a cave diver underwater in Britain, November 1959. Phil Davies wears back-mounted air cylinders and a Normalair full face mask, in the sump between the First and Second chambers. This apparatus was being tested by Dr. Allan Rogers at Bristol University for sump rescue and foul air recovery work by the Mendip Rescue Organization at the time.
Photo by Luke Devenish

Few can have won their spurs quite so rapidly as Mike Thompson, and after only four training dives, all in Wookey Hole Caves. Three more trips were recorded before the end of 1959. Over the weekend of 7th and 8th November 1959, yet another television programme about cave diving was filmed for the old Television Wales and West company. It gave a chance for him to get to Eight between shooting scenes in the tourist chambers. Another noteworthy photographic session was organised by Luke Devenish on 27th and 28th November; the very first attempt at underwater photography in cave diving[16]. A lot of shots were taken and several of the prints gave the lie to those who had cruelly dubbed it 'Operation Photo-Phlop'. Posing for cave photographers could be a miserable business, especially when hanging around underwater waiting for suspended sediments to clear to obtain sharper shots. On this occasion Phil Davies was the 'victim'.

Fig. 9.4
After taking his historic photograph, Luke Devenish is helped ashore by Howard Kenney whilst Jim Hanwell completes the log, 1959.
Photo by Edward W. Dyer, Wells

Just before Christmas on 12th and 13th December, Mike Thompson was taken to the Ninth Chamber by Brian de Graaf, a fitting present and on his ninth visit to the cave within little over a year. In a letter to Oliver Wells written early in the New Year, 1960, Brian noted:

Mike Thompson has had his first trip to Nine in the fastest time yet, the water positively steamed around him.

This was the first spare moment that Brian de Graaf had found to keep Oliver in touch with CDG news. He and his wife, Valerie, had been hectically busy since moving to a large neglected property at Llangorse which virtually needed to be rebuilt. Career opportunities had taken the Chairman to a new life in America, and contemporaries in the Group were now faced with putting jobs and family affairs first back home. Inevitably, there were fewer occasions for the small number of qualified divers to train the newcomers. As Brian, again, remarks in his New Year letter:

I think we could easily accommodate all the CDG...without overcrowding.

[16] Luke Devenish died on 24 February 1994, aged 73.

Fig. 9.5
Mike Thompson returns from his first dive to the Ninth Chamber on his ninth trip upstream in mid-December 1959.
Photo by John Cornwell

Diving equipment now occupies a large room all to itself instead of being heaped up in one corner.

By 1960, serviceable equipment was also in short supply. Job lots of diving gear surplus to the Navy often produced mountains of spares, but few complete sets. But it remained important to keep up good contacts with RN establishments so that no bounty was overlooked. Luke Devenish managed to sell unwanted aluminium bottles and brass frames for scrap at £5 a hundredweight to boost meagre CDG funds and clear some space at Crooks' Rest. Meanwhile, John Bevan and Charlie George, both Navy men from South Wales, were useful new recruits to cave diving by Brian de Graaf. Charlie was able to get complete new UBAs for £20, though it must be remembered that this was most of a month's wages at the time. As Brian put it:

Charles Owen George can be the first of the new generation of CDG divers born and bred on the UBA.

There was also news of the latest Admiralty aqualung with 3,600 lbs per square inch working pressure bottles; but these were scarce and priced well beyond most divers' means then. Most made do with modified SGAMTU kits created from spare parts.

At the May Annual General Meeting of the CDG in Wells in 1960, John Buxton took over as Chairman since the post had been vacated by Oliver Wells. Not only that, John had also devoted much energy to acquiring and making up mixture sets at Bedford for the deep push in Wookey Hole past the Fifteenth Chamber. Weekend training sessions at HMS *Vernon* were arranged to try out both equipment and techniques. The Group relied and centred on this approach so far as those on Mendip were concerned. The signal from Bedford that all was ready was eagerly awaited. Mike Thompson and the others carried on training, with Brian de Graaf taking the brunt of the sessions. As Brian confessed in a letter to Wells in April:

I'm a little fed up of 3 to 4, 3 to 5, 3 to 7, 3 to 9, 1 to 3, etc., etc., ad nauseum and back.

Fig. 9.6
Jack Waddon makes final adjustments to his kit before diving upstream, 1960.
Photo by John Cornwell

*Fig. 9.7
Jack Waddon prepares to
dive upstream, 1960.
Photo by John Cornwell*

But it was all worthwhile, of course. Mike Thompson records that the trips early in 1960 led to a more leisurely tour of the Ninth Chamber with Phil Davies, Jack Waddon, Ken Dawe and Brian de Graaf once again. There was also an opportunity for 'a wallow' with a P-Party 70/30 mixture set on his tenth trip to Wookey Hole. The next occasion there was the one on which Mike Thompson joined John Buxton and Charlie George when the lowest parts of Fifteen were reached.

Charlie George came from Cardiff where he was articled to solicitors. He had previously been a Royal Navy diver and expert at using UBA sets. Only Oliver Wells in cave diving circles had such knowledge of mixture breathing, but he was in America. Mike Thompson's diary takes up the story thus:

Saturday 17th December 1960. Wookey Hole.

Having been issued with a P-Party this was to be the first occasion on which I would go below 30 feet...We entered the water at 7.45, John leading... [from the Ninth Chamber]...The three of us with mixture dived straightaway. We went down the mud bank and then proceeded down a slope for about 100 feet to the top of a steep drop. We descended gradually. I was second in line and could see John's light down below my legs. Once at the bottom we were on a level sandy floor. The pressure was noticeable. This was chamber Fourteen. At the other side a narrow slot descended at an angle of about 80 degrees. I didn't realize it at the time but this was the terminal point reached by Oliver and John. John went on down the hole. I followed with Charles and we came to a drop about 20 feet down. We then returned to the top of the slot. Both depth gauges gave a reading of 55-60 feet. John belayed himself to the line and dived again. He reached the bottom after 20 feet when he found that the floor had levelled out. He was able to cross a small chamber and see upwards into what appeared to be a large mass of water. Could it be that the deepest point has now been reached! John attained a total depth of about 80 feet. As soon as he had returned we set off back to Nine. With a full bag the ascent was very easy if a little uncontrolled at times... Presumably the next trip will be an attempt to push on although the depth is about the maximum for 60/40 O_2/N_2 mixture.

*Fig 9.8
John Buxton, the new Chairman of the CDG, helped by Dan Hasell, the new President, dives wearing a new wet-suit. Mike Boon (left) and Steve Wynne-Roberts (right) observe as aspiring cave divers, 1960.
Photo by John Cornwell*

Fig. 9.9
Mike Thompson sets out to investigate the deep Fifteenth Chamber on 17 December 1960, with Charlie George and John Buxton.
Photo by John Cornwell

Thus the River Axe in Wookey Hole Caves began to ascend again and the turning point had been reached. But the risks of going to over 70 feet deep on the way upstream, then having to return to the same depth through inevitably murky water on the way back were considered to be unacceptable. It would have been possible to adjust the oxygen-nitrogen mixture on the P-Party respirators to go on, but these sets were already fifteen years old, and it was a tribute to their technology that they had done so much over that time. In any case, closed-circuit equipment was already being replaced by aqualungs in the Navy and the Fire Services. And new aqualungs could be bought in the sports stores at reasonable cost. It made sense to make the transition.

Let John Buxton have the last words on this era for he led the final push with mixture sets. His memories of these deep dives to Fourteen and Fifteen remain as crystal clear as 'the clear green water' that attracted him to Wookey Hole:

If my memory can be relied upon at this length of time, I don't think I had seen the drop below the 30 foot level down to Fourteen until Bob Davies's aqualung dive. As fervent O_2 bottom walking divers we were prepared for an awful abyss! Earlier when I reached the drop off, with a tight wire from me to Nine, Bob had already stirred up the surroundings somewhat but I was able to see bits of rock below the edge and a definite slope before the mud blotted everything out. Later, when I had fished Oliver Wells on his recce and he had described the dive on mixture to 50 feet it had sounded quite exciting, but when we eventually went together and kept ahead of our own mud cloud, my impression was that the dreaded abyss hardly existed. Nearly all the way was a steady steep walk with only an occasional hand needed to steady us. I think that dive was the clearest water I saw. When the three of us dived to Fifteen on 17th December 1960 using mixtures sets, I remember we had a large triangular lead weight with eyelets. Michael Thompson carried this with him as a belay for any further steep extensions downwards. The third diver was Charlie George. When we got to Oliver's final belay I gave my wire end to one of the other two and popped through the hole laying out the wire as I went. I remember that it went down still deeper as I descended feet first with great care! By this time the depth gauge on my Aflo was 'off the clock'. It had been calibrated by either Bob Davies or Graham Balcombe to 60 feet; which, in the 'olden days' represented a hundred percent margin on the 30 feet limit. A subsequent check on a dead weight calibration gave 70 feet depth. The bottom looked as though it might go a little deeper, but was nearly horizontal and more muddy than the coarse sand in the slot. The water was murky so that the chamber could not be seen at all clearly. It looked as though it went off somewhat to the right upstream. I realised that this was as far as we were likely to go at that moment in time, so returned by coiling the wire on my hand. The other two, seeing the light and a large cloud of mud started their retreat, so I sprinted back to the lead weight belay and caught them up. Together the lead weight was removed from the snaplink or whatever held it and my wire was tied to the end of Oliver's. I then popped back into Fifteen and, in the middle of the more open muddy floor, cut my wire and tied on the triangular lead weight to mark the limit explored. I understand this point was not passed for another six years. Wookey water had got into my blood and I returned there to dive on any excuse: archaeological; B.B.C. TV filming; training, and practice with mixtures. I was very sorry when financial cramp forced me to stop!

John Buxton's diving companion, Oliver Wells, remembers:

At the time when this dive took place, John was already more interested in family affairs. In September of that year, Audrey Buxton wrote to us that their daughter 'has 4 teeth and can stand up, holding on with one hand'. John took to diving in the sea with an aqualung, catching scallops and lobsters. The very large lobster claw which so impresses visitors to his house dates from this period.

So it was that the active centre of the Cave Diving Group returned largely to Mendip and focused upon a new generation whose main interest was in pushing Swildon's Hole further under Priddy, maybe even bound for Wookey Hole itself. In the shallower sumps of this swallet system, the need to pursue closed-circuit mixture diving simply did not arise.

Despite the ominously low gradient of the streamway, there was a theory that older Swildon's drainage could have gone to Cheddar but had subsequently lowered and changed direction when intercepted or captured by the River Axe during the development of Wookey Hole Caves. Derek Ford's significant detailed study of cavern formation for his doctorate thesis at the School of Geography, Oxford University, in 1963, cites evidence for underground river diversions in Swildon's. Thus the debate among geologists and geomorphologists on Mendip raged no less than that among divers as to which equipment to use. Indeed, each fuelled the other's dream of 'following the stream from Priddy to Wookey Hole through caverns measureless to man'. Such ambitions were immortalised by Mendip's very own bard, 'Alfie' Collins of the BEC, in a particularly apt 'Speleode' entitled: 'The Nautical Narrative of Percy Pound'. The full text will be found in *Reflections* and the saga ends thus:

So let us draw a modest veil
Upon this portion of our tale
And follow Percy, as with a crash
One final sump the waters smash

To carry Percy, still afloat
To Wookey Hole, complete with boat.
As mid the waters wildly pitching
He landed in the Witches Kitchen

Confounding every caving bloke
Who thought he'd get to Rodney Stoke.
And thus today, our Percy Pound,
Runs steamer trips beneath the ground

And navigates the Mendip Queen
From Wookey Hole to Priddy Green
A life upon the ocean wave
Is lived by Percy down a cave.

Fig. 9.10
Drawing from 'Alfie' Collins' Reflections by 'Jok' Orr

There can have been few other periods during which the blend of theory and practice proved to be so fruitful to both facets of caving: the science and the sport. Herbert Balch and Ernest Baker would have approved of the times.

The following chapters will reveal why these were lean years for diving in Wookey Hole Caves. If we move on almost four years in Mike Thompson's diaries, after the deep dive to the bottom of Fifteen in 1960, the final rites ending some twenty years of closed-circuit diving were appropriately enacted in Wookey Hole. The principal local divers were Mike Thompson and Steve Wynne-Roberts with friends from the north including Ken Pearce, Alan Clegg, Bob Jarman and Gordon Nolan. Others present included John Cornwell, Phil Romford, Dave Causer, Fred Davies and Mike Boon. Olive Hodgkinson also put in an appearance, no doubt true to her uncanny instinct for historic events in her caves. All the divers except Steve Wynne-Roberts were using new aqualungs. Steve used an old trusted UBA Mark II respirator with a Mark I soda-lime canister; a modification which was to lead to some excitement. Undetected ageing cracks on the connections were subsequently found!

16th February 1964. Wookey Hole.

Steve and I planned to revisit the dry passages linking up the avens from Wookey Eleven and Fourteen discovered by R.E.Davies after he got lost in Thirteen. I had read his proposals for rediscovering this passage but they appeared to be hopelessly impractical. We decided to swim particularly in the light of our last dive at Wookey, and to lay a new line from Nine-Two to Thirteen. With Ken Pearce, we swam uneventfully to Nine-One. I took two spare 40 cu.ft. cylinders using them to Nine and keeping my tank (50 cu.ft. at 160 atmospheres) for the onward journey. After discussing reserves Ken decided not to go on. Steve and I then started our dive from Nine-Two. We swam down the mud bank and then into the downward sloping passage. I was leading following the existing line whilst Steve followed laying a new line. Shortly after starting I realised that Steve was no longer with me. I turned back and then found myself on top of him. He, afterwards, told me that his feet

Fig. 9.11
Steve Wynne-Roberts gears up to re-examine the higher level avens, discovered by Bob Davies, for possible dry ways on. Thus, his assembly includes a NiFe battery (on his left hip) and cap lamp (trailing in the water), 16 February 1964.
Photo by John Cornwell

were buoyant and he had been unable to stand up! After this little contretemps we swam on presently emerging from the passage into Eleven and the edge of an amphitheatre. Last time I visited this spot, the only time, I was second of three bottom walking divers and was unable to see anything of any note. This is a most impressive place with the bottom sloping steeply down. We swam down under a wide low-roofed arch. Passing beneath this across a sandy floor the roof soared away and we were obviously at the foot of Bob's rift and the beginning of the Fourteenth Chamber. We made signs at each other and began the ascent. Afterwards, Steve told me that he thought I had hold of Bob's line, in fact I didn't, I got hold of the line he was laying and started to follow him up. He then dropped back to let me lead as I was following the mythical line. The result was comical; a battle as to who shouldn't be in front. We started up on the far right, facing Fourteen, of the rift which sloped away to the left, the result being that we constantly banged against the ceiling; during the course of the ascent Steve 'knocked on' (the by-pass) twice with unpleasant results which did not manifest themselves immediately. After what appeared to be an age we surfaced in a narrow rift, with muddy sloping edges. To our left a passage sloped upwards and to the right and curved away. After a struggle we both got out of the water and ditched our kit. The passage to the left was almost 3 feet in diameter. Bob's wire was belayed to a boulder, but the towel he had left there in 1955 was gone. As we started the passage sloped upwards for 20 feet emerging into a small chamber whose floor consisted of a boulder wedged into a rift at right angles to the passage. The counterpart part of the rift existed in the roof in the shape of a narrow arch, possibly climbable. It was not choked as described by Davies, but there were freshly broken boulders on the floor. We now turned back. We straddled the rift and turned the corner only to find that the water continued and joined the roof. No go. We decided to return and kitted up. As we were about ready Steve swore; when 'blowing up' he had partially pushed the soda lime can out of the breathing bag. He pushed it back, tightened the clamp and off we went swimming vertically downwards. The water had now cleared and we seemed to be between towering cliffs. The descent seemed very quick and we were soon passing under the arch into Eleven and up the slope. In Nine we rejoined Ken and climbed across the Pinnacles. I dived first followed by Ken and reached Three followed shortly by Ken, but no Steve. After a pause Ken started back. I changed bottles and followed. We found him still in Nine. He had started and realised that he had about a quart of water in the bag. He had returned to empty it. We now started again. I stopped every so often to wait for Steve, the last time in Five. I then emerged into Three and blow me Ken was the next to appear. He went back, but a second or two later Steve appeared. He had had to go up into Four to clear CO_2! Thus ended a trip over which I have mixed feelings; the grandeur of the submerged cave contrasting with the worrying defects of lashed up mixture kits.

By now, Mike Thompson had used all three types of respirator in Wookey Hole Caves; first, closed-circuit oxygen, then a mixture set and, lastly an open-circuit aqualung. It was only the second time that air had been used in the farthest sumps then known in Wookey Hole. The ghost that had haunted since Bob Davies' lucky 'Night of the Thirteenth' was now friendly if not laid. Steve Wynne-Roberts had coolly overcome three mishaps on this dive, a measure of his competence and

Fig. 9.12
Steve Wynne-Roberts and Mike Thompson return to base with 'mixed feelings' about the operation on 16 February 1964.
Photo by John Cornwell

commitment in awkward situations. Whilst many would have taken the hint and reverted to less hazardous pursuits, Steve, in fact, went on to become a professional diver gaining great experience around the world's offshore oil fields and earning a reputation for always being at the forefront of those engineering advances essential to underwater technology[17]. He quietly and reliably got on with the job as part of the family-like contingent of divers from the Shepton Mallet Caving Club who pushed the Swildon's streamway in the early sixties. Also, at Stoke Lane Slocker, on eastern Mendip, Steve Wynne-Roberts had been the first to get through the constricted Sump 2 on 16th September 1962 to discover and explore the open stream passage beyond. Earlier attempts by John Buxton in 1956 and Phil Davies in 1959 had been thwarted because their old style respirators were simply too bulky. Jack Waddon was also cheated by the unaccustomed tightness of this sump. Modifications had been made by Steve Wynne-Roberts, Mike Thompson and Fred Davies to suit their work in the constricted sumps of Swildon's, but more of this in the following chapter. Also, in August 1963, Steve Wynne-Roberts and Mike Boon had accompanied Ken Pearce and others on an expedition to dive the terminal sump at the bottom of the celebrated Gouffre Berger near Grenoble, France, at a great depth of 3,680 feet below the entrance. Although they did not get far, the new lightweight cave diving equipment had successfully stood the test of a major expedition.

Tragically, Jack Waddon was not so lucky. On 3rd November 1962 he had planned to get through the recently pushed Sump 2 in Stoke Lane having re-rigged his own gear. In the event, heavy rain made the trip impossible and Jack decided to practise in the Priddy Mineries pond that evening instead. Something went badly wrong underwater for he experienced an acute lack of oxygen, passed out and failed to surface. In the gloom it took Mike Boon and Fred Davies over an hour to locate him. They towed him ashore, even managed to resuscitate him, dashed to the hospital in Wells and hoped. But Jack died there about ninety minutes later without regaining consciousness. He was a keen member of both the BEC and Wessex Cave Clubs having caved on Mendip since his school days at Taunton. His main interests in the scientific aspects of caving led him to most limestone areas in Britain and to Germany and Austria during his National Service days ten years earlier. On the day that he hoped to see the new passages in Stoke Lane, however, his methodical approach and natural caution were somehow not enough. Ironically, Jack Waddon died, aged 30, in the same pond where the 1935 Wookey Hole Cave Divers had learnt their craft.

When Gordon Marriott died in Wookey Hole Caves in 1949, almost five years elapsed before cave diving revived on Mendip. When Jack Waddon died, therefore, everyone hoped that history would not repeat itself. It was a time for steadying hands such as Mike Thompson's to keep things together. And so they did with much rethinking of both equipment, techniques and new training procedures. Friends from other regions rallied to the CDG, a new constitution and outlook followed at the Annual General Meeting of the CDG in Wells on 14th May 1963. The *Letter to Members* later in August tells the full story and lists the few involved with cave diving at that time. Thus, it was doubly cruel when, with many Mendip cavers in the Yorkshire Dales over Easter 1964, Alan Clegg drowned while diving the Master Cave Sump in Lancaster Hole. Alan was the second cave diver to lose his life in action underground. Once again Mike Boon had the unenviable task of retrieving a lost cave diver and trying to resuscitate him. From that moment, Mike Thompson handed over the reins and only dived once more; a last look at the far reaches of Stoke Lane Slocker with Bob Pyke. But he had had a good five years, and his diaries have much more to tell one day[18].

If ever a symbolic, if belated, act was needed to confirm the final passing of the first half of the fifty year period celebrated by this book then it must be at the end of 1967. Olive Hodgkinson wrote a letter to Oliver Lloyd on 8th November saying:

> Last summer I was in desperate need of extra car parking space and I very much regret that I shall have to pull down the old diving hut to give us this additional room apart from having to give up half of my kitchen garden and take back the allotments I had let people have. I do not know what equipment is in the diving hut or who holds the key but I should be most grateful if you would contact whoever is responsible.

---

[17] Steve Wynne-Roberts died in 2001, aged 63.   [18] Mike Thompson died in 2007; his diaries were bequeathed to the UBSS.

So, times were changing at Wookey Hole Caves too. By Christmas, Crooks' Rest and its cache of old diving paraphernalia were gone. Some of the treasures eventually went into the display of cave diving equipment in the little museum at Wookey Hole, the rest who knows where? Only the written records survive and memories linger.

Fig. 9.13
John Cornwell diving at the Mineries Pond, Priddy. Photo from the John Cornwell collection.

Fig. 9.14
After handing over the reins of the CDG, Mike Thompson takes on cave digging projects on Mendip. Seen here during blasting open the Twin Titties entrance shaft at Priddy, and wearing Luke Devenish's ear defenders, 1969.
Photo by Jim Hanwell

Chapter Ten
Changing Gear
based on Fred Davies

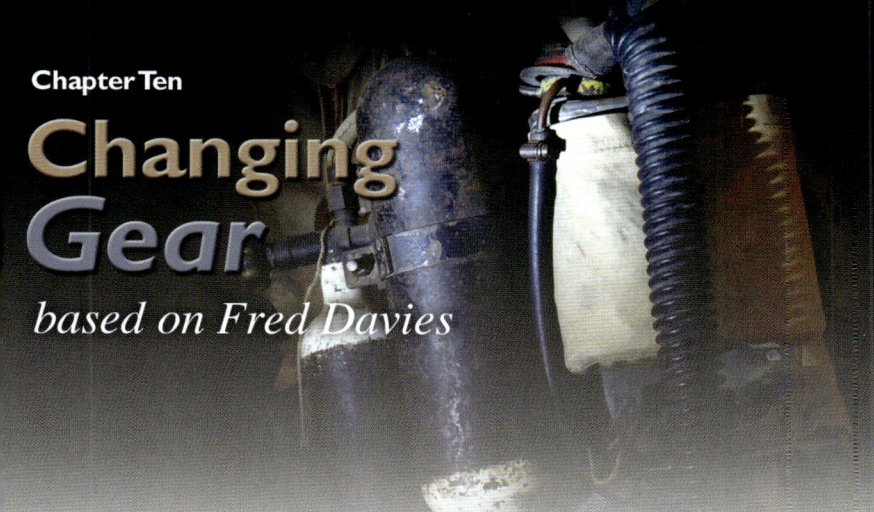

There are many similarities between the start of the next twenty-five years of cave diving in the early sixties and the pioneer days of the mid-thirties. Once again, Swildon's Hole captured most attention. The maxim in vogue was: practise in Wookey Hole but push in Swildon's. Just as well, too, for headway along the Swildon's streamway required slimming down both the number of people involved and the equipment used. Although no one really imagined so at the time, new foundations were laid for the subsequent return to Wookey Hole. Much as the first success with standard helmets and air lines had led to Graham Balcombe and Jack Sheppard developing equipment such as the slim-line 'bicycle respirator' and returning to Swildon's, so the proven and trusted rebreathers were trimmed and miniaturised. And compressed air made its comeback unexpectedly.

After the Oxford University student, Neil Moss, became stuck fast down a narrow rift in Peak Cavern and died owing to excess carbon dioxide in March 1959, there was concern among cave rescuers about recovering people from passages where the air was foul. Bad breathing conditions had been experienced by the large sherpa parties humping packs of diving equipment through the fossil Paradise Regained Series to Swildon's Four. Dr. Oliver Lloyd, then Secretary and Treasurer of the Mendip Rescue Organization, took on the problem of what would be done on Mendip in the event of an incident involving foul air or, of course, recovering an injured caver underwater through a sump. He enlisted the help of Dr. Allan Rogers, a colleague in the School of Medicine at Bristol University. Allan had just returned from Antarctica as the physiologist on Vivian Fuchs' epic crossing of that continent. He had a particular interest in respiration and was a seasoned Mendip caver as a member of the University of Bristol Spelæological Society.

Within little time, Oliver and Allan had assembled an apparatus comprising compressed air bottles with extended high pressure hoses to rubber masks that covered the entire face. Suitable full face masks and their sophisticated demand valves were being manufactured by Normalair, an engineering company based at Yeovil nearby. This firm had developed a reputation for its oxygen equipment for high altitude aircraft and supplied many Himalayan expeditions, notably the successful assault on Everest in 1953. They kindly donated components to the MRO to evaluate and make up the first Sump Rescue Apparatus

Fig. 10.1
Getting things done with the bare essentials: an approach championed by Fred Davies, 1971.
Photo by Jim Hanwell

by the summer of 1959. A 16 mm film of the equipment in use was shot down Swildon's Hole at Sump 1 and shown by Oliver Lloyd to a large and eminent gathering of physicians attending the Edward Long Fox Memorial Lecture at Bristol University in 1960. The cast of cavers packed the gallery by invitation and Oliver's lecture and film were simply entitled 'Cave Rescue'. It was quite an occasion. Many MRO Wardens, even those without cave diving experience, became familiar with and happy at using the Normalair Sump Rescue Apparatus. Phil Davies dived with a Normalair mask and back-mounted air bottle in Wookey Hole Caves before Christmas in 1959. Compressed air had staked a claim in one area of cave diving at least.

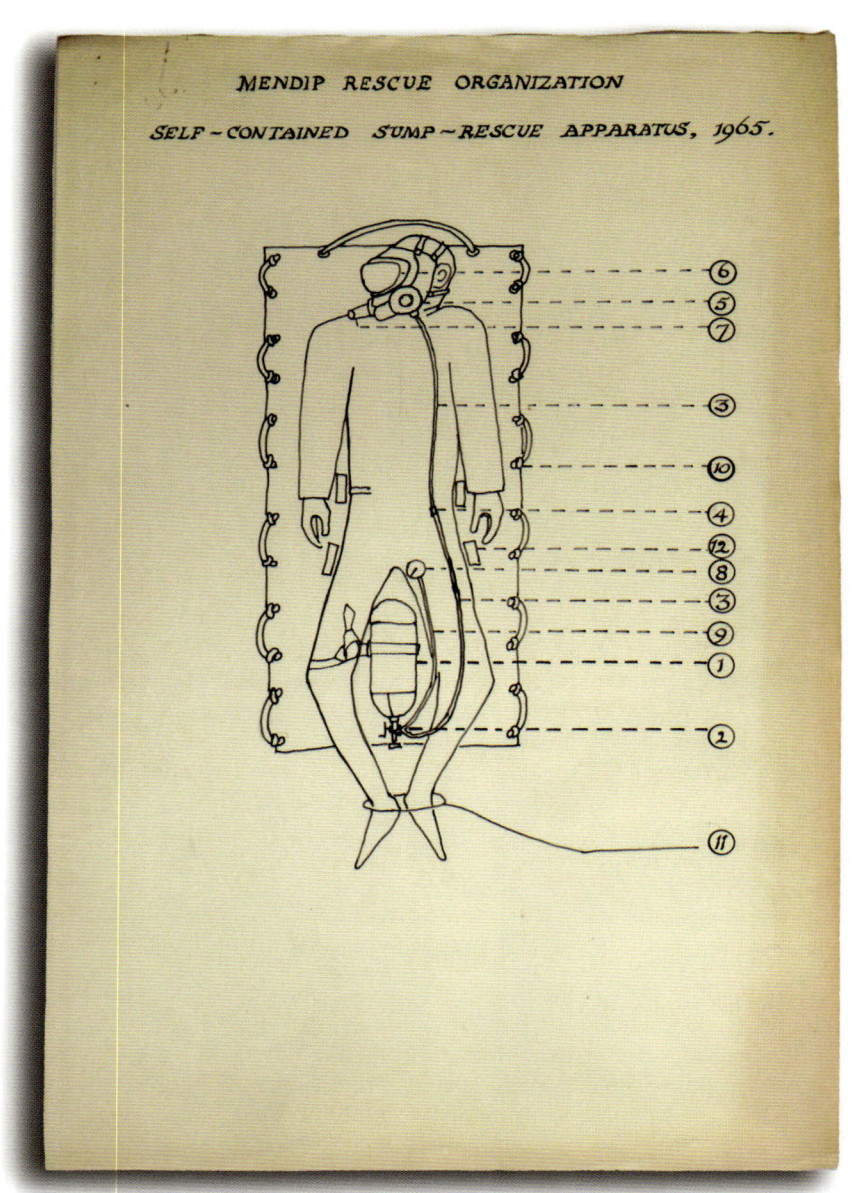

Fig. 10.2 The Normalair Sump Rescue Apparatus and MRO Carrying Sheet set-up of the 1960s is illustrated by Oliver Lloyd, 1965

The moment was also ripe for air to be used in exploring new sumps. Someone naturally inclined to believe that the minimum of personal possessions was the best way to get the maximum amount of caving done led to the simplest and most economic solution. John Michael Boon practised this creed to perfection! Already a master at borrowing equipment and putting it to good use, it fell to Mike more than anyone to make the change of gear. The least amount of kit to get him by, or rather, through, was enough.

That things should turn out for the best in the direction chosen by Mike Boon does not cast any doubts on earlier divers, of course, nor even his contemporaries who tried out different systems. The previous generation had pushed everything to its limits, and even beyond in many instances. The only way forward was through experimenting rather than in conforming. A pursuit that did not lack practical endeavour was given the extra effort that tipped its fortunes. Whilst such chance may favour the prepared individual in the end, it also requires the group to stick to the task hopefully. At the time the Shepton Mallet Caving Club's pragmatic outlook was particularly suitable: one which is appropriately summed up by Fred Davies' now immortal adage:

Caves be where you find 'em!

Another Mendip story, maybe a little apocryphal, is told about Mike Boon's arrival at the Shepton Hut. One dark evening in 1958 there was a knock on the door. As it was opened, the cosy glow inside lit an elf-like face, eager and boyish even if the figure was hunched in the gloom. It said:

I'm Mike Boon and I'd like to go caving.

Fig. 10.3
Another believer in the least amount of kit! Mike Boon enters the Bristol Exploration Club's old wooden hut 'The Belfry', Priddy, greeted by John Cornwell, 1960. Photo by Jim Hanwell

Once inside and warmed, the enthusiasm grew and plans were made.

Mike has written from his home in Canada and recalled how his debut actually arose:

As to how I started caving; did you know I was brought up near Stoke-on-Trent in Staffordshire? Although we moved to St. Albans, Hertfordshire, after the War ended when I was five, I still looked on Staffordshire as home and used to go up there to stay with my dear old Auntie Gladys. I invited one of my school chums on one of these jaunts and we passed several cave entrances on a walking tour of North Staffs and Derbyshire. The 'caves' I most remember were an extensive series of mine workings near Matlock. One of the locals took us down these; a fabulous maze. When we got back we formed an informal group at St.Albans School. After a foray to Mendip in which we did Upper Swildon's and Swildon's down to Sump 2, we eventually bottomed August Hole and got into the Paradise Regained Series in Swildon's. On leaving school the group stayed together for a bit, then broke up. I first came into organised caving thus: I had taken a bus from Bristol and, either on the bus or getting off it, had fallen into step with Jerry Wright heading for the Shepton Hut. After some negotiation with Ken Dawe, the Shepton took me to Swildon's Four where my light went out 60 feet up an aven!

Luckily for caving, Mike Boon was not to be deterred. Indeed, he was always exhorting himself and others to 'follow the stream'; a Shepton battle-cry invented and popularised by Mike Thompson but inspired by lines from Herbert Balch's favourite poem, 'A glow, a gleam'. This verse appeared in all the early Wookey Hole books written by Balch. Like Balch, Mike Boon had a way with words and liked to write up his ventures. *Down to a Sunless Sea* by J.M. Boon was published by Mike in Edmonton, Canada in 1977. The first two chapters are classic accounts of the exploration of the Swildon's streamway. Only Mike could tell the story so vividly and lucidly. At the time, if anyone could be said to live up to his name, it was surely Boon.

Fred Davies came from Street where Mendip's impressive southern slopes can just be seen across the flat moors or Levels past Glastonbury. Maybe it was because the town was supplied by Mendip groundwater from the Rodney Stoke Rising along the Cheddar Road from Wookey Hole that caving got into his blood from an early age! A love of cycling, cross country running, climbing, Scouts and the hard life certainly helped. He joined the Shepton Mallet Caving Club before going to Exeter University to read physics. There he made a fortuitous bar-room encounter with Kenneth Dawe from Bodmin in Cornwall. Weekends exploring and crossing Dartmoor together were punctuated by frequent visits to Mendip caves, and Ken too joined the Shepton Mallet Caving Club. National Service was kind to Fred and took him to Catterick within reach of the Yorkshire Dales. Ken was posted to Cyprus but the caving partnership survived, as they usually do. They were among the most enthusiastic and tireless of the sherpas on Swildon's diving operations during the latter years of the fifties.

Oliver Wells' successful passing of Sump 5 in November 1958 was the prompt for Ken and Fred to get in on the act and start diving themselves. By that time, however, Fred was living and working in North Wales whilst Ken was more conveniently living in Bath. Fred returned in 1961 with his family to Street as a local teacher and the opportunity to cave and train regularly with the others was assured. Wookey Hole's deeper sumps were to cheat him out of progress beyond the Deeps on the other hand, because he invariably found trouble when clearing his ears. Most of his considerable energy could thus be directed at the Swildon's streamway. After John Buxton's deep dive to the bottom of Fifteen in Wookey Hole, Fred Davies recalls the prevailing attitude among Mendip cave divers as:

Wookey – yes, of course it's interesting, it's the other end of Swildon's!

The prophets were already predicting a connection that had yet to be proven for sure. A more cynical view heard now and then was that diving at Wookey Hole had become little more than, 'Diving for diving's sake'.

The first big diving operation in Swildon's Hole was set up during the summer of 1961 with the objective of getting through Sump 5, digging away the gravel and mud bank beyond, as suggested by Oliver Wells, and so opening an air space when the pool drained. Lower water levels would also improve the approach along the maze of ducks in Swildon's Five known as Buxton's Horror; a place that John Buxton first passed in September 1958 and is unlikely to forget. This time, Mike Thompson, Phil Davies, Jerry Wright and Ken Dawe were the principal divers using their trusted oxygen rebreathers. Ken master-minded the organisation and, no doubt, with memories of the earlier operation when Wallington died, covered every detail. June 17th would be the day for the main push.

Intent on making the most of his supporting role, Mike Boon set about getting himself equipped and trained through the Watford Underwater Club back home. So it was that he turned to air. Although contrary to prevailing CDG guidelines which deterred the use of air after Bob Davies's lucky escape in the Thirteenth Chamber of Wookey Hole Caves, the sumps in Swildon's were

Fig. 10.4
Fred Davies completes his training in Wookey Hole Cave using a conventional SCAMTU closed-circuit respirator, 1962
Photo by John Cornwell

clearly a different proposition. Moreover, Mike Boon's winsome charm and winning ways were enough to make the necesssary allowances for him. And the current generation of active divers, aware of what was needed, sensibly recognised the value of someone prepared to have a fresh look at cave diving itself. Phil Davies as the most active of the senior divers on Mendip agreed that Boon could go, and the family-like contingent of Shepton Mallet cavers did the rest. No one said 'No' anyway. They encouraged and enabled Mike to accelerate through the ranks in his own way.

As might be expected from someone so single-minded, Mike Boon devised his own arrangement of the gear and used the smallest air bottles possible. He was greatly assisted in all this by Denis Mead of the Watford Underwater Club and fondly recalls:

Thanks are due to Denis Mead for endless patience in discussing the problems involved as well as practical help in making the kit.

The air bottles used were ex-War Department (W.D.) aircraft oxygen bottles known as Tadpoles or 'Taddies'. Mike slung them between a waist belt and a shoulder strap (the latter soon found to be unnecessary) on his side and within easy reach, rather than wearing them on his back as with conventional aqualungs. Not to be outdone by the available SGAMTU, SEBA, DSEA and ATEA kits worn by the others, he called

Fig. 10.5
Mike Boon wears a chest-entry ex-W.D. Submarine Escape Suit (called a 'Goon Suit' by Mendip cavers), probably over his pyjamas! 1960.
Photo by John Cornwell

his own gear 'Nyphargus' after the tiny blind shrimp-like creature *Niphargus* that lives underground in rock pools and streams, especially in Swildon's. If this minute troglobyte was at home there maybe it could be emulated. The term was apt.

Mike Boon's account of the epic all-day trip that cracked Sump 5 on Saturday 17th June 1961 takes pride of place in *Down to a Sunless Sea*. On the following Monday he wrote triumphantly to Oliver Wells in the United States:

> *Well we made it! After about six hours digging by the diving party, Sump 5 gave a giant despairing gloop and before long the support party could join the divers in Six. The whole operation went very much to plan, good telephone communications from Four to Five to Six, excellent kitchen in Four run by Fred and Oliver Lloyd and only minor mishaps with diving gear.*
>
> *Six itself is almost 250 feet long, consisting of the gently curving streamway you explored in '58, then a sharp bend to the left where the passage becomes a large inclined rift, followed by a second sharp bend to the right into a large sump chamber. Here a big muddy passage bored upwards, leading into extensive crawls. One ends in a blind pitch, the other continues upwards to a considerable height above the stream until a slippery rift is reached which cannot be climbed without mechanical assistance. The tributary stream you noticed leads back towards Paradise Regained (as do the other side passages) and ends in a static sump. Your suspected oxbow was in fact an oxbow, though there may be a passage entering it by a difficult mud slope. On the first big bend there is an aven discharging a fair stream.*
>
> *The final sump is deep and dirty, with a bobbing raft of blackened candle ends on the surface. Mike (Thompson) and Ken (Dawe) will have a preliminary dabble on July 7th, and the best of luck. There is a dig over the sump which looks hopeful. I hope that your aqualung is going well this summer. Mike ... is going to France this year on one of the diving holidays run by Triton. I do a little myself, but only in cold and rather inhospitable gravel pits!*

The last paragraph in this letter reveals that the use of air by some cave divers was gradually becoming accepted. Given that the times were receptive to new ideas, it was now more a matter of personal preference rather than hard and fast rules favouring closed-circuit oxygen sets. Much thought and training was going on to overcome the problems faced earlier when using open-circuit aqualungs in caves. So it was that Mike Boon had been allowed to use his 'Nyphargus' kit on his dive through Sump 5; a sump that was successfully turned into just a duck the same day. It had served a significant purpose so far as Mike was concerned. A 'Promised Land' had been reached and another step forward in cave diving had taken place. It was the first of many more fresh exploration accounts that Mike Boon was destined to write.

Fig. 10.6
Survey of Swildon's Five and Six by Derek Ford and Oliver Wells. December 1958

Fig. 10.7
Personal recollection of his dives in Down to a Sunless Sea by Mike Boon (1977)

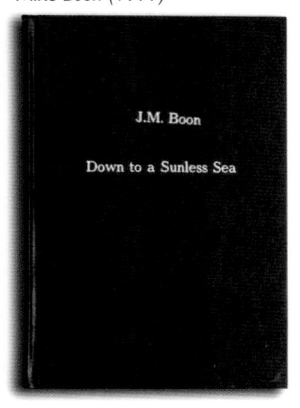

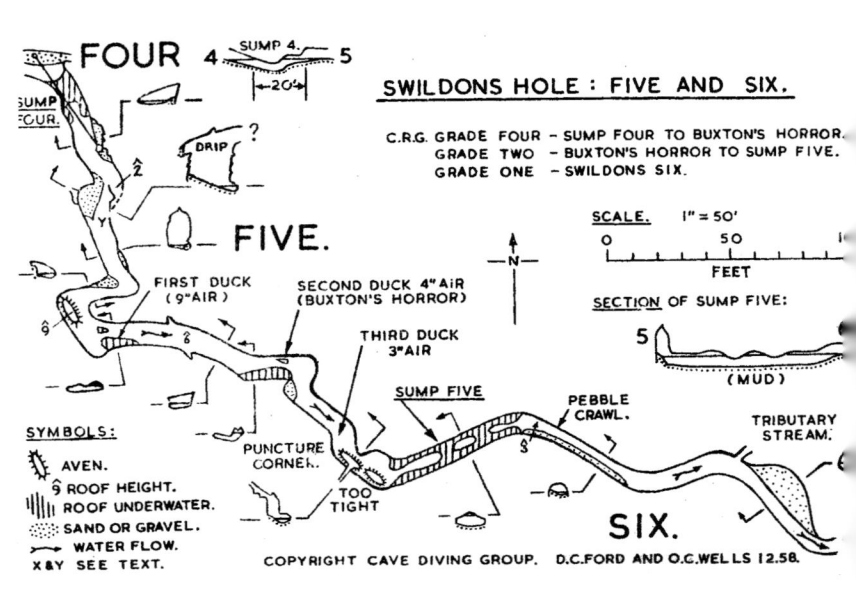

The first hand description of Mike's very first cave dive through Sump 5 is given in *Down to a Sunless Sea*. It is typically eloquent:

> It was now my turn. This was my first real cave dive; I was frightened at the prospect but glad to be getting it over with. After that unreal moment when one's head dips under water I was filled with a curious sense of detachment as if I was watching someone else dive from the inside. I felt the telephone wire with my right hand and pulled myself along seeing only the greyish water slip by a few inches below the roof. In what seemed no time at all I was aware that the waters were no longer swirling over my head and I entered the Promised Land, Swildon's Six.

Three weeks later on 7th July 1961 he was in the lead through the virgin Sump 6, experiencing the darker side of cave diving, but being successful, nevertheless, through sheer determination:

> I submerged and crawled along the bed of the sump, feeling along the lefthand wall. Sure enough, the wall was undercut further along. After a moment's hesitation I tried to move in feet first. But the passage was only twenty inches or so high, and the blocks of lead strung on the length of wire round my waist jammed immediately. I came out and this time tried headfirst, squeezing along a few inches at a time, my back tight against an irregular rock roof, my chest against a mud floor. At times I had to wriggle sideways to free the bottle from the rocks. It was pitch black and very frightening but the chance of breaking the sump urged me on. In any case there was no longer much likelihood of being able to turn around and certainly no hope of backing out. At one point I thought I felt an increasing resistance to breathing and for a dreadful moment thought the air was giving out. No, there were those solid, comforting gusts of air again, thank God. Suddenly, the water thinned to a light grey and in a moment of agonising hope I rose upwards. Could this be the end of the sump? Damn! Still rock overhead. Damn! At that moment my head broke surface in Swildon's Seven. Hardly able to believe my eyes, I peered into the passage ahead, a clean washed triangular archway in the rock stretching to an abrupt bend to the right. I tried to get out of the water and gave three tugs on the line, the signal for more rope. The line didn't budge; not an inch more could I get. I was pinioned in the tiny sump pool with only my head above water. I took a last longing look at the six feet high passage stretching away with the intriguing sound of the trickling stream beyond the corner, then prepared to dive back. With mask back on, I paused to gain the resolution to face the slot again. If I wanted to see daylight again that was the way it would have to be ...

And it was, of course. This kind of trip belied Boon's wry claim to be:

> By nature a pessimist!

What is more certain is that diving was a means to find new caves and not an end in itself for Mike Boon. Foremost a caver, he downplays his role in the development of cave diving, much as Graham Balcombe and Jack Sheppard took for granted their affair with the Swildon's streamway long before Mike was born. Never one to accept pessimistic reports from others either, Boon always looked for himself before dismissing the prospects for new finds. With less bulky equipment, this approach was bound to pay off. Thus it was that on 12th October 1961 he made the first passage of Sump 3 by diving upstream from Swildon's Four and arriving in the complex of air bells that Graham Balcombe had first entered from the other end with his 'bicycle respirator' twenty-five years earlier in November 1936. It was near enough to be a fitting meeting of the ways.

Many months passed before the new route was to be used in earnest, for such was the care taken in those days over planning and preparing cave dives. The first trip along the entire length of the Swildon's streamway through the sumps from One to Six was on 2nd June 1962 by Mike Boon, Mike Thompson, Steve Wynne-Roberts and Fred Davies. By avoiding the tortuous route through Paradise Regained from One to Four, subsequent diving operations became quite different. The days were over for large sherpa parties ferrying equipment through this difficult route, particularly the notorious Blue Pencil Passage that had been blasted open by Dennis Kemp in 1957. The first shots had been fired by Oliver Wells and Keith Chambers back in the spring of 1955, using gelignite provided by Frank Frost and electrical wire obtained by Oliver Wells from the Engineering Laboratories at Cambridge University. The gentleman in charge of the lab stores there was rather angry when the wire did not come back! It took two years of determined blasting by Dennis Kemp and Len Dawes to force a way through this apparently unpromising passage. But Blue Pencil became the back door

*Fig. 10.8
Steve Wynne-Roberts outside the old Shepton Mallet Caving Club Hut at 'The Beeches', Priddy, wearing a full wet-suit and his newly modified slim-line breathing apparatus, 1962.
Photo by John Cornwell*

*Fig. 10.9
Fred Davies dons his own side-slung respirator for a training dive in the St. Cuthbert's Minery Pond, Priddy, 1962.
Photo by John Cornwell*

for divers to regain the main streamway beyond Sump 3. Five years later it had served its purpose and only the most nostalgic old hand could mourn the end of the marathon expeditions to Swildon's Four which had to be planned like their mountaineering counterparts. A new era of light-weight, self-supported cave divers had begun.

By the beginning of 1962 Fred Davies and Steve Wynne-Roberts had boosted the number of trained divers. As well as their exceptional caving experience, both had the necessary blend of theoretical knowledge and practical skills to fashion and forge the new approach. Their refinements of available closed-circuit oxygen equipment, however, still did not equal that of 'Nyphargus' for tight sump work. But proof that the latter had the edge for the new style cave diver had to be earned the hard way in the end. Fred Davies explains:

When 1962 started Steve Wynne-Roberts and I had joined the ranks of divers. The statutory visits to Wookey Hole had been made and hours had been spent wallowing in the Mineries pond wearing SGAMTU sets. But all this seemed too far from the constricted passages of Swildon's. I personally opted for a Davis Submerged Escape Apparatus (DSEA) as the simplest form of light-weight kit. An oxygen bottle was slung across the bottom of the chest-mounted breathing bag, a face mask used in place of the original open gag and I had a rebreathing apparatus that would last close to an hour on one fill of soda lime and gas.

Using this gear I worked with Mike Boon when placing bolts and a good reliable guide line through the Swildon's Two and Three complex on 5th May 1962. Then on 2nd June we mounted an attack on Sump 6 - Mike Boon, Mike Thompson, Steve Wynne-Roberts and I. The other three soon passed into the new territory of Swildon's Seven, first glimpsed by Boon the previous summer, but I found my DSEA chest-mounted kit too restricting in tight places and was forced to stay in Six to await their return. They reached Sump 7 having explored another 300 feet of streamway.

Steve had made himself a side-mounted rebreathing apparatus by combining components from the Amphibious Tank Escape Apparatus (ATEA) and the Submarine Escape Breathing Apparatus (SEBA), which had recently become available as Government Surplus. With the full bulk of the apparatus tucked under his right armpit, Steve easily passed the constricted Sump 6.

I went home to gather the components for a similar ATEA/SEBA kit, but was still using my familiar DSEA on 9th June on a solo dive through the sumps to bring out the maypoling pipes used on an earlier push up the Cowsh Avens above the roof of Swildon's Four. By 23rd June, I had built and trained with my new respirator and went with Mike Boon and Steve Wynne-Roberts to push Sump 7. Bob Pyke and Dave Turner joined us, having travelled via Blue Pencil, and free-dived Sump 4. The rest of us had taken all the gear along the streamway and dived through the sumps with our slimmed down respirators. This time I easily passed Sump 6.

Bob set up a kitchen at the sump to provide hot drinks on our return. Steve and I left our respirators and lead at the beginning of Swildon's Seven and helped Mike to carry his air set and spare bottle to Sump 7. After a few initial ducks, it involved Mike lying flat in about 2 feet of water as I remember it. His legs gradually went from view; there was a lot of kicking and splashing, then silence ...

Steve and I were beginning to chill and get alarmed when a violent splashing announced Mike's return. I have never seen anyone quite so distressed. He had passed an upward and slightly to the right underwater squeeze by removing his bottle, pushing it ahead and following with his body. In this way he had discovered Swildon's Eight, but the return journey through the new sump had been even more difficult and whilst fighting to get back Mike lost his face mask. After a brief discussion he was persuaded to leave whilst Steve and I stayed on for a while. He left the cave straightaway having had enough for that day! Although Mike left a fully charged bottle strapped to the roof in Eight for his next attack, he never returned for it. The rusty cylinder was retrieved almost three years later by Mike Wooding when he discovered Swildon's Nine. It now has pride of place among the 'trophies' that hang on the wall at the Wessex Cave Club's headquarters behind Eastwater Farm.

Mike Boon's own graphic account of his dive to discover Swildon's Eight in *Down to a Sunless Sea* ends his obsession with the place. He was to go farther afield and fulfil his commitment to find new caves. On that night in June 1962, a mud-splattered and drained Mike Boon had drinks with Jim Hanwell in the New Inn on Priddy Green and was quietly relieved to be there at all. Nearly 400 feet beneath them Fred Davies and Steve Wynne-Roberts were discovering the audacity and brilliance of Mike's effort:

Fig. 10.10
Mike Boon's trusted 'Tadpole' air bottle. Now a trophy at the Wessex Cave Club's headquarters, Upper Pitts, Priddy. It was recovered from Swildon's Nine in 1972.
Photo by Mark 'Gonzo' Lumley

Steve and I decided to at least 'have a go', so we went back for our respirators and lead and returned to Sump 7. Each in turn tried the sump. I kept my right hand firmly on the oxygen valve, pushed underwater until I could feel rock all round. I then filled my lungs but emptied the breathing bag until I could make myself no smaller. I pushed hard yet only progressed a further three or four inches. Carefully, I backed out, reinflated the counter lung and then felt a sudden rush of water into my left arm. I had rippped the cuff of my dry suit and had to call off my attempt.

When Steve tried, he had a similar experience and ripped the neck seal of his suit. We were a sad and miserable pair on the long journey out of the cave with so many sumps to pass.

The advantage of 'Nyphargus' in constricted, short sumps had stood its sternest test yet. Closed-circuit oxygen was successfully used for a couple more years in longer duration dives, but the tide had turned against it. A different outlook and approach to cave diving naturally followed.

News of Mike Boon's epic discovery of Swildon's Eight travelled fast that summer. Jack Waddon was particularly interested in the progress being made with miniature BAs for he was planning to dive the not too roomy entrance sump in Threaplands Cave, a rising beneath Cracoe Fell in Wharfedale, Yorkshire. Fred Davies recalls:

Jack had heard reports of our progress in Swildon's and invited Steve and me to join him on a dive in Threaplands. We travelled together in my car and met Jack and his wife, Dorothy, late Friday night on the first weekend in September 1962 in the field outside the entrance. There is a 60 feet long sump just inside the entrance leading to lots of cave. We dived the next day using the SEBA/ATEA system that Steve had invented. Jack used his larger SGAMTU apparatus.

I led and followed the line to the far side of the sump. Steve arrived shortly afterwards and then we sat and waited for Jack. He didn't arrive, so I went back through to find out the reason for the delay. His face mask had a leak, but Jack also found the route underwater too tight for his equipment. After I had relaid the line into a roomier part of the sump and his face mask was clamped, he was able to join us in exploring the cave beyond. However, Jack had no caving lamp, helmet or proper caving boots for he had been more accustomed to a dive being an end in itself.

Fig. 10.11
Fred Davies with his own slim-line SEBA/ATEA respirator about to dive the constricted Sump 2 in Stoke Lane Slocker, 1962.
Photo by John Cornwell

A major push on the second sump in Stoke Lane Slocker back on Mendip was planned for a fortnight later and Jack Waddon was invited to join the party. As this sump was even more constricted, he was given the chance to try Fred's equipment. In addition to the SEBA/ATEA sets, an extra SEBA bottle and Normalair demand valve were carried. Steve Wynne-Roberts successfully pushed through this sump for the first time and returned to tell those waiting. He swapped his BA for the Normalair one on his second attempt so that Fred Davies could use Steve's SEBA/ATEA and Jack Waddon borrow Fred's set as arranged. But although Steve and Fred saw Jack's bubbles on the far side of the sump, he was unable to follow them through. Mike Thompson, whose gear that day included an old milking coat borrowed from Stanley Stock at the farm, then had a turn with the kit and succeeded in passing the underwater squeeze that had foiled Jack. Steve, Fred and Mike entered several hundred feet of new streamway with decorated chambers above, bypassed another sump and finally turned back at Sump 4.

Jack Waddon was determined to put a similar SEBA/ATEA set together for himself and practice for the next Stoke Lane Slocker push planned for early November 1962. At the time, he was the most experienced diver in the CDG to maintain an active interest in the sumps down Mendip's swallet systems rather than those in Wookey Hole Caves. He was especially helpful and encouraging to the younger divers. Other senior members of the Group were busy elsewhere and a do-it-yourself attitude gathered momentum.

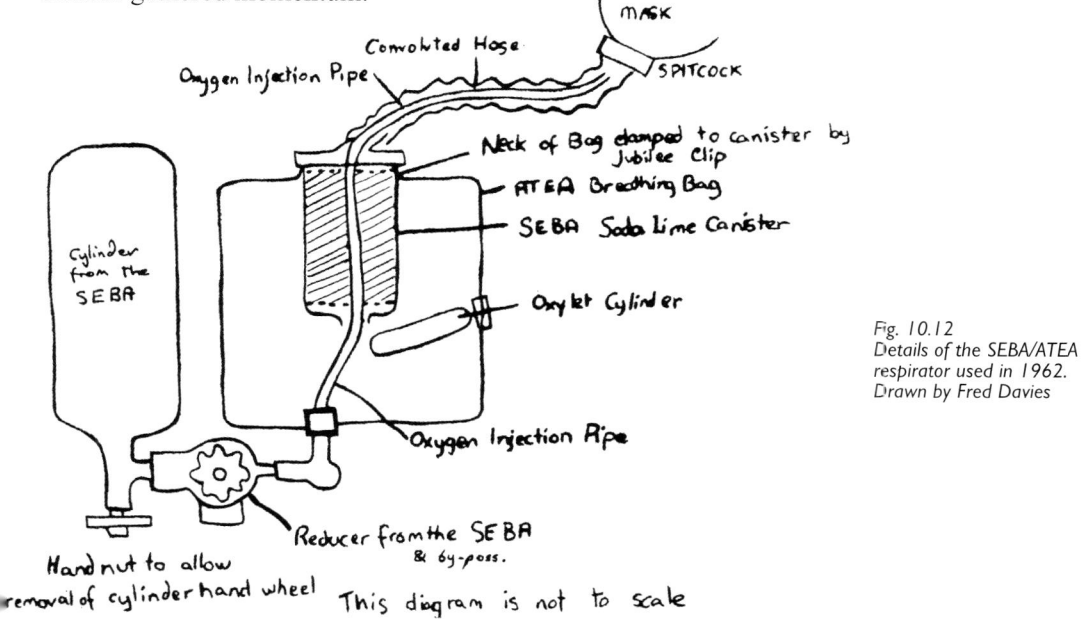

Fig. 10.12
Details of the SEBA/ATEA respirator used in 1962.
Drawn by Fred Davies

Bad weather forced a postponement of the Stoke Lane trip on Saturday 3rd November. Jack still had some finishing touches to do on his new SEBA/ATEA set and borrowed Fred's for a practice dive in the Priddy Minery. He fitted his own reducing valve to this set and, in the absence of a meter, altered and compared the flow rates to his own by ear. After dressing into full caving kit, which included woollens, immersion suit, boiler suit and nailed boots, he did a trial dip to adjust his weights and set off underwater just before 6 p.m. By then it was dark and the other divers were back in the Shepton Mallet Caving Club Hut unaware of Jack's intentions but content that he was more expert and experienced. It was a great shock when Dave Turner, one of the two cavers left on the bank, rushed to the hut fifteen minutes later very concerned that there was no sign of Jack in the pool. Fred Davies donned another kit and set out on fruitless searches of the murky water, and then handed over to Mike Boon at 7 p.m. Jack was located after ten minutes, towed ashore and given artificial respiration until he breathed spontaneously but with difficulty. This assistance was continued until he reached the hospital in Wells at 7.40 p.m. But Jack died there an hour later from a pulmonary oedema.

A full report of the incident was published by the CDG in August 1963. It appeared that Jack had lost his bearings and, unaccustomed to the set, had caused a carbon dioxide build up in his efforts to get out and release his weight belt belatedly. Anoxia caused the fatal oedema, although there was sufficient oxygen from the set to maintain his life during the hour or so he was probably unconscious underwater. There was some criticism of the SEBA/ATEA kit by naval experts used to respirators with larger capacities and flow rates. But their sets were no good for diving constricted sumps in caves, of course. Again, the differences between essentially open-water divers and cave divers became all too clear. On the night in question, those troubled by the events came to the conclusion that cave divers had to do things for themselves. In the following year, as mentioned in the previous chapter, the old regional sections of the Cave Diving Group devised by Graham Balcombe were strengthened; a new constitution to match was agreed and revised Diving Qualifications for both trainers and trainees established. As Charles Wyndham Harris put it eloquently, these changes were the Group's lasting memorial to Jack Waddon.

*Fig. 10.13
Jack Waddon's last dive at Wookey Hole Cave in 1961, before losing his life in the St. Cuthbert's Minery Pond on Saturday 3 November 1962. Photo by John Cornwell*

In a letter to Oliver Wells dated April 1963, Fred Davies was able to review better days and prospects:

A stir in the CDG ... We used the SEBA/ATEA kit again at the Easter weekend. Agen Allwedd at Llangattock, S.Wales, has produced a sump so far from daylight that a fast moving, unloaded, caver is expected to take 3½ hrs to reach it.

The lead weights were taken in a fortnight earlier and Steve Wynne-Roberts and I carried our own BA's to the site of the dive, both wearing wet suits. We dived individually in turn using ordinary cap lamps as lighting and laying a line from a reel holding 300ft.

We between us reached a point at least 150ft, probably nearer 200ft, into the sump by which time we were about 15 ft below water and on the edge of an underwater pot. The sump passage was at least 6 ft by 6 ft in cross section, a fine gravel bed down the centre and some soft mud at the edges. Visibility zero.

Not a bad performance for the BA we reckon. We earlier in the year did a dry dive and tried to induce CO_2 excess by climbing up and down a ladder wearing a 35 lb weight belt. Forced to give up by over heating of the breathing gases without any CO_2 symptoms. The canister seems capable of permitting a fair amount of work.

In the summer of 1963, similar approaches enabled Ken Pearce of Derbyshire to dive the sump at the bottom of the Gouffre Berger, backed-up by Steve Wynne-Roberts and Mike Boon, the main proponents of light-weight diving at the time. By February 1964 following Mike Thompson and Steve's return to the Thirteenth Chamber of Wookey Hole described in the previous chapter Fred Davies wrote again to Oliver Wells in the United States:

The situation with regard to cave diving is now very different in this country. Compressed air has been clearly demonstrated, again and again, as being a perfectly satisfactory cave diving tool, especially for short preliminary probes. Almost all keen cavers are well equipped with wet suits and the impulse to just step out and buy a cylinder and demand valve is very high. In fact there are many people beginning to attack sumps who have not made any effort to join the CDG...

Recently Mike Thompson, Steve Wynne-Roberts and I were invited to join the Northern Section in an attack on Langstroth Cave. There were eight divers operating, I was the only one using oxygen (UBA), all others were using compressed air BA's (their own property)...

I am sure that compressed air is going to be used more and more; especially on the very tight, short sumps and the deeper dives. Mike Thompson and Steve Wynne-Roberts recently reached Wookey Thirteen ... Steve used mixture (UBA) whilst Mike used air and had ample reserves. Both were swimming and laying a line. All instruments on the wrist, miners' cap lamps (about 3 watts) as lighting on caving helmets, i.e. no Aflo, just a light line reel paying out courlene line which floats and is less likely to become silted over. With the advent of cylinders with higher working pressures, the advantages so far held by rebreathing sets are being whittled away.

Long after these events Fred Davies reflects that the change had a lot to do with the different attitudes and equipment needed when diving in Wookey Hole Caves compared with the sumps in Swildon's Hole. It was a matter of supply and demand, too, regarding the availablity of equipment, including clothing:

It seems to me that the mental attitude of a cave diver before the sixties was moulded by many years of emphasis upon Wookey Hole Caves as a diving site. His entire equipment was designed for big roomy spaces, and he did not expect to get out of the water for long so clothing did not have to be suitable for normal caving as well. In 1958, for example, we had carried heavy rubber and twill dry suits into Swildon's Four for the divers to don before entering the water. By 1963, on the other hand, Steve Wynne-Roberts, Noel Cleave and I revisited Swildon's Seven to make the survey, accompanied by a few other cavers as far as Sump 2, but were on our own after that for about ten hours. Our diving experiences, in Swildon's especially, led us to expect to carry all the gear for ourselves and be able to remove it when out of the water without the assistance of dressers. Each diver had to maintain a guard upon his own consumption of gas without a controller remaining behind to keep a log as in earlier days. Whilst we found a companion diver was a security when base feeding a line through a new sump, he would invariably follow if the sump was passed. Solo diving was the ultimate outcome following such changes.

We soon learnt to settle for the lightest clothing available, such as the yellow submarine escape suits with an umbilical entry and neck seal. These so-called goon suits had come onto the market late in the fifties. Some divers modified them by fitting a dry hood, but my preference was always to use one of the new-fangled wet suit hoods independently. In those days, the wet suit itself was not considered to give enough protection for long trips! Only thin, single skinned neoprene versions were available. However, I did find it an advantage to cut off the uncomfortable feet of the goon suit, fit ankle seals and wear wet suit socks instead. Thus topped and tailed with the fabric of the future, I wrapped my middle warmly as of old. Wearing this, I found little difficulty in passing all the usual caving obstacles and still kept warm under water. For me it was the most comfortable dress I ever used.

The standard CDG respirator was the SGAMTU which supported life for one and a half to two hours when filled. Although beautifully comfortable when supported by water it was bulky and difficult to don single-handed. Once fully weighted, it turned the diver into a lumbering hulk when out of water. The modified DSEA that I used gave me a duration of about an hour and was substantially lighter and portable. Its smaller chest-mounted breathing bag, CO_2 absorber and oxygen cylinder made it easy to put on and take off. Steve Wynne-Roberts' brilliant ATEA/SEBA adaptation was even more manoeuvrable. He slung the 5 litre breathing bag, CO_2 absorber, oxygen cylinder and reducing valve on a shoulder strap, haversack fashion, under his right arm. A waist belt provided extra support. It was the small size of this arrangement that enabled Steve to get through the awkward second sump in Stoke Lane Slocker. My DSEA would never have made it. Mike Boon's use of small 'taddies' became popular and they were pumped to rather more than Home Office recommendations when duration was at a premium. When gas supplies were urgently needed they were even filled by decanting oxygen from the welding bottles of a friendly local garage! Single hose regulators with the second stage carried on the gag were used with a face mask covering the eyes and nose. Although Mike's air kit showed even greater advantages in his daring dive of Sump 7 in Swildon's, one feature in favour of oxygen rebreathing was the smaller heat losses involved. In fact, it supplied you with warm air. I have not seen such a feature stressed by other writers; yet, nowadays, this principle has been developed in the life-saving hot air devices used to resuscitate hypothermia cases.

Why then was the rebreather rare in cave diving by the mid-sixties? I believe that it has more to do with the decline of Government Surplus supplies of equipment against the rapid growth of the commercial market for open-water skin divers rather than something within cave diving as such. Costs dictated the change as much as anything. A modern cave diver wishing to buy a rebreather will have to spend a lot of money to acquire a good safe set. Almost the reverse was true around 1960. There is also the question of straightforward practicalities, of course. Oliver Wells recognised this first when last diving Sump 5 in Swildon's in 1958. He knew that his AFLOLAUN was unnecessary and carried ordinary lighting, for example. We too always wore normal caving helmets and lamps over our face masks. Travelling downstream with one's own disturbed silt from the time you entered a sump made the water thicker than pea soup. I have never actually seen anything in Sump 6, for instance, but navigated by feel alone. You couldn't do this in Wookey Hole's spacious sumps.

This brings me, finally, to the 'bottom-walking' versus 'finning' debate that had arisen in Wookey Hole Caves. Gas economy and safe navigation were the main arguments favouring the bottom-walkers with oxygen rebreathers. But it was also the distances they had to dive and the conditions encountered underwater. It is of interest that Mike Boon's 'Nyphargus' air kit was first designed for bottom-walking rather than fin swimming, whilst I found on several dives that my ATEA/SEBA oxygen gear was fully capable of supporting my breathing needs when finning. In diving the shorter downstream sumps in Swildon's Hole, the difference was simply irrelevant. In tight, constricted underwater passages when you are fighting with a squeeze, the question of walking or swimming is meaningless. With the return to Wookey Hole Caves, on the other hand, it was obvious and correct to fin as this was the only way to cover the greater distances to the Fifteenth Chamber and beyond. And the depth required that air was used.

In a letter to Oliver Wells dated 19th July 1961, Mike Boon wrote:

Swildon's really is an amazing cave, the further you push it the more there is to explore.

He could have said much the same about this crucial experimental period in the development of cave diving attitudes and equipment. The next generation to return to Wookey Hole Caves were the beneficiaries and so were bound to tackle things differently.

Fig. 10.14
The way on upstream from the Third Chamber in Wookey Hole Cave continues to beckon a new generation of cave divers, with different gear and methods. Meanwhile, The Crocodile remains 'on guard', as ever, 1950.
Photo by Luke Devenish

Chapter Eleven

An Independent Air

based on Oliver Lloyd

No history of cave exploration on Mendip, and in many British caving regions for that matter, would be complete without mentioning Dr. Oliver Cromwell Lloyd's close involvement and influence. Since his arrival on the scene in the early fifties, 'OCL' touched upon almost every facet of both the sport and the science at some time. From the mid-sixties he even fashioned modern cave diving.

Although Oliver did not take to caving seriously until over forty years of age, his love of the good things about the past helped him to make close links with the active cavers of the day and the older generation. With Luke Devenish, Howard and Richard Kenney, for example, Oliver helped to bear Herbert Balch to his final rest following the great man's death on 27th May 1958. By then he had already taken over from Howard as Secretary and Treasurer of the Mendip Rescue Organization; a post he occupied until Jim Hanwell was appointed in March 1972.

Like Balch, Oliver was to die in his sleep. And it was twenty-seven years later, almost to the day. Preparations to celebrate the Jubilee Year of cave diving at Wookey Hole Caves had already begun; so, sadly, he missed the fun of the reunion of cave divers spanning the years there on 4th October 1985. Had he lived on there would have been more stories to tell about cave diving in both Swildon's and Wookey Hole, especially from 1965. They were his favourite caves. But this account must suffice with a few choice snippets from Oliver's records; the ones unlikely to be remembered or published elsewhere.

The numerous obituaries and notices, from *The Times* and *British Medical Association Journal* to tributes in every major caving publication, tell stories of a cultured and complex personality who was a notably charismatic figure both at work and play. Oliver Lloyd was appointed as Reader in Pathology at Bristol University in 1949. Both his Professor, Tom Fewer, and the Professor of Medicine, Bruce Perry, had been active members of the University of Bristol Spelæological Society for years. There were strong links between academic staff and students in 'the spelaeos'; a happy and fruitful tradition that Oliver himself was to lead later. Dr. Donald Thomson, then a medical student, recalls OCL's conversion to caving after the first ritual trip to the spelaeos' own find at Charterhouse-on-Mendip, G.B. Cave:

Fig. 11.1
Oliver Lloyd, on the first of many visits to Swildon's Hole, climbs the Twenty Foot Pot, 1953. Photo by Don Thomson

Life was difficult for him in the early 1950s. His wife was ill and at times he felt overwhelmed. He used to cycle out to Mendip, but somehow did not at first fall in with the caving community. Eventually, he did ask if we would take him caving again, a request which resulted in a trip to Sump 2 down Swildon's one Saturday in 1953. It was in the days before wet suits or even dry suits, and he could hardly be blamed for stating firmly the following Monday that he would never go again. By the following weekend, the bruises were fading, and he was caving nearly every weekend for the next twenty years.

Within the year, OCL had the novel idea of organizing so-called 'scavenging trips' to clean-up caves by scrubbing off unsightly stains of spent carbide and removing any litter. The editor's first encounter with Oliver was on such a trip at Sump 1. Whilst dutifully sticking to his allotted post and scrubbing away alone at a particularly stubborn crust of carbide, OCL breezily arrived complete with wickerwork basket. This was ceremoniously laid on a boulder and unwrapped to reveal a welcome flask of hot coffee and magnificent chicken sandwiches. We had an unlikely picnic and chat.

Oliver Lloyd was not a dull man and he liked to enliven those who were. He did not wish to be much involved in writing-up accounts for this commemorative book for fear of it being too difficult to convey both the facts and the fun when so much was weighted towards the former. Or was it because he sensed there was little time left for the job? More than once he claimed a preference to live for the moment. In a letter to Jim Hanwell on this subject dated 9th February 1985, Oliver warned:

Since you are undertaking this job, I hope you will follow the precepts of Tacitus! For me, all that I have done or watched being done is already on record. As I have nothing to add, perhaps you can use the available material without asking me to do anything more. Whatever you do, avoid being dull and worthy. Dullness is not one of your faults, but I'm thinking of the stuff you may get from your correspondents.

Looking back over some old Diving Reviews I am reminded that there was virtually no serious diving in Wookey Hole from 1962 to 1966. I think that the only important contribution made by me was to go and chat up Mrs. Hodgkinson in 1965. But don't go putting that in!

Ever yours,

Oliver.

To this, the reply was that it was too good to miss out, especially since Tacitus himself had argued that history was chiefly:

To rescue virtuous actions from the oblivion to which a want of records would consign them...

On the rest of this quote, we were both aptly silent!

Fig. 11.2
From the mid-1950s, Oliver Lloyd gets involved with most aspects of Mendip caving and cave diving. He is seen here with Dan Hasell.
Photo by Tony Morrison

We met finally only a couple of days before he died. Oliver last visited Priddy to launch the British Cave Rescue Council Conference on Mendip over the weekend following Friday 17th May 1985. He screened his 'Cave Rescue' film made by the Mendip Rescue Organization in 1960. Historic shots of sump rescue work at Sump 1 in Swildon's took everyone back to where cave diving had been conceived before hard-hat diving began at Wookey Hole Caves. They also recorded the advent of air as described in the previous chapter. Oliver was there at the beginning of this last phase of cave diving and his film show to many caving friends was a fitting finale. He was on form to the end.

As recently as 8th May 1985, too, Oliver had come to second thoughts about contributing to this book himself. He typed out the following as a short vignette 'to liven things up.' His accompanying letter concluded:

If you like it and want any more of this sort of thing, send me along the other chapters, as they come, and I'll put the Lloyd gloss on them.

Fig. 11.3
Oliver Lloyd undertakes the 'boring occupation' of helping divers. At Sump 2 in Stoke Lane Slocker with John Buxton, watched by Struan Robertson, 1 September 1956.
Photo by Tony Morrison

This was not to be, alas. Here is the only such highlight concerning Wookey Hole Caves:

Assisting cave divers in the nineteen fifties was a boring occupation. They took ages. First assembling their kit, then testing for leaks, then putting it all on and then doing their breathing drill, before they were ready to set out. Nowadays kitting up in Wookey Hole takes about twenty minutes. No wonder the dives made by the Old Men took half the night. So it is not surprising that the sherpas pushed off and went exploring in the upper parts of the cave. I got to know these very well and still take student parties up there, when I tell them the following tale.

Right at the front, just over the main entrance, was a small opening, through which jackdaws came and built a communal nest about the size of a double bed. I had been inspecting this one day (Wells gives this as the 30th June, 1956) and on descending met two of my party of assistants going up. They were using 'Balch's Dependables', in other words candles, for lights. The name is derived from a famous remark by Balch that 'by common consent the candle is the most dependable illuminant, because it casts no treacherous shadows.' On reaching the jackdaws' nest one of them had the misfortune to upset his candle into it, setting the twigs alight. They tried beating it out with a slipper and thought they had succeeded. But no. Fanned by a stiff breeze from outside, the whole nest took fire and burnt fiercely, emitting about the foulest smoke I have ever smelt. Jackdaws are not careful about the disposal of their unwanted products. There was nothing we could do about it; the approach to the fire was a squeeze, into which the flames were being blown. When Oliver Wells passed us on his way out from his dive he said, 'I think you might have left us a line through all this.'

By about four in the morning the fire had nearly burnt itself out and we were able to crush the embers and bury them in mud. The cave still stank. We had to leave the front door open for ventilation and I was volunteered to guard it until the guides arrived in the morning, after which I was expected to go and apologise on behalf of the Cave Diving Group to Wing Commander Hodgkinson.

The first guide to arrive was Colin Bristow, one of our students. When I had told him the story he said, 'When the Wing Co hears about it, you won't half catch it!' Very encouraging. But soon after a wonderful thing happened. Professor E.K. Tratman turned up. He had heard about two of our people getting into trouble and came specially to help me apologise. Now Trat is well known for the many kindnesses he has done, but I think this beats all. So we faced the Wing Co together. The Wing Co took it quite well, considering what a nuisance we had been, and sent in to Wells for all the Airwicks he could get to purify the cave atmosphere. His only stipulation was that neither of the culprits was to be allowed to enter the cave again. This was not followed. One of them was a trainee diver. He finally had the misfortune to spend a lot of time upside down in the water, because all the air in his baggy suit had gone into his legs. On righting himself he gave up diving.

Then Trat and I went to see Mrs. Hodgkinson, who was in the kitchen kneading a cake mixture. Trat at once spotted that she was wearing the Bronze Age gold bracelet, that had recently been dug up at Ebbor.

'But', protested he, 'It is priceless; and yet you wear it when you are cooking.'

'I like to think the thoughts of its original wearer, when I am making a cake, as she must have done.'

'But the Wing Commander has had a copy made of it. You could wear that.' She drew herself up to her full height. 'Me! Wear an imitation!'

As the French would say, 'la portage était tout à fait folklorique.'

Fig. 11.4
Olive Hodgkinson and Oliver Lloyd celebrate the long awaited breakthrough in Wookey Hole Cave, 1970 – see Chapter 13. Photo by Norman Heal, Cheddar, given by Olive Hodgkinson to Jim Hanwell

Fig. 11.5
Olive Hodgkinson's Bronze Age gold bracelet, originally found in Bracelet Cave, at Hope Wood, Ebbor, by Ted Mason, 1955. Photo by Luke Devenish

So we added another verse to the caving song.

> We never help cave divers, And as the smoke rolls outwards
> They are a frightful bore, We cough with all our might,
> We set fire to the bat shit 'Per igne via asbestos,
> And sit outside the door. Blow you Jack, I'm alight.'

At the end of October 1963 Oliver Lloyd twice visited Stoke Lane Slocker just downstream from the village of Stoke St. Michael. Of all people, he picked up an acute hepatitis infection, known as Weil's Disease, which attacks the liver, may even damage the kidneys and could be fatal. Rats are the source of the offending *Leptospira* organism, and at the time they were infesting the banks of the polluted stream near the sink. Oliver was very poorly indeed, but was on the mend by Christmas and well on the road to a complete recovery in the New Year; a process he described himself:

The first thing to do is to get out of hospital as early as possible. After that you eat and sleep as much as you can and go in for 'graduated exercises'...The turning point will be reached when you feel you can start caving again. This will be a lot earlier than they think. Tiring and rather slow at first, you will easily get out of breath. Last of all, back comes your wind and the spring in your step, and you can draw a line under what has gone before.

Oliver was obviously delighted with his speedy recovery for his general fitness had stood the test. Lesser mortals at 53 years of age might well have succumbed. The line he had crossed by the New Year saw a new and even younger man in many ways.

As OCL got older his particular fancy focused more and more upon cave diving. He had helped on most diving operations on Mendip for almost a decade and, of course, had practised often with the MRO's Normalair Sump Rescue Apparatus in pools, though never through a sump. Then, at the age when most seek a quiet life, Oliver took up cave diving himself. Go-ahead young cavers studying at Bristol University seemed to provide him with the impetus. Their enthusiasm was just the tonic OCL needed. Even if they were happy to get on with it alone, Oliver was always there to share the action and fun. He was to grasp the reins of cave diving and earn his spurs as a diver. His moment probably came over Easter 1964 in Yorkshire.

Oliver Lloyd had taken over as the scribe of 'Mendip Notes' in the Wessex Cave Club *Journals* from Oliver Wells in 1955. He adopted the pseudonym 'Cheramodytes', meaning 'one who creeps into holes'. The style was unmistakable. He was at his most whimsical describing the customary trip to Yorkshire over the Easter weekend from 27th to 30th March 1964:

The Jolliest Party Ever

Members of the UBSS came as guests, bringing with them a guitarist. They also brought with them an ugly rumour about a diving accident in Lancaster Hole. Such rumours have an unpleasant way of turning out to be true, so Mike Thompson went and phoned Bob Leakey. No sooner had we got our three guitars in tune than he returned to say that Alan Clegg had been drowned while diving in the Master Cave Sump in Lancaster Hole.

We heard the details later from Mike Boon. At the conclusion of the diving operation the float line got stuck when being drawn in, and Alan Clegg went in to free it. In the process of doing so he got tangled up in the line himself, lost his gag and drowned. Mike Boon immediately went in to free him, and although resuscitation was started within two minutes of the accident he never recovered. Clegg had been aqualung diving for some years and had done a number of cave dives with the CDG. He was regarded as one of their most steady and reliable members. The others are now asking themselves, if this can happen to him, who can't it happen to?

The party languished and not a song was heard...

Mike Thompson and Oliver Lloyd went to see the authorities in Lancaster the following day to sort out any legal and pathological issues arising from the incident. These were straightforward enough, but the lesson about the dangers of sorting out tangled lines underwater had been learnt in the harshest way. Alan Clegg had become the mainstay of the Northern Section when Mike Thompson revived the lapsed regional framework of the Cave Diving Group. His death was a bitter blow. When Mike stepped down from organizing cave diving that day, Oliver was there to fill the breach. Indeed, from that fateful Easter in 1964, OCL was always part of the action in CDG affairs. He was to live up to his forenames and a New Model CDG was in the making.

Back on Mendip, the stage shifted accordingly from Priddy to Bristol. Enter Mike Wooding, Dave Drew and Dave Savage with double neoprene wetsuits, a couple of Scubair regulators between them, but more than their share of the improvising spirit that freed the young in the sixties. Another free spirit of the times was Mike Jeanmaire, a young Axbridge Caving Group tearaway who already rejoiced in the nickname 'Fish'. Mike's daring free-diving trip through Swildon's Sumps 1, 2 and 3 and back on 27th September 1964 was a feat that fascinated OCL. And well it might because he had done a lot to provoke the challenge.

After a couple of adventurous cavers, Russell Morley and Robin Sims, mistakenly free-dived Sump 2 in May 1962, and just about managed to return safely, Oliver had been in two minds about the foolhardiness of others who planned to repeat the venture. Morley's hair-raising account of the escapade, whimsically entitled 'Sumping for Beginners', is probably one of the most incredible of its sort and appeared in the Wessex Cave Club *Journal* 85 (July 1962). News of this new dimension in free-diving sumps travelled fast and was largely greeted with incredulity. For some it was the next logical step. Wearing his MRO hat, Oliver counselled against repeating the performance. But he might as well have been goading so far as 'Fish' was concerned. For the first time in print Mike Jeanmaire recalls his epic free-dive of Sump 3 in particular.

Fig. 11.6
Dave Drew, 1966.
Photo by Tony Morrison

Fig. 11.7
Mike Wooding (left) and Dave Savage in Wookey Nine, 1985.
Photo by Tony Morrison

> In the early sixties, the Cave Diving Group seemed to me to be at the top of the tribal structure of Mendip clubs; romantic, mysterious and only attainable after an endless apprenticeship. As a very young member of the Axbridge Caving Group, our trips were led by older experienced cavers on a rota basis. Only those over 18 years old were encouraged below the Forty Foot Pot in Swildon's and no one in the group had been through Sump 2. I was anxious to do both but did not feel that I could do my own thing until 'coming of age'.
>
> But changes were clearly on the way, if only because of the greater numbers of cavers of my age group joining in; we were the post-war baby boom. In 1960 I had a chance meeting with Mike Boon and, on a couple of trips with him down Swildon's, discovered that it was possible to make your own rules. We went as far as Sump 2 on one trip and experienced an August thunderstorm flood in the Upper Series on the next visit. This caused a rescue call-out and, afterwards, I felt like an outcast. Mike's remark that older cavers often see clearly but don't hear too well was a crumb of comfort. On a caving pilgrimage to South Wales the following year, I became impressed by Bill Little's outlook in the South Wales Caving Club. Not only had he adapted the wet suit for ordinary caving but also advocated a 'new view' of speed and efficiency; if you felt you could do something, then get on with it! Bill very kindly helped me to make my very first wet suit in 1961.
>
> One day in the spring of 1962, whilst walking to Nordrach, I met Oliver Lloyd. He startled me by leaping out of the roadside ditch at Charterhouse. He told me of the adventures of two Wessex Cave Club members who had mistakenly free-dived Sump 2. Whilst glad that they had got away with it, of course, Oliver was highly critical and referred to them as 'brave fools'. I walked on contemplating what seemed to be an illogical judgement. Thinking about all those who had free-dived known sumps on Mendip such as Len Dawes through the awkward Sump 4, I concluded that it was not just the sump that mattered but the person

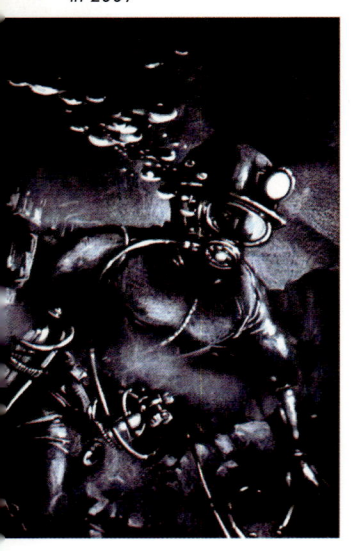

who chose to dive it. Hadn't Sir John Hunt prefaced a book on British climbers, and written about Graham Balcombe's exploits: 'that the impossible becomes the possible; the possible becomes the probable; one must always contemplate the impossible'. By the time I had reached the Axbridge Caving Club Hut, I had vowed to free-dive Sump 2. And why not Sumps 3, 4 and 6 as well.

I caved with a small band of like-minded cavers. We had developed a method of free-diving which was as safe as possible and they did not dismiss my plans. My approach was to gain as much information as possible from diving reports, then reconnoitre the sump concerned using a base fed 20 foot waist line. We dispensed with boiler suits to reduce drag underwater, and found this preferable above water too. Hyperventilation was used despite advice to the contrary. We also learnt to rest on the final approaches to a dive to reduce CO_2. I even inserted a weak link into the chinstrap of my helmet and ensured that the headlamp cable was detached from the back and fed across the chest. These precautions would prevent me getting trapped by my cable. On the dive, I would use fins and strap a spare torch to my ankle. Thus we practised in the sump pools down Swildon's until a modicum of discipline was achieved.

In September 1963, Colin Graham, Colin Edmunds, Chris Smale and I set off to tackle Sump 2. I dived and was surprised to surface in the Little Bell almost immediately. I was even more surprised to be able to be overheard at base via a tiny airspace. Colin Graham came through to join me, and then I set off again for the Great Bell. Just before gaining this airspace I reached the wall exactly as described by Morley and Sims. I knew that all I had to do was to duck under this. And so I arrived in the Great Bell. Another short duck led me to St. John's Bell. Here I sat on the mud bank in triumph! Contemplating the line that disappeared into Sump 3, I removed my torch, attached it to the line and let it sink to the bottom of the pool to light up the way on. After a while I dived down to retrieve it and beyond saw a passage I knew well from previous trips upstream from Swildon's Four. But this day I returned to rejoin my friends still waiting in Swildon's Two.

When I read Graham Balcombe's account of his feelings after diving Sump 2 in 1936, I recall that my emotions were exactly the same. No doubt these were triggered by the chilling effects of not using a hood and the high buzz of adrenalin trailing into anti-climax. Sump 3 was not free-dived until a year later on 27th September 1964. I was accompanied by Colin Graham, Colin Edmunds and Mike Ferraro. By then, tales of my exploits had spread and were met with disapproval from the 'establishment'. Everything was done secretly. Colin Graham followed me to St. John's Bell. I then dived to the bottom of the Sump 3 pool, made a slight move to the right into an easier path and soon arrived in Swildon's Four. It was complete and straightforward, but next time we would use lead weights. When the news broke, incredulity increased and my sanity was questioned. Soon afterwards I had to leave Mendip to work elsewhere, and so free-diving the other sumps had to wait. During that time, attitudes changed and such exploits became accepted. I became the Cave Diving Group's youngest member on 20th December 1965.

The dive on 27th September 1964 had been on my mother's birthday. When I explained to her the resistance we had met, she told me that both she and her brother had been jailed in Switzerland when very young. Their crime was skiing, then illegal. Now everyone skis in Switzerland!

The only mention of Mike Jeanmaire's achievement was a brief report by Tony Oldham in *The Mendip Caver* (No. 10, February 1965) and a couple of sentences in Peter Johnson's book on *The History of Mendip Caving* published in 1967. Significantly, perhaps, 'Cheramodytes' was silent, and certainly in awe of this feat like most Mendip cavers. It is just as well that not all our virtues are only written 'in water', for Mike

Fig. 11.8
a. Mike 'Fish' Jeanmaire in 1968, who subsequently became chairman of the CDG.
Photo from Malcolm Foyle

b. Painting of 'Fish' in his element by Mark 'Gonzo' Lumley presented to him by the CDG upon his retirement as Chairman in 2007

became a long serving Chairman of the CDG.

During the autumn of 1964, the Bristol team of Wooding, Drew and Savage set about training themselves in the Abbot's Pool, Bristol, the Mineries on Mendip, the short sump in Stoke Lane Slocker and, of course, the longer challenges down the Swildon's streamway. Getting their new equipment right was crucial, especially the balance between good insulation and the correct buoyancy and trim for finning beneath submerged rock.

On Saturday 5th December 1964 the trio successfully dived through Sumps 1, 2 and 3 in Swildon's. Because they were sharing only two valves and cylinders, however, some free-diving took place too. From this they concluded that, with thicker handlines, they would also be able to dive all three sumps in this way. Sump 1 posed no problem; Sump 2 was free-dived with 'comparative ease', and even Sump 3 was only 'a strain on one's lungs.' Sump 4 was awkward but already well known to experienced cavers whilst Sump 5 was no more than a duck since its downstream dam had been dug away. There was nothing to prevent a determined caver reaching Swildon's Six without breathing apparatus. And there was always Blue Pencil Passage and Paradise Regained to use if needs be.

The seven-hour trip through Swildon's sumps had been seminal. Confidence ran high and it was done without a support party; that is, until Oliver Lloyd met them on their return to Swildon's Two! On Priddy Green that evening the triumvirate founded the Independent Cave Diving Group. OCL became a self-appointed steward and sage so that they were not alone. He kept the divers' logs among other records[19]. Cave exploration on Mendip particularly flourished over the following year. Many discoveries were made and lots of changes in caving were afoot once more. Each fuelled the other.

The first quarter of 1965 sets a hectic pace. Barely had the New Year hangovers cleared when the ICDG trio dived through to probe Sump 7 in Swildon's. Although failing to get through, they learnt enough about the place to hasten their return. On surfacing just before midnight on 3rd January after over twelve hours underground, their log of the trip concludes:

Found OCL by the fireside in Main's Barn, and was he pleased.

The following weekend saw the inauguration of the Council of Southern Caving Clubs at Bristol University in the Geography Department above the old 'Spelaeo Rooms'. The thirty-five caving organizations present from the South of England elected Oliver Lloyd as its first Honorary Secretary and Treasurer. Oliver already held a similar post in the Mendip Rescue Organization, of course. His well known philosophy of caving matched the mood of everyone on Mendip, and his suggested motto for the new CSCC was popular enough:

Live and let live.

Individual cavers were essential and the enthusiasm of young explorers such as Wooding, Drew and Savage should never be suppressed by heavy-handed organizations. Mendip cavers even felt alone in cherishing a tradition that they thought was becoming a thing of the past in other

Fig. 11.9
'Live and let live'! Oliver Lloyd addresses the Inaugural Meeting of the Council of Southern Caving Clubs, as the first Honorary Secretary and Treasurer, with Jim Hanwell, the new Chairman, January 1965.
Photo by Tony Oldham

[19] These included a short run of journals circulated among ICDG members, now in the CDG library.

parts of the country where there was more support for a national organization to govern the sport of caving. Many would refer to the CSCC as the 'Southern Confederacy' and Oliver was likened to one of its aristocratic Generals. Whilst he might have been more of a Jefferson Davis in appearance and position, OCL also had much of Thomas J. 'Stonewall' Jackson's conspicuous eccentricity so far as caving affairs were concerned; the hapless Confederate General was often compared with Oliver Cromwell, no less. And OCL always held that new discoveries relied upon the few prepared to question and get on with it, not the many who thought they knew what the answers were. A thinking caver would achieve more than a committee, however knowledgeable the latter.

Jim Hanwell took the chair of the CSCC, being secretary of the largest caving club in the country at the time. This was to Oliver's liking, too, for the Wessex Cave Club held similar views. Its first President, Herbert Balch, always encouraged young cavers and Frank Frost who took over this office certainly ensured that the old traditions lived on. Frank had dived a bit at Wookey Hole with the pioneer 'hard-hatters' in 1935, and the Wessex itself had formed the November before. Even then the young enthusiasts for the sport had wanted to look after themselves.

By January 1965, the Wessex was just thirty years old and their 99th *Journal* issued at the end of the month captured the prevailing mood well. Derek Ford, by then Professor of Geography at McMaster University in Canada, reviewed the development of Swildon's Hole as a way of thanking all who had helped his research. The lowest parts of the streamway were the newest and so likely to become more restricted further downstream with longer, deeper sumps. Willie Stanton had thought much the same after his detailed survey of the system during the early fifties. Yet both were only too aware that cavers were always finding more to add to the map. Only a surveyor who had spent hours in the cave could have written an account with the somewhat plaintive title of:

No end to the darn place.

Wooding, Drew and Savage intended to put the pundits to the test.

The 100th *Journal* of the Wessex Cave Club dated March 1965 was something special again. It included an article by Dave Drew and Mike Wooding on their discovery of Stoke Lane Five, Six and Seven on 8th February. (They found the terminal stretch called Eight later that year). And to top it all, readers were treated to accounts of their even more recent discoveries of Swildon's Eight to Twelve on successive weekends with Dave Savage. They had reached the end of what was then the deepest cave in the country.

Despite persistent attempts over the past twenty years to push beyond the terminal sumps reached in 1965 no more breakthroughs have been made in either system. This stalemate is also a lasting tribute to the extraordinary achievements of the Wooding-Drew-Savage team at the time. Let each one tell part of the Swildon's saga for there is no more known of the streamway to Wookey Hole. The extracts are from the Wessex Cave Club's 100th *Journal* dated March 1965:

Swildon's Eight and Nine

Mike Wooding

The last chapter in the story of the exploration of the Swildon's streamway was written in June 1962 when Steve Wynne-Roberts, Mike Boon, Mike Thompson and Fred Davies of the Cave Diving Group returned to the cave for an attempt on Sump 7. In an epic dive Mike Boon alone was able to get through despite losing his facemask and at one stage his gag. He left his breathing apparatus by the sump and set off down the streamway at a trot until brought to a halt by the inevitable sump Sump 8. This, he said, looked easy.

In January an ICDG party consisting of Dave Drew, Dave Savage and myself entered Swildon's Seven and had a look at Sump 7. Partly due to inexperience and partly due to a misleading description of the sump we did not get through, although two of us had several attempts. Correspondence and conversation with members of the CDG party led us to the conclusion that some blasting was necessary in the entrance to the sump. Mike Boon kindly supplied a sketch of the sump showing that the first section of 15ft. led to an air bell from which the second section, also of 15ft., led off. It was this second part which caused Mike's desperate struggles. A month later Dave Savage and I carefully planted 5lb. of bang and retired a respectful distance before setting it off. Fumes were still thick after an hour, carpeting the approach to the sump in knee-deep fluffy white vapour.

On the weekend of the 27th February the other members of the team were otherwise engaged, so I decided to have a look at the effects of the bang. I was lucky enough to be offered the use of their ladders by an Imperial College party en route to free-dive Sump 4. As I approached Sump 7 I sensed that something was different about the place. What was originally an awkward duck to enter the sump pool was now a gaping hole. Not only had the massive boulder blocking the sump disappeared but one wall had detached itself from the roof. It was easy to float into the half-way bell, and here the water was neck deep. I thought that the way on would have been obvious but it took four dives to find the correct hole. It was too tight to get through with the side mounted cylinder, so in approved fashion I took it off and pushed it in front of me. Despite not wearing a weight belt I found this section tight and the flexes of my twin headlamps kept snagging. Once in Swildon's Eight I hurried downstream through two ducks and was stopped by a miserable pool, heavily silted Sump 8. A rusting air cylinder was jammed in the sump and I pulled this out and examined the thing. A block of wood was protecting the tap and as I pulled this off the tap started to hiss violently. Very much startled by this I tossed the cylinder away and dived into the widest part of the sump. It was short (about 10 ft.) but tight, and it was a very relieved diver who surfaced in Swildon's Nine. By now I had become concerned about my reserve of air so left the breathing apparatus on a shingle bank and headed downstream. Less than 100 ft. further, the spacious passage turned sharp left. I scrambled over huge boulders and found myself on the brink of a large pool – Sump 9. It is an evil black place but appears to be a good diving site!

Fig. 11.10
Mike Wooding dives in Wookey Hole Cave, with Dan Hasell as Controller, 1966.
Photo by Tony Morrison

Fig. 11.11
Dave Savage prepares to dive at Wookey Hole Cave, 1966.
Photo by Tony Morrison

Swildon's Ten and Eleven

Dave Savage

On the 6th March, M.J. Wooding, Dave Drew and Dave Savage again entered Swildon's and made a relatively uneventful trip through Sump 7, though the place is still awkward and tight, despite the removal of the first portion a few weeks ago. With one person in Eight it was quickly established that there was a talking connection from Seven to Eight, via large boulders to the left of the sump pool. We quickly passed Sump 8 and began preparing to dive Sump 9. Dave Drew dived first on the end of a 150 ft. Courlene line, and returned after 50 ft. had been paid out. He reported a large steeply descending passage with a very soft muddy floor, but failed to clear his ears and had to return. Dave Savage dived next, and after dropping about 15ft., the large passage levelled out and it was possible to reach an air surface about 80ft. from base. On his return Mike Wooding then dived and passed beneath the air space at a depth of 15ft-20ft, traversed another 60ft. following gravel on the floor, and found several smaller air spaces (one of these may be below 'water level') eventually reaching dry land in Swildon's Ten.

After 20ft. a short (3 ft.) sump led to Swildon's Eleven and 100 ft. of roomy stream passage to Sump 11. Dave Savage dived next into Ten and Eleven, after Mike Wooding's return, and poked a short distance into Sump 11 which looked easy. More interesting though, to the right of Sump 11 is a steep, slippery climb of about 30ft. at the top of which is an opening, 3ft. by 4ft., through which can be heard the sound of a large stream falling about half a dozen feet. This opening obviously provides a bypass to Sump 11 and the noise is coming from Swildon's Twelve.

Swildon's Twelve

Dave Drew

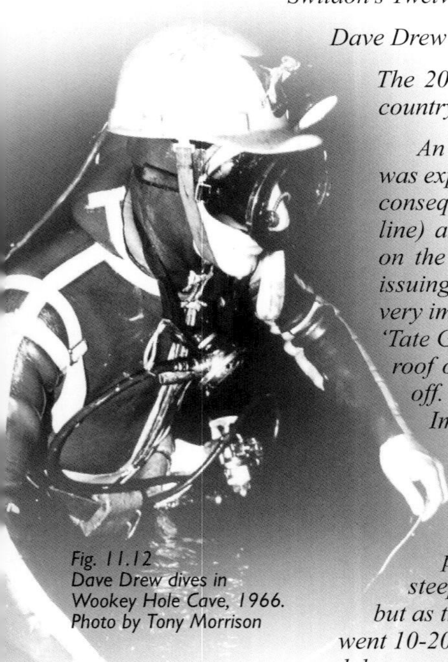

The 20th March 1965 found the trio back in Swildon's - deepest cave in the country!

An uneventful trip through the sumps brought us to Swildon's Eleven which was explored carefully (high-level by-pass to a duck found); no side passages of consequence discovered. The steep climb by Sump 11 was tackled (permanent line) and proved to be very awkward 30ft. high and muddy. A gentler slope on the far side led down to the streamway of Swildon's Twelve; the stream issuing from Sump 11 some 30 ft. away. Swildon's Twelve proved to be 110 ft. of very impressive passage some 10 ft. wide and up to 60-70 ft. high with fantastic 'Tate Gallery' effects and 'windows' splitting the passage at several levels. The roof of the passage was not reached but many muddy tubes and avens lead off. Sump 12 is an unprepossessing looking pool developed along the strike. Immediately downstream of Sump 11 the gradient of the floor steepened abruptly for a 60ft. stretch and included one 4 ft. pot (a feature not seen in Swildon's since Four). Sump 12 was then attempted. Dave Savage dived first, and reported a steeply descending passage (45 degrees) extending for 50 ft. at which point he reached the edge of an underwater pot and returned. Dave Drew dived and appeared to spiral down very steeply, descended the pot, and continued steeply down with no obstruction, but as the passage showed no signs of rising he returned to base. Mike Wooding went 10-20 ft. beyond this point and reported that the passage became choked with mud, however, he thinks there is an alternative route to be tried. Around 90 ft. plus, of rope was paid out and with the passage descending at such a gradient the depth attained was 40 ft. at least. The only reason for not pushing the sump further was the high rate of air consumption at this depth which would have left us with an inadequate safety margin for the return trip.

Fig. 11.12
Dave Drew dives in Wookey Hole Cave, 1966.
Photo by Tony Morrison

News of these discoveries soon spread and crossed the Atlantic to Derek Ford in Canada. Dave Drew sent him a first hand description with a survey of the new streamway to Sump 12. And Derek passed the word straightaway to Oliver Wells in New York in a letter dated 14th April 1965:

At the limit of exploration in Sump 12, the cave is 535 to 540 feet deep, the British record. Bully for Mendip!

> At first sight of the section my face split into a broad grin. Swildon's is behaving exactly as per my theory - sumps getting more frequent and deeper, just as they should, bless their souls. I think that the divers' limit cannot be far off, perhaps it has been already reached at Twelve. Digging in the abandoned routes, such as Shatter Passage, is much more promising.

Whilst gleeful to be proven right, of course, Derek strikes a somehow sad note with wistful hope for further progress along higher routes. But his theory for Mendip sumps has stood the test of time only too well. He has also contributed much to geomorphological studies of cave formation world-wide since then, and was, for a time, President of the Union International de Spéléologie. The sport of caving and science of speleology that mixed and matured on Mendip with devotees such as Derek Ford have travelled far over the years.

In June 1965 Dave Drew was responsible for introducing a major water tracing programme to Mendip as part of his research in the Department of Geography, Bristol University. He studied the sources and hydrology of the groundwater rising from the St. Dunstan's Well and Ashwick Grove Springs into the River Mells near Oakhill. His doctorate thesis was accepted in 1967. A second phase of the so-called Mendip Karst Hydrology Research Project moved to central Mendip in January that year and delimited the Cheddar and Wookey Hole groundwater catchments at last. A third phase concentrated on the Burrington Combe area in April 1968.

The MKHP helped put Bristol University Geography Department to the fore and the researchers involved became known internationally as the 'Karst Police'. Their supervisor was the enthusiastic David Ingle Smith. It was 'Dingle' and Dave Drew who compiled *Limestones and Caves of the Mendip Hills*, eventually published in 1975; a key to the sources of information about karst studies in the area. Dave also contributed the chapter 'Cave Diving' to the *Manual of Caving Techniques* published by the old Cave Research Group in 1969; a review of how equipment, techniques and training had radically changed in the CDG. But, back to their beginnings and the sport that precedes the science and the books.

Oliver Lloyd's first cave dive was on 31st January 1965 through the short sump in Ludwell Cave, near Weston-Super-Mare. He was 54 years old! His companion that Sunday was Dave Savage and their log records:

> OCL and DS explored the sump. This was OCL's first cave dive with apparatus and it felt very strange.

He had problems with both his mask and hat whilst underwater, but spent several minutes simply sitting there and:

> Getting used to it.

Years of serious swimming and free-diving sumps, especially in Swildon's Hole, gave him the necessary confidence. Only the previous weekend, for instance, he had joined Mike Wooding and Dave Savage on a free-diving trip along the streamway to Swildon's Six. It fulfilled an ambition that had captured Oliver's imagination ever since the adventurous free-dives of Sump 2 and Sump 3 described earlier. 'Fish's' bold breakthrough in free-diving fascinated OCL in particular. After all, he had effectively laid down the challenge in the first place. His own pretext to follow suit was a survey of the air bells comprising Swildon's Three with Mike Garrett. Oliver wrote:

> We had planned to do a survey, on air, of Swildon's Three, but owing to the indisposition of DPD's, OCL's and DS's demand valves we went without... Sump 2 was passed without incident. MG declared himself delighted. Halfway through Sump 3, however, OCL hit his lamp, which lost its lid and went out. On emerging in Four he had trouble because there were so many holes in his boiler suit, that it was difficult to find the one which led to the pocket, where his spare lamp was. On our way through Swildon's Four we met some of the Cambridge University party, including Clive Westlake. 'What', he said, 'was it that you said about those two chaps, who dived Sump 2 last year?' OCL replied that 'when you bring it off people say, "Jolly good show", but if you fail, they say, "Bloody fools for trying!"'

'Mendip Notes' by Cheramodytes and his 'colleague' known as 'Cave Beetle' in the 100th Wessex *Journal* concludes:

A New Death Cult, or Cousteau Lloyd? What shall we do with him? People that don't conform are always a nuisance and how easy it is to sit in your regulated little world and say 'fool' to those who step outside. Oliver has free-dived Sumps One, Two, Three, Four and Five in order and then back again. He is mighty pleased with himself; and well he may be and many of us are pretty sore, we hide it under a lot of talk about safety. Me? I'm just scared at the thought of it. Ever since Morley and Sims made a mistake in May '62 Oliver has been working up to this. He said then, 'For my part I am thankful that such 'bloody fools' exist. Life would be dull without them' [20].

Fig. 11.13
Oliver Cromwell Lloyd in Wookey Hole. 1970.
Photo by Norman Heal, Wells

Oliver was not alone. During an all night vigil to collect water samples from the far reaches of Stoke Lane Slocker during the following week, Dave Drew and Mike Wooding took the very exceptional and risky step of taking stimulants on the trip. Their ICDG log records:

Four grains of Caffeine Citrate were taken by each diver and withdrawal commenced.

MJW 'heard' a party of schoolgirls coming down the Three streamway! Once through Sump 2 more soup was consumed. DPD had four more grains of CC and with considerable difficulty all kit save two bottles was taken out of the cave. During the trip out both divers experienced loss of consciousness a number of times when they stopped, and auditory and visual hallucinations. The presence of a 'brass band' in the Muddy Oxbow seems to have been overlooked by previous visitors to the cave, but was clearly heard by DPD who also perceived flashing lights and loud, irregular heartbeats resulted from the slightest effort.

They concluded that both were 'at the limit of endurance' despite the use of stimulants. Another new lesson had been learnt; one unlikely to have been tried in any other circumstances.

Yet another ICDG first was Mike Wooding's seven-hour solo diving trip described by him earlier to probe Sump 9 on 27th February 1965. The air of independence had reached its ultimate expression and no one else was in contention. Back in January 1935 when a young and enthusiastic cave explorer, Eric Hensler, had written to Herbert Balch about the prospects of pushing Sump 1, the reply had been:

Very serious efforts have been made to pass the trap at the bottom of Swildon's Hole, and great attempts at blasting a way in, in vain. The place is too constricted to wear a diving suit. Floating charges have been sent through and exploded in vain. The hill above was shaken by the discharges. I do not think therefore that your efforts would succeed. If you join the new Caving Club, and the enclosed circular will enable you to get in touch with London members, from them you will learn all you need to know.

Yours faithfully,

H.E. Balch

The 'new Caving Club' then was the Wessex Cave Club. When Graham Balcombe got to hear of Eric Hensler's bid he wrote from Buckland St. Mary, Chard, to establish his claim on 6th February 1935:

Jumbo happened to mention that you were interested in Swildon's Pool operations, and that the MNRC reports on the subject had been sent to you. May I mention that the reports are inaccurate in this respect, that work there has not been abandoned; on the contrary, negotiations have almost been concluded for a fresh expedition in the near future, with improved apparatus. If you should yourself be contemplating an attack on the barrier, we should esteem it a gracious act on your part, if you could see your way to deferring operations

[20] Clive Westlake's logbook records his words as 'People can only say "You bloody fools" when you've failed and we haven't!'

until our next attempt. Should you wish to accompany the forthcoming expedition we will be pleased to arrange for this. My address after this weekend will be c/o The Engineer-in-Chief, G.P.O., Radio Section, Armour House, St.Martin's le Grand, E.C.1.

Best caving wishes,

Graham Balcombe.

Competition then was thus dealt with graciously. And so it would be once again thirty years later. After his solo trip, Mike Wooding wrote:

The diver had a welcome drink in the Victoria, and was given a lift to Wells where he called on Mike Thompson and made peace with CDG. Mike eventually suggested a pint, so they went to the Hunters' and saw, among others, Fred Davies and Steve Wynne-Roberts; the latter provided transport back to Bristol.

On 6th February 1965, Oliver Lloyd had dived for the first time in open water inside Wookey Hole Caves under Mike Thompson's control. By early March the ICDG contingent continued their work within the established Cave Diving Group. All submitted to the standard CDG tests during March under the watchful eyes of Mike Thompson and Luke Devenish. Only Oliver was referred and the notes of 13th March record:

OCL used too much air and must practise using less. The others performed satisfactorily, though DS is believed to have lost some weights. At a meeting of the Cave Diving Group, Somerset Section, the same evening at the New Inn, Priddy, all four were admitted as Diving Members. OCL was required to fulfil two conditions; first, to practise using less air and, second, complete the remaining three of the five cave dives required by the rules.

Needless to say, he soon did both.

Although it was not known for sure yet, it became pretty clear that the Swildon's stream headed towards Wookey Hole beyond Sump 12. The Eastwater and St. Cuthbert's streams certainly did so. In the words of the contemporary cave diggers' song about the prospects for Tankard Hole near the Hunters' Lodge Inn:

Taking Swildon's as a feeder and St. Cuthbert's as a drip!

Only the CDG could get leave to dive upstream in Wookey Hole Caves, of course.

Apart from Mike Thompson's and Steve Wynne-Roberts's dive to explore dry passage in the Thirteenth Chamber the previous year, there had been no attempt to push beyond the 70 foot deep Fifteenth Chamber since the dive by John Buxton on 17th December 1960. All had used mixture rebreathers and walked. Now it was up to those who used open circuit air sets and finned. Fast swimming would compensate for less gas supplies.

Coincidentally, there had been an identical five year hiatus at Wookey Hole since Wing Commander Gerard Hodgkinson's death in 1960 gave his widow, Olive, grounds to develop the area's tourist attractions in her own way. During February 1965 she was taken to the lying in state of Sir Winston Churchill and had been struck by a newspaper suggestion that a fitting national memorial to such a great statesman would be:

A piece of English countryside preserved in its natural state ... as a permanent reminder of something we nearly lost.

From that moment, Olive set about handing over 126 acres of nearby Ebbor Gorge to the National Trust. Plans to diversify exhibits at the caves followed. A revival of cave exploration with the wide publicity that accompanied every discovery was welcomed. It was back to business at the Great Cave.

Open-water training for the return of cave divers to Wookey Hole occupied the summer months and, on Saturday 6th November 1965, the Wooding-Drew-Savage team arrived there with Oliver Lloyd, Bob Pyke, Noel Cleave and Paul Allen. But, still unaccustomed to giving up their independence to OCL as the required Controller for such Wookey Hole ops, all except Noel and

Bob walked out on the set-up, even after signing the equipment log. Later that evening, OCL led his depleted team from the Third to the Ninth Chamber after a trip from the First to Third following Bob Pyke. Oliver found two pennies, a camera cap and Noel Cleave's torch on the river bed. His chief reward, however, must have been the 15 minutes spent in the majestic Ninth Chamber for the first time; its oldest visitor. On all subsequent trips, Oliver was also at pains to record how much of the so-called 'Crocodile', in the Third Chamber lake, was above water level; a useful tradition that has since lapsed because most dives now start from the Ninth Chamber.

Christmas cheer and goodwill soon solved the organizational difficulties and all were back in force on Saturday 15th January 1966. George Pointing was their Controller. The objective was to get divers to the Fifteenth Chamber and the end of John Buxton's line. Mike Wooding and Ken Pearce did so and a fine time was had by everyone. They were back again the next weekend with Dan Hasell in control. This time Oliver followed Dave Savage down to 70 feet and both were enticed by what they saw. Dave reported:

The slit is about 6 feet high and 15 feet wide. Floor levels out after last weight and then goes down about 60 degrees to vertical. Could see no bottom or roof.

A big well-planned push on Saturday 26th February was thwarted, however, by muddy water following heavy rain on Mendip. Eager press men were fed with a heroic story by Mrs. Olive Hodgkinson and the archaeological potential of the Fourth Chamber as a burial site made a newsworthy diversion. Another attempt by Dave Savage and Oliver Lloyd on 26th March also ended when a tangle of lines was encountered and removed from the Fifteenth Chamber.

At long last, their persistence paid off on Saturday 30th April 1966. So confident had both divers become owing to their familiarity with the dive to the Fifteenth that they were unsupported. Once again OCL followed Dave Savage and progress forwards was made. Moreover, it was upwards. Their report ended:

Afterwards Lady Wing Co was so delighted that she embraced each of the divers, Dan on both cheeks, and then rang up the Press.

Oliver, however, confessed to feeling 'rather far from home', and future pushes were to depend upon Dave Savage's considerable initiative. Indeed, Dave was to hold the distinction of having dived farthest down the Swildon's streamway and along the underground River Axe in Wookey Hole Caves for the next few years. He tells his own story about Wookey Hole in the following chapter.

*Fig. 11.14
From the left: Bob Drake, Jeff Price, Dany Bradshaw and Bob Cork carry out a sump rescue practice at Wookey Hole with Alan Mills in the stretcher, 1985.
Photo by Rich West*

Oliver, meanwhile, was to turn his hand to mastering the extreme difficulties of underwater cave surveying, editing CDG *Newsletters* and, above all, training up-and-coming cave divers in the students' swimming baths at Bristol University. Friday night sessions there became the high spot of every week for him. In due course, Oliver also took over organizing all Wookey Hole diving operations and, somehow, had a hand in every subsequent breakthrough. Each aspect would be another story, and all who knew Oliver would have something different to contribute. This chapter, after all, has focused upon little more than one year in over thirty during which OCL went caving, the one in which he probably exerted his greatest influence and did the most. That this is appreciated by cave divers is evidenced by their naming the first big discovery in the underground river at Cheddar in 1986 'Lloyd Hall' while a distant find in Notts Pot, Lancashire, has been called 'Oliver Lloyd Aven'. One day there may be another en route from Swildon's to Wookey Hole. He would like that.

This account will close with mention of the beginning and end of Oliver Lloyd's caving days on Mendip. When Luke Devenish blasted the stalagmite boss beyond Balch's Forbidden Grotto in Swildon's, Oliver although a comparative newcomer, was invited to join the exploration party. It was 26th January 1953 and his diary also showed it to be the Feast of the Conversion of St. Paul. Birthday Squeeze at the far end of the crucial Double Trouble Series, in fact, commemorates Oliver Lloyd's birthday later that year when he was the first one through on 4th August 1962. This created the Short Round Trip. Caving was his own road from Tarsus in many ways, and he once tried to convince me that diving allowed him to go caving without so much pain from an increasingly arthritic hip! After his death, he was found by two of his divers; Bob Drake, then Secretary of the Wessex Cave Club, and Marco Paganuzzi, a University 'spelaeo'. On 4th August 1985, a large gathering of cavers was on Mendip to celebrate what would have been Oliver's 74th birthday by visiting Swildon's Hole. It lashed with rain all day, the cave flooded and those present had to abandon the wine and fruit cake to carry out a rescue! Guess who was chuckling somewhere?

Fig. 11.15
Last summer.
Photo by Martyn Farr

Fig. 11.16
A fitting legacy of Oliver Lloyd's long involvement in cave diving and the Mendip Rescue Organization (now Mendip Cave Rescue). Today's improved Sump Rescue Apparatus is demonstrated in the Wookey Hole Resurgence at the 2005 British Cave Rescue Council Conference by divers Martin Grass and Keith Savory with Carol Tapley in the Short Slix stretcher.
Photo by Mark 'Gonzo' Lumley

*Fig. 12.1
Contact prints from January 1966, a peak period for the Independent Cave Diving Group.
Photos by Tony Morrison*

Chapter Twelve
Up to the Eighteenth
by Dave Savage

After John Buxton had entered the Fifteenth Chamber on 17th December 1960 no one else appears to have visited it again until 1966 when the fourth phase of diving at Wookey Hole began. In this phase the divers were a completely new generation, all of whom used the aqualung and swam with fins, rather than oxygen rebreathing equipment and bottom walking techniques. Previously open-circuit air equipment and finning in Wookey Hole Caves had been viewed with deep suspicion, especially after Bob Davies had swum away into the Thirteenth Chamber in December 1955. The availability and use of wet suits rather than the very much more bulky and cumbersome dry suits also gave the new divers a greater mobility without losing too much thermal insulation.

Several of the new generation of divers, including myself, had been inspired by accounts of diving at Wookey Hole particularly by John Buxton's description of passing through the 'slot' down onto the sandy floor of the Fifteenth Chamber, and his enticing description of a possible way on. Our diving up to that time, however, had been in much smaller underwater passages such as the sumps in Swildon's Hole and Stoke Lane Slocker, and we felt frustrated because there seemed to be no easy prospect of diving beyond Buxton's limit. Consequently when the chance arose to dive in Wookey Hole, as described in the previous chapter, we eagerly took the opportunity to acclimatize ourselves to the clear and spacious underwater passages.

On 15th January 1966 a group of seven divers explored down as far as the Fifteenth Chamber for the first time since John Buxton's last visit in December 1960. They examined the state of the old guide lines and considered tactics for further exploration. Eventually, on 22nd January 1966, I ventured for the first time down to the end of the old heavy black line in the Fourteenth Chamber first laid by Oliver Wells. The slot could be seen through the dull yellow glow of my single miner's cap lamp It didn't appear to be particularly tight, and I judged it to be about 6 feet wide by about 2 feet high and sloping quite steeply downwards into blackness. Tying a short length of orange Courlene onto the weight at the end of the old line I paid it out slowly and passed tentatively into the slot. After about 25 feet of narrow passage, the steeply sloping muddy floor gave way to a yellow sandy floor with an apparent way on through the dark grey gloom to the right, just as described by John Buxton.

A little over a month later on 26th February Mike Wooding and Ken Pearce supported by other members of the Cave Diving Group made a first attempt to push past Buxton's limit in the Fifteenth Chamber. Wooding and Pearce went to the end of the Fourteenth Chamber together

Fig. 12.2
Dave Savage with a conventional wet-suit, Texolex helmet and miners' NiFe cell lights, about to dive upstream from the Third Chamber following a six-year lull in exploration at Wookey Hole Cave. He wears back-mounted twin-set compressed air cylinders, and spare side-mounted reserve set, 1966.
Photo by Tony Morrison

181

*Fig. 12.3
Dave Savage, assisted by Rich West, enters the water to reconnoitre the slot beyond the Fourteenth Chamber, January 1966.
Photo by Tony Morrison*

and then, whilst Ken sat at a depth of 60 feet paying out line from a reel, Mike passed through the slot and attempted to find the way on from the Fifteenth Chamber. Unfortunately, the water was still very cloudy after recent heavy rain, and with such poor visibility he was unable to find a way upstream beyond the Fifteenth Chamber. Instead he completed a loop and finished up where he had started, at Pearce's feet!

The next attempt to push forward was on 26th March 1966 when Oliver Lloyd and I swam down to the Fourteenth Chamber. I tied the line from my reel onto the old line once again and started off towards the slot. Once again fortune was not entirely on our side because after only a few feet the reel jammed and I had to return. I made another attempt using the line left by Mike Wooding on 26th February, but as this was completely tangled and quite unusable, we swam back to the Ninth Chamber to sort out the jammed reel. Once again I swam down to the end of the old line, but yet again, our luck remained out, for this time one of my two Scubair demand valves became faulty forcing us to abandon the attempt.

Five weeks later on 30th April, we tried again, and this time luck was with us. The water was very clear and we felt confident. I dived first taking a line reel containing 200 feet of Courlene and, having tied it on to the end of the old black line, went into the slot and kept to the floor reaching the bottom of the Fifteenth Chamber at a depth of approximately 70 feet. The way on appeared to bear right over a level sandy floor in a wide passage[21] and I followed the right hand wall for a short distance until abruptly an elliptical shaped passage led off upwards at an angle of over 30 degrees[22]. This was a very exciting development for it meant that at last there was a chance of another air space, possibly even a major one. Swimming up the passage following the clean bare rock with its large shallow scallop markings was a pleasant change from the ubiquitous sand and yellow silt of the rest of Wookey. The slope is such that no significant amount of fine sediment has accumulated, and it was an odd sensation to touch the floor without stirring up dense clouds of silt. I followed the passage up at the same slope for about 150 feet and then the left hand wall abruptly vanished with the floor continuing

[21] Dubbed the Sixteenth Chamber, a term no longer in use.

[22] Referred to at the time as Chamber Seventeen

sloping upwards at the same rate. At this point the floor and roof are only 4 feet apart but since there was no sign of any walls a clear idea of direction was impossible. Instead I followed the floor along the line of maximum slope until the line from the reel ran out. The depth at this point was obviously relatively little and I judged it to be about 20 feet. Consequently it was frustrating to return to base in the Ninth Chamber over 400 feet away.

A diving team consisting of Dave Drew, Oliver Lloyd, Ken Pearce, Mike Wooding and myself assembled for the next push on the evening of Thursday 26th May. The plan was for one diver to take a reel containing 200 feet of line and, after fixing it to the end of the one laid previously, continue upwards along the bedding plane passage for as far as possible. After this attempt another diver was to push further with another line reel or explore any air spaces if luck was on our side.

I was soon swimming up to the end of the line laid on the previous occasion. The new line was quickly fixed and I proceeded hopefully upwards following the scalloped rock floor still sloping upwards at 30 degrees. After nearly 100 feet of new line had been laid the passage became narrower and suddenly I broke surface in a very tight sloping rift which had so little room above the water surface that it was impossible to turn my head. Above me the rift became even narrower though I could see for at least 10 feet. Horizontally, the rift also continued tight for as far as I could see. Obviously it was impossible to get out of the water into an air surface of this nature so I jammed the reel into what seemed a good crack and started back for the Ninth Chamber. The way back was a rather painful journey since one of my ears refused to clear on the way down to the bottom of the Fifteenth Chamber; fortunately it did not burst the eardrum and I reached Nine with nothing worse than a badly bruised eardrum.

Mike Wooding and Ken Pearce dived together next, but Ken soon returned having been unable to clear his ears on the way down to Fifteen; however, Mike continued alone and soon reached the air space by now dubbed the Eighteenth Chamber. After searching for a while he wasn't any more successful at finding a useful airspace, and after having a bit of trouble with the line which was now becoming loose, he returned to base.

The next obvious step was to thoroughly explore the bedding plane passage at the highest point of which we had found Eighteen. Unfortunately the line laid from Fourteen had become too loose to follow without a danger of getting entangled in it and so, before any more pushing could be done, this line had to be removed and relaid. Initial attempts to retrieve it by pulling from the other end in Fourteen were completely unsuccessful; the only effect being to create large loose coils of Courlene in the Fifteenth Chamber which were a deadly snare for the unwary diver. There were several trips made specifically to remove this loose line, but none of them was successful straightaway. For example, an attempt by Steve Wynne-Roberts and Mike Wooding on 30th June to pull the line downstream to the Fourteenth Chamber failed when the line became jammed somewhere upstream. Yet another attempt which I made on 24th September failed when the line was found to have been mysteriously cut about 20 feet upstream of its belay point with the rest of the tangled mass well out of reach. This line blocked the way on for nearly two years, and it wasn't until 3rd February 1968 that it was eventually removed after three further attempts by Chris Gilmore and myself.

Although no progress upstream was made during these line removing operations it provided excellent training at working underwater in difficult conditions. The technique that we eventually found successful was for two divers to go to Fourteen; the first tied himself on to the end of the old line there and the second went ahead into the Fifteenth Chamber with an empty line reel whilst being lifelined by the first diver with a line from a full reel. The second diver untangled the coils of loose line by winding them on to his empty reel and then following the loose line up towards the Eighteenth Chamber, all the time taking out line from the first diver's reel. The first two attempts had to be abandoned after communication difficulties, but the third went perfectly and nearly 400 feet of loose line was removed in a dive lasting 25 minutes.

The way on was now clear for further pushing operations and a trip was planned for 17th February 1968. This time a reel containing 500 feet of brand new line was used with a powerful beam gun to give a better appreciation of the large underwater passage near the Eighteenth Chamber. I had the first dive and was soon following the steeply sloping passage up towards Eighteen having belayed the line I was carrying in the Fourteenth Chamber. On reaching the right-angle bend, instead of going up to Eighteen I followed the lower part of the bedding passage with the intention of going well past the airspace and possibly reaching a more convenient air surface. Somewhat unexpectedly the floor continued to rise when I had expected it to be horizontal at this point. And then, about 250 feet beyond the slot a mud bank effectively blocked the way on upwards since it left only a 6-inch gap between it and the roof. A way to the right was tried, just below the mudbank, but this too was abandoned when the line pulled across into a very narrow part of the passage. I retreated some 50 feet and then tried a new direction moving around the right hand side of the tight section. Very soon the passage widened slightly and then, at a point nearly 280 feet from Fourteen, a larger space could be seen upstream. At this point, for no apparent reason, my left ear began to ache and attempts to clear it by conventional methods proved useless. It was obvious that the pain was not going to go quickly, so I started to return to Fourteen, winding up the line until a convenient resting place for the reel was found in the form of a small flat ledge at a depth of 30 feet, nearly 230 feet from the Fourteenth Chamber. Until the line eventually began to rise again after going down to 70 feet in the Fifteenth Chamber, I had visions of a burst ear drum and all its consequences. As it was I had a few drops of blood in the bottom of my mask when I eventually removed it in Nine and a very painful left ear. This was the second time that I had experienced the same problem, and it was rather difficult to find a satisfactory explanation since Mike Wooding had suffered exactly the same trouble when he visited the Eighteenth Chamber. Whatever the explanation, it certainly seemed to be a problem associated with going deep twice in rapid succession.

The next visit to these distant parts of Wookey was made more or less accidentally by Phil Collett on a 'look and see dive'. His original intention had been to inspect the Fifteenth Chamber, but never having been there before he only realised where he was when he came to the end of the line fixed to a large reel. He didn't stop to look round but returned immediately to Nine. On 18th May 1968, Phil and myself had the next real attempt to push on. The plan was for Phil to return to the reel he had glimpsed briefly on his previous dive and then push forwards trying to find the way on. I was to dive second, continuing the push if necessary. Phil reached the reel and swam slowly along the inclined bedding plane, found a boulder around which he wound a loop of the line, and then continued for another 40 feet. At this point the depth was only about 10 feet but the passage itself had become tight and visibilty had become poor; consequently he returned to Nine after leaving the reel at the furthest point reached. After waiting for 20 minutes following Phil's return, in order to give the water a chance to clear a little, I dived next and went to the boulder that Phil had noted. Beyond this the line was very loose and the passage rapidly became tighter and, not very reluctantly, I too returned to base.

We appeared to be tantalizingly close to new air surfaces in 1968 but the final exits from the deepest sumps so far encountered were apparently too tight. However, the underwater passages near the present limit of exploration were very extensive as can be seen from the accompanying survey, and it was not inconceivable that a passable way on existed. With more sophisticated equipment and the divers that were bound to come forward within the next few years, who was to know what might be found?

All the serious diving we had undertaken at Wookey Hole since 1965 was done with self-contained open-circuit air breathing equipment, wet suits, fins and other conventional accessories. The first cave dive using air at Wookey Hole was made by Bob Davies in 1955 with a conventional twin hose aqualung; but now our preference was for two stage single hose regulators since these are much better suited to cave diving in all respects.

In 1966 we dived the shorter sumps in Swildon's and Stoke Lane using single air cylinders, usually with a capacity of 26 cubic feet, fitted with a Scubair single hose demand valve and pressure gauge. We wore 3/16-inch thick neoprene suits, usually with an extra layer over the trunk, and fins. We only used single 26 cubic foot bottles once on a reconnaissance in Wookey Hole, however, since they gave less than 15 minutes of air. For the first pushing dives beyond the Fifteenth Chamber, a twin set fitted with a single Scubair valve was used, the air cylinders being either 40 cubic feet bottles or 26 cubic feet 'tadpoles'. Sometimes a spare 26 cubic foot bottle fitted with an independent valve was also used. For the later dives the equipment consisted of a twin set made up of two 40 cubic foot cylinders and fitted with a two-stage, single hose valve and pressure gauge. The types of valves used with this arrangement have included the Orca, Malibu, Snark and Sea Bee air regulators. Invariably, a spare 40 cubic foot cylinder fitted with a separate valve was also used and fitted to the diver's side. This obviously gave a greater safety margin and also, equally important, gave the diver a tremendous psychological advantage in a long dive. The unfortunate disadvantage of this equipment was its uncomfortable weight out of the water even with lead weights reduced to 4 or 5 pounds. This became especially noticeable when the diver had to climb almost vertically for 15 feet into the Ninth Chamber wearing fins; a very trying operation.

Lighting was almost always a single NiFe cell and this appeared quite satisfactory. We also tried using a powerful beam gun, but the the extra light that this gave wasn't a tremendous advantage.

Reflecting on those dives which we made at Wookey Hole during the mid-sixties, it now seems incredible that we used to trust our lives to such crude equipment. I remember that invariably my equipment was leaking from a homemade manifold and that the spare valve and bottle, if any was carried at all, was usually one which had given trouble on a previous dive. Equally alarming in retrospect was the complete trust we placed in a single ex-National Coal Board miner's cap lamp and bulb which was never designed to be operated underwater let alone down to depths of 60 to 70 feet. In all the dives we made, however, I cannot recall one occasion when either lamp or bulb failed, even though the cap lamp was completely filled with water and the bulb itself was totally immersed in dirty cave water. I have occasionally been asked if our relatively crude equipment hindered our progress, particularly when we were a long way from base. I was never conscious of it at the time, but realise now that it was probably a subconscious inhibiting factor which always caused us to pull out of a dive before over-committing ourselves.

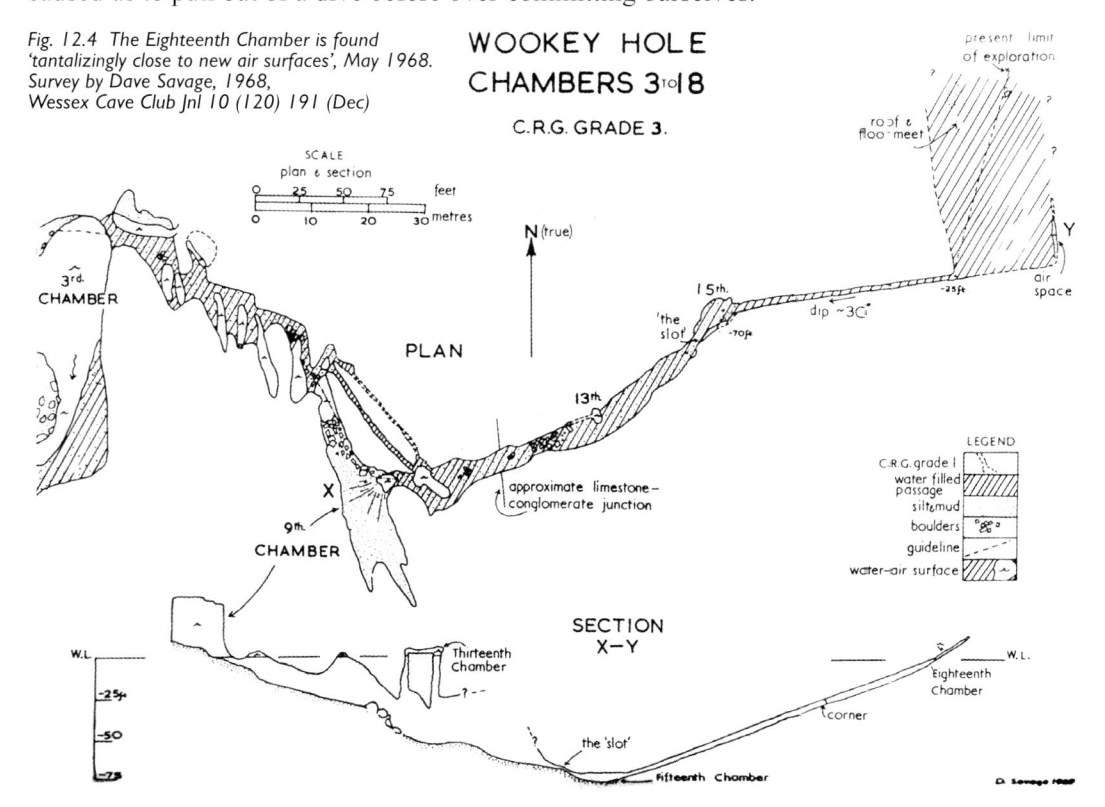

Fig. 12.4 The Eighteenth Chamber is found 'tantalizingly close to new air surfaces', May 1968. Survey by Dave Savage, 1968, Wessex Cave Club Jnl 10 (120) 191 (Dec)

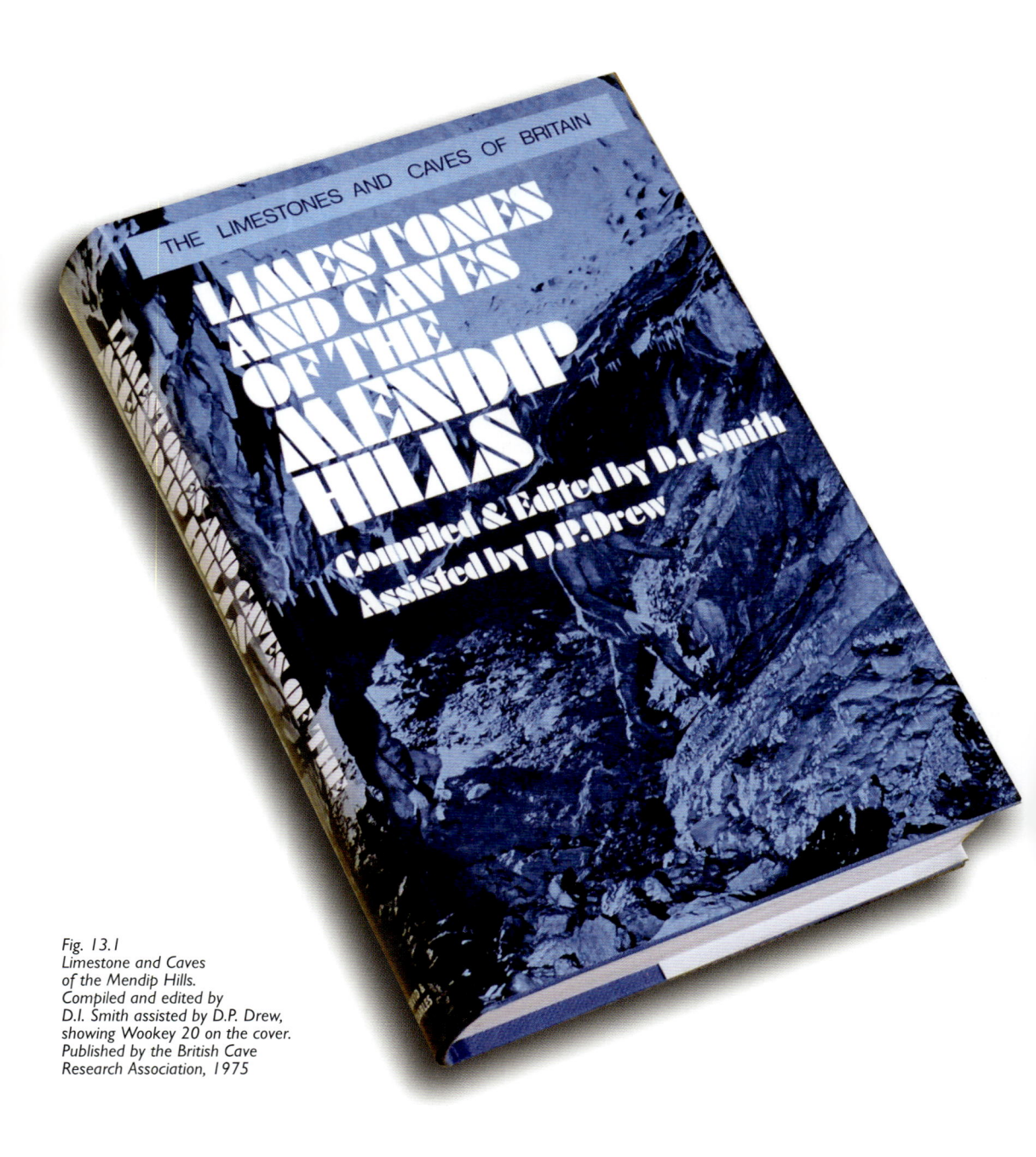

Fig. 13.1
Limestone and Caves of the Mendip Hills. Compiled and edited by D.I. Smith assisted by D.P. Drew, showing Wookey 20 on the cover. Published by the British Cave Research Association, 1975

Chapter Thirteen
Success in the Seventies
by Brian Woodward

For almost thirty-five years the underground River Axe had gradually yielded its secrets to cave divers in Wookey Hole, but only chamber by chamber. Each one had its own lake surface which could only be reached underwater through deep sumps. Only the old tourist route from the entrance to the Third Chamber followed a definite dry passage, formed in the geological past when the river flowed at a higher level. Dave Savage's discovery of a tiny airspace in the Eighteenth Chamber, and the barely submerged inclined bedding plane beyond by 1968, revived hopes that more such dry passages lay ahead waiting to be explored. At worst, new air spaces and more bell chambers seemed likely.

Success came with the arrival of the 1970s. The hard work and lessons of the earlier divers paid off handsomely at last. Rather like coming-of-age, the new cave took on a different character. More lofty and lengthy passages were discovered in quick succession. Soon, 'Wookey Twenty', or simply 'Twenty', and so on would be used to announce each new discovery. By the end of the decade the numbers would roll off the tongue; 20, 21, 22, 23, 24 and 25. And from 23 the River Axe was found to be a raging torrent; a real river, rather than limpid lakes. Its dimensions deserved the rarely used term 'Master Cave'. All the streams off the hills around Priddy that entered swallets such as Swildon's Hole, Eastwater and St. Cuthbert's joined together under Mendip and contributed to the flow through Wookey Hole Caves.

In keeping with the tradition of cave diving in the Great Cave, the breakthrough followed pushes led by one enthusiast, for the stalemate between Wookey Hole and the Cave Diving Group which had lasted throughout the sixties was broken by the arrival of John Parker from Pontypool. John started training with the Somerset Section CDG in 1968. At the time he set his sights on Wookey he had already built a reputation for himself in South Wales by making significant discoveries in White Lady Cave, Porth-yr-Ogof and the Hepste Resurgences. John worked on the basis of looking for himself, never being psyched out by pessimistic reports of earlier divers; so, he more than anyone else was able to tap the potential of Wookey Hole. He set a new standard in cave diving which was not surpassed until Oliver Statham, Geoff Yeadon and Martyn Farr took over as the leading divers using improved techniques and equipment in the mid-seventies.

Being modest about his achievements John Parker does not recall too much about the discovery of Wookey 20. He was surprised that the sump appeared to be wide open and that it went so easily. The date was 3rd January 1970. Six divers plus assorted controllers and assistants entered the cave in the early evening, as was usual at that time. The cave management were not too keen to get divers and visitors mixed up, hence the late start. Chris Harvey and Andrew Brooks were having a training dive, going as far as Nine and being guided by Mike Jeanmaire. Meanwhile, James Cobbett, Brian Woodward and John Parker were intending to push on beyond the limit reached by Dave Savage and Phil Collett. It was Brian's and John's first real experience of diving at any depth in a cave. The

Fig. 13.2 Brian Woodward and John Parker in Chamber 3, January 1970.
Photo by Norman Heal, Wells

three pushing divers used two bottles each which was unusual at that time but this was considered necessary as Dave Savage had been using similar equipment on his dives. James Cobbett had twin 50 cubic foot bottles mounted as a back pack, which he was later to regret, while John and Brian used two side mounted 40 cubic foot bottles each. Lighting consisted of one NiFe cell each. It is fair to say that John, with 950 feet of line on his reel, was the optimist, Brian was not too sure about what to do with all the gear around his waist, while James displayed his usual confidence.

Andrew Brooks had been delegated to keep log in Wookey Nine as he was the only one of the divers with a watch; Oliver Lloyd was acting as Controller in the Show Cave.

At 5.33pm John and James dived together from the Ninth Chamber with James in the lead as the only diver present to have been beyond Wookey 9; in fact, on his first dive in the cave earlier, he had reached the remote and famous Thirteenth Chamber. Visibility was excellent; however, on this occasion the slot in 15 was silted up quite badly, preventing any further progress by James with his back-mounted set. Reluctantly he had to hand the line reel to John who then continued alone. Not far beyond 15, John found Dave Savage's line where it emerged from the silt. He belayed his line reel to this and followed the old line in crystal clear water eventually reaching the previous limits of exploration. Picking up the old line reel, he continued on for about 50 feet seeing no sign of the mud bank which had halted Dave Savage's progress. Presumably the July 1968 flood had removed this. John eventually reached a pile of boulders which was passable; but, to his left up an inclined bedding, he was attracted by the glistening of an air surface. Heading for this, he surfaced in a narrow rift. De-kitting in a confined space in the water, he was able to leave his gear on a small ledge, negotiate a narrow channel and swim into the Wookey 20 sump pool proper[23]. The water was

[23] Strictly speaking, this is the Nineteenth Chamber and the subsequent side passage is Wookey Twenty.

deep and the climb out on his left up some awkward boulders was made worse by the presence of a heavy deposit of mud and his slippery wet suit socks.

Fig. 13.3
John Parker in Wookey 20, 1970.
Photo by Brian Woodward

After scrambling over some large fallen blocks John found the passage eventually turned to the left and he was rewarded with the magnificent sight of the large Twentieth Chamber. On his right was a superb fluted rock wall reaching up to the roof while in the centre of the chamber a sandy floor led down to a clear lake with a greenish hue. The privilege of being the first person to enter this place would have been sufficient reward for anyone, but there was more to be found, much more. At the far end of the chamber on the right a strike passage led off. John followed this for about 2,000 feet, gaining an estimated 300 feet in altitude until no easy way on could be found.

Fig. 13.4
Christine Grosart by the fluting in Chamber 20.
Photo by Clive Westlake

Meanwhile James Cobbett had returned to the Ninth Chamber. After waiting for about 15 minutes he made a second dive to the slot which again barred his progress thus forcing his return. Half-an-hour had now passed since John had last been seen and there was some concern over his safety.

Brian then dived, and with side mounted bottles he was able to proceed beyond James' limit but he was rather apprehensive on finding John's line reel not far beyond the slot. Visibility was still excellent, allowing the line to be followed easily up the large ascending passage round a prominent left hand bend. It was with some relief that the air surface in 20 eventually came into view. At least John had passed the sump safely. As Brian was de-kitting he heard John's approach and shouted out. John's cheery Welsh voice came back: 'Come and look what I've found, there's an H-U-G-E chamber, bigger than anything else in Wookey'. Even allowing for the 'Parker Factor', the tone in John's voice did indicate something big had been discovered. The two divers then set off to retrace John's steps. True enough the chamber was larger than anything else in the cave and progress along the wide ascending passage beyond only required one to crawl on one or two occasions. At its farthest point the passage divided; a narrow ascending rift went off to the right, while the larger left hand fork dropped down over some boulders. On returning John pointed out where he had scratched his initials on the wall 'J.P. 1970'; to this were added the initials 'B.W'.

By this time there was a tremendous amount of activity between Wookey 9 and the entrance as the two divers were long overdue. Mike Jeanmaire dived as far as the slot and returned to 9 while James Cobbett returned to base in 3 to ask for more air should an underwater search be required. Oliver Lloyd sent out for more bottles and also called out Tim Reynolds to form a rescue team if required. Spare bottles were eventually ferried through from 3 to 9 by James and Andrew. Chris Harvey and Andrew then left 9 whilst James and Mike kept vigil.

After exploring all of the main passage in 20, and oblivious of the activity in other parts of the cave, John and Brian returned with bruised feet to the 20 sump pool. Kitting up in the water was eventually completed without losing any gear. The two divers then dived together stopping at the deep part of the sump where John had left his line reel. There were one or two old lines partly buried in the silt, these were tidied up a bit and John's line reel was reclaimed. By this time visibility was quite bad. Brian indicated to John the outward line and John moved off. Unfortunately Brian could not follow immediately as one of his valves had become entangled in the line; but, after a couple of minutes this was sorted out and the two returned to 9. The twitching of the line in 9 came as a great relief to Mike and James. On hearing of the progress, James described his back-pack in the most colourful and explicit language imaginable. John then led out to 3 to announce the good news.

Fig. 13.5
Tim Reynolds and Phil Collett in Wookey 20, 1970.
Photo by Brian Woodward

Everyone was keen to go down again as soon as possible, but Mrs. Hodgkinson, the owner of the cave was not one to miss some valuable publicity. Consequently the next dive was delayed until 24th January when the press and BBC had been organised to come along in addition to the divers. On this occasion John, Brian and James were joined by Maire Urwin, Phil Collett and Tim Reynolds. Maire held the distinction of being the first woman to dive at Wookey Hole since Penelope Powell in 1935. Andrew Brooks had also had a training dive to 9. In view of the length of dry passage in 20, all of the divers carried a pair of boots in addition to their diving gear. The main aim of the dive was to carry out further exploration and to try and survey the new discoveries. There was a possibility that the steeply ascending passage might reach close to the surface. An inlier of limestone at the head of a little combe known as Smokham Wood, about 1,500 feet due north of the entrance to the cave, seemed to be a likely spot above the end of 20. But conventional surveying beyond the deep sumps could not locate the end accurately enough. It was decided to use magnetic induction techniques whereby a party on the surface with an aerial could home-in on a 'bleep' signal transmitted by those underground. Such a technique had been used first in Britain by Norman Brooks in 1955 down Longwood Swallet, a cave system close to Cheddar Gorge. This so-called radio location equipment was further developed, notably by South Wales cavers, and used successfully in mapping the very complex Ogof Ffynnon Ddu system. A full account of such techniques appears in *Surveying Caves* by Bryan Ellis (1976). Now radio-location was back in use on Mendip.

Russell Pope's radio location equipment was brought from South Wales by Mel Davies, a Cwmbran caver who was trying to organise Welsh divers much as Oliver Lloyd had done in Somerset. Mel had travelled across in his VW Camper which was his pride and joy. He was often seen inside this spotlessly clean van eating his sandwiches and drinking flasks of hot tea, while cold muddy cavers, who had been pushing nasty holes in the ground so that he could write about them, were left outside hoping for a few crumbs, or a mouthful of warm liquid to come their way. But alas, no; Dickens could have written a book about it. Since both Mel and Oliver regarded John Parker as hot property as far as finding new caves was concerned, the organisation of Wookey Hole diving operations was one of enforced tolerance between both controllers. James Cobbett did have the audacity to step inside Mel's van on a later dive at Porth-yr-Ogof in South Wales, which did not do the atmosphere much good on the trip. Not long after, Mel moved up to North Wales away from the action.

The dive to 20 on 24th January proved to be quite eventful for one or two of the divers. With six of us in the sump the visibility was not good. This, coupled with buoyancy problems (lack of) in the deeper part of the 450 ft long sump meant that, instead of swimming gracefully up the ascending passage on the far side, some divers had to crawl up the first part using the guide line to pull on. This resulted in the line being pulled into the low inclined bedding on the left near the far end of the sump. There were some very relieved expressions on the faces of some of the party when they surfaced in 20; even James lost some of his cool confident outward appearance.

Once in 20 the party donned boots and proceeded to the large lake chamber. Here, the transmission coil was spread out in order to start the radio location process. Unfortunately, the underground party were behind time, so the surface team (described as 'The Mountain Party' in CDG *Newsletter* No. 14 and consisting of Mel Davies, Jim Hanwell and Janet Woodward) who had been wandering around in the pouring rain and darkness were surprised when the first weak transmission stopped 5 minutes after they had first picked it up. The reason for this was that the underground party decided to try and make up for lost time and get back on schedule. The transmission coil was moved to a point about a third of the way along the main passage and the next transmission commenced at roughly the right time. This second transmission was picked up loud and clear and a good fix on its position made. In the rarified atmosphere 500 feet above sea level a small cairn was built by the 'Mountain Party' to mark the spot. During this second transmission, which lasted for an hour, the underground party went to the end of the main passage and extended the left hand terminal passage by 200-300 feet. A rough compass survey was made on the return journey to the sump. Back at the sump pool, the line was relaid from its original position and belayed at the point where the climb out of the water is now made. This avoided the low bedding which had caused some problems on the way in.

On the return dive numerous boots were temporarily misplaced in the sump (some are still missing) and James Cobbett managed to get caught up in the line which he had to cut; fortunately no one was behind him!

Fig. 13.6 Maire Urwin, James Cobbett, Brian Woodward, Tim Reynolds, John Parker and Phil Collett celebrate the discovery of Wookey 20, January 1970.
Photo by Norman Heal, Wells

Surfacing in 3 the party were met by the press and Mrs. Hodgkinson who provided some celebratory drinks. The following day James Cobbett was interviewed for both BBC radio and television in Wookey 3. John and Brian appeared on ITV and BBC television news programmes respectively later in the week.

The press were very keen to get some exclusive photographs of the new discovery and some lucrative offers were made. However, Mrs. Hodgkinson favoured the idea, not surprisingly, of getting some photographs taken and letting all the newspapers carry the story thus maximizing the free publicity. Following this the number of visitors to the cave increased tremendously and a small donation of fifty pounds was made to CDG funds.

On the 14th February 1970, John Parker and Tim Reynolds tidied up the line between 9 and 20 and found a short passage near the 20 sump pool which eventually led back to water level. In Nine, Colin Priddle and Brian Woodward removed some of the old line from Coase's Loop. Colin also climbed the aven at the south end of 9:2 and tried to push a tight bedding plane, but this was too small to make any significant progress. It was on a later attempt to get into this passage that the high-level extension in 9:2 was found by John Parker and Brian Woodward.

Three weeks after this dive, on 7th March, a trip was organised to take some photographs in the new passage. In addition, extra bottles were carried through to 20 in order to examine the lake in the large chamber and to make the first exploratory dive upstream of the 20 sump pool. The party consisted of Tim Reynolds, Andrew Brooks, John Parker, Phil Collett and Brian Woodward.

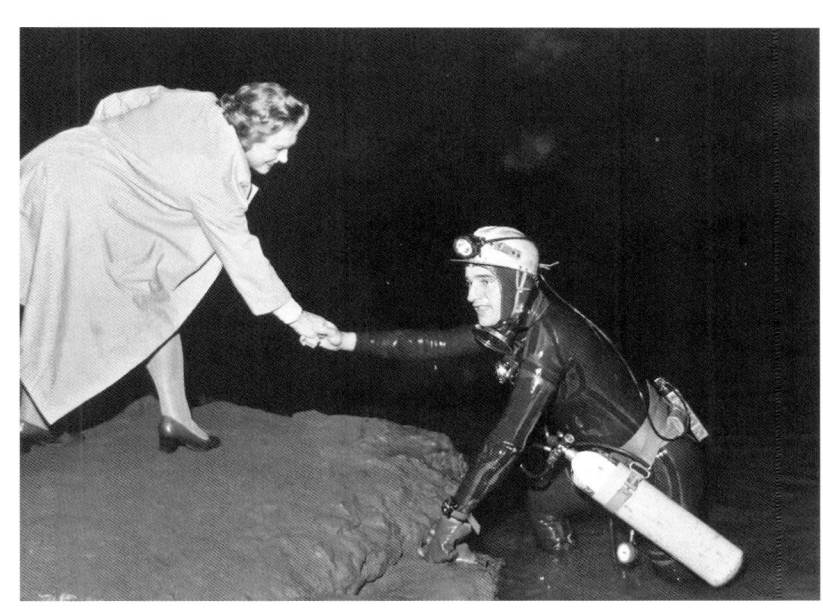

*Fig. 13.7
Clive Hodgkinson greets
James Cobbett after the discovery
of Chamber 20, 1970.
Photo by Norman Heal, Wells*

*Fig. 13.8
Colin Priddle, Brian Woodward,
Oliver Lloyd, John Parker,
Tim Reynolds, Chris Harvey
and Andrew Brooks, 1970.
Photo by Norman Heal, Wells*

By this time divers were getting familiar with the buoyancy problems in the deeper parts of the dive to 20; yet despite this, a large amount of silt was still stirred up resulting in poor visibility for the later divers, hence the speed at which people kitted up increased. The first diver would usually have good visibility which led to a race to be in pole position. The exception to this was Phil Collett, affectionately referred to as 'Cod Fillet' by his friends. Phil gained a reputation for being unbelievably slow in getting changed. Why he was so slow was never clear. His wet suit did have a nasty hole in the crutch and this tended to dampen his enthusiasm. Some thought that he waited for the water to warm up due to the presence of other divers! On some occasions he was so delayed getting into the sump that the water must have cleared again before he dived. One major disadvantage of Phil's approach was that the selection of boots which had been left in 20 was not too good. Any half decent boots were snapped up by the first divers leaving the dregs for the latecomers; but Phil did not seem to worry about this.

Tim Reynolds dived the lake in 20 to a depth of about 45 feet where it funnelled down to a muddy bottom with no obvious way on. This was subsequently confirmed by two later dives; John Parker in November 1974 and Chris Milne in October 1984. Chris did report seeing dark spaces between the rocks but no progress could be made.

Some photographs were taken in the new passage by John and Brian, with Phil, Tim and Andy acting as models. Interestingly, one of these photographs showing Tim and Phil at the far end of the new passage appeared on the dust cover of the book *Limestones and Caves of the Mendip Hills* compiled by David Ingle Smith and David Drew (1975). The same photograph was also reproduced on page 299 of the book. Even more interesting was the fact that this was credited to Nicholas Barrington who, although well known for his photography and guide books on Mendip caves, had never been beyond the limits of the Third Chamber at the time!

The most significant event of the dive on 7th March 1970 was John Parker's first exploratory push 300 feet upstream of Wookey 20 into what is now Wookey 21. After negotiating the boulders near the start of the new sump, John entered a wide passage up to six feet high which gradually descended. He eventually arrived at the lip of a large underwater pitch which he descended for about 50 feet. At this point the passage was very large and he started to lose his sense of direction so had to return. This demonstrated the inadequacy of a single NiFe cell as a source of lighting. After this John made a bracket which was capable of holding two NiFe headpieces on his helmet. This resulted in a better light as well as improved safety and contributed a lot to the next significant find in Wookey.

Fig. 13.9 Brian Woodward in Wookey 20, 1970. Photo from Olive Hodgkinson

On 11th April 1970 another trip was organised to get a better radio location in Wookey 20. Tim Reynolds and John Parker were to set up the radio location transmitter in 20 while Mel Davies and Jim Hanwell made up the surface party. Unfortunately Tim's light went out on the way from 9 to 20 and he had to stop in order to sort it out. While doing this in the absolute blackness of the sump he misplaced the two ammunition tins he was carrying. Despite both divers making searches for the missing tins, they could not be found. Consequently, the original purpose of the dive had to be abandoned. All this was unbeknown to the surface party, of course, and they had to be content with the signals from the Pen Hill television mast that towers above one of Mendip's highest points.

On the same day Oliver Lloyd was also experimenting with the development of underwater cave surveying between 3 and 9. As the sumps which were being dived got longer, it was important to improve the accuracy of the sump surveys, and this was something that Oliver had become very concerned about. He surveyed a closed loop traverse from 3 to 9 and back again from 9 to 3. Unfortunately, some of the tags on the line which served as survey stations were missing and this caused a poor closing error in his survey.

The loss of the radio location equipment resulted in a second dive by Tim and John a week later in order to search for the missing ammo boxes. On this occasion both wore two NiFe cells. In addition, they also had a beam gun and a 250 foot line reel to aid their underwater search. The main purpose of the dive was not achieved, but there was a silver lining cast by the improved lighting. While carrying out the search John noticed a large passage on the left of the stream at Wookey 15. They attached their line reel to the main line and entered the new passage. Before long the base of a 40 foot underwater pitch was reached. This was ascended with some difficulty as both divers were overweight. At the top of the pitch a shallow horizontal passage was entered, this was followed past a large slab on the left of the stream until the line reel eventually ran out after 250 feet. On their return John and Tim literally fell down the 40 foot pot, landing on top of each other in a cloud of silt and mud. The problem of being overweight in the deeper parts of the passage from 9 to 20 was quite common, particularly if other equipment such as batteries for the radio location gear was also being carried. There was a delicate balance to be achieved between being either underweighted at the start of the dive and sticking to the roof, or being overweight at the deepest parts of the sump. The consequences of being overweight were that the diver had to crawl up the first part of the ascent until his buoyancy increased with expansion of the neoprene of the wet suit. This enforced underwater crawl, needless to say, resulted in very bad visibility. The use of bulky Adjustable Buoyancy Life Jackets, which preceded the modern dry suits used by later divers, were not considered at the time, as they would have made some of the lower sections of the sumps more difficult.

This new find by Tim and John was very exciting as it was thought that they had possibly entered a completely separate streamway from the one which had been followed to 20.

Two weeks later on 2nd May 1970 a trip was organised to continue the exploration of the new underwater passage, now called 15a. Some training was also carried out as Martin Webster, John Elliot and Colin Graham were having their first dives in Wookey Hole. Maire Urwin, Andrew Brooks, Brian Woodward and John Parker were the other divers and everyone went to the Ninth Chamber. From there John dived using a single 40 cubic foot bottle and a line reel to continue the exploration of 15a. This dive was significant in the fact that it was the last pushing dive done in Wookey using a single bottle and valve. The diving techniques which Mike Boon had pioneered in the early sixties had reached their limits in the hands of John Parker. In Wookey Nine everyone watched patiently for the twitching line which would signal John's safe return. After he had been gone for about 20 minutes, an air of tension started to build up. A 40 cubic foot bottle should last for about 40 minutes on the surface but much less at the 50 feet maximum depth of the sump. Also, without any back-up equipment, John must have been getting beyond his safety limit. In the sump John had proceeded to the previous limit reached by Tim Reynolds and himself two weeks earlier. After tying on his new line reel, he continued forward in the shallow passage and eventually met the main line from 9 to 20 near the 20 sump pool. Thus the discovery was simply an oxbow off the main route from 9 to 20. But the new route to 20 proved to be about 100 feet shorter than the original route and, with a maximum depth of 50 feet was also shallower; hence, it became known as the Shallow Route. After surfacing in 20, John returned via his original Deep Route to complete a round trip of about 900 feet, much to the relief of the party in the Ninth Chamber. This dive, plus the 200 feet dive from 3 to 9 had all been done on a single bottle and left no room for error. After this time it became less and less common to see divers using single bottles even whilst training. Two independent bottles and valves then became the order of the day.

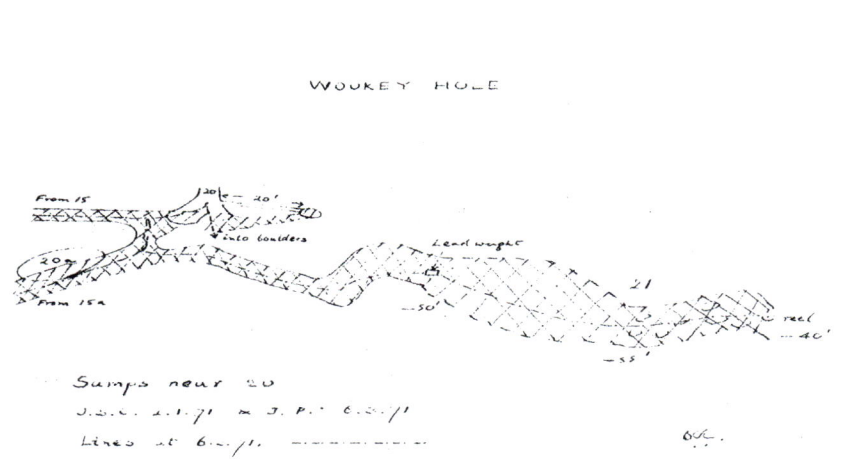

Fig. 13.11
High level passages above 9, surveyed in 1971.
Survey by John Parker

Fig. 13.10 Sumps near Wookey 20 sketched by Oliver Lloyd, 1971

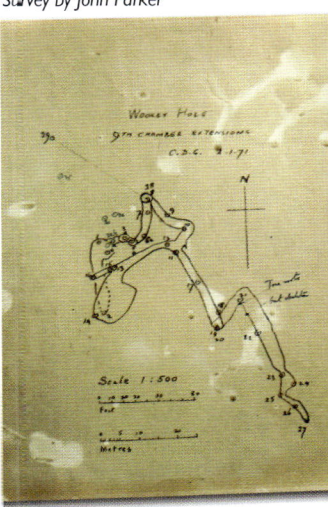

Not satisfied with his new discovery underwater, John was keen to climb and try entering the tight bedding plane which Colin Priddle had looked at half-way up the south end of the Ninth Chamber on 14th February earlier that year. Maire Urwin, Colin and John Elliot returned to 3 while John started to climb. On reaching the bedding plane, John entered it head first but soon became stuck. Eventually his cries for help were answered by Brian. On reaching John, all that could be seen were two legs sticking out of the tight descending passage. With a bit of assistance John was retrieved and confirmed that the passage was a bit tight! At this point it was noticed that the rift which had been climbed continued on up and, despite only having wet suit boots, both managed to continue up until a horizontal passage was reached.

John went one way and Brian the other. John's passage closed down while Brian's progress was halted by some fine 2 feet long white stalagmites. But the way on could be seen so John was called back. 'Don't damage the stal', he said, 'I might be able to squeeze past'. With remarkable efficiency John demolished the offending stalagmites and, sure enough, he was able to 'squeeze past' followed immediately by Brian. They entered a fine balcony which looked down on the Nine sump pool some 80 to 90 feet below. A further short climb led into another passage but before entering this they called down to Martin Webster and Andrew Brooks to come on up. The new passage climbed gradually, but the way on was eventually blocked by a fallen slab. The significance of the new passage became apparent when it was noticed that there were roots coming down from the roof, obviously they were very close to the surface. The total length of this new extension above the Ninth Chamber roof was about 500 feet. On 28th November 1970 this discovery was pushed by John Parker, Andrew Brooks, Colin Priddle and Paul Esser to gain access to a further 10 feet of passage. At the farthest point reached there was a good inward draught and bat droppings on the floor. Oliver Lloyd was also on this trip but decided that he preferred the sumps to the exposed climb up into the new extension. On 2nd January 1971 Phil Collett, Martin Bishop and Simon Howes surveyed the new passage accurately and recorded a total length of 659 feet including the climb out of Nine. At the same time James Cobbett and Paul Esser went to 20. James tried to follow John Parker's line into 21 but could not proceed very far as the line had apparently been pulled into a tight section of the passage. On a second attempt at sump 21 and using a line reel left by John Parker, James managed to enter an alcove and rediscover Wookey 20 via another route through the boulders at the start of the sump. This passage is shown on the survey in CDG *Newsletter* No. 19 (April 1971) facing page 21. The length of Paul Esser's account of this dive is inversely proportional to the length of the discovery. They did, however, manage to retrieve the ammunition boxes and radio location gear which had been lost nine months earlier by John Parker and Tim Reynolds.

At this time the underwater passage between 9 and 20 had still not been surveyed with any degree of accuracy. It was to Oliver Lloyd's great relief that Paul Esser, his new but ill-fated protégé offered to carry out this survey using methods Oliver had gradually been improving. On 23rd January 1971 Paul surveyed from 9 to 20 via the Shallow Route. Unfortunately, just before reaching 20 he lost the slates on which he was recording his survey data. But luckily the sump was not very complex and Paul provided a useful provisional survey from the data he had remembered. This survey was reproduced in CDG *Newsletter* No. 19 (1971) facing page 20. Tragically it was Paul's last contribution to the story of diving at Wookey Hole Caves, for he lost his way and drowned on a dive into Porth-yr-Ogof, South Wales, only three weeks later on 13th February 1971. However, Oliver Lloyd helped to institute annual Paul Esser lectures at Bristol University as a more lasting memorial to this young student; at only 21 years old, the youngest trained cave diver in Britain to die in action.

Now it is a well know fact that the best way to discover a new cave is to make a survey of the existing passage. This invariably guarantees that the next person to look seriously at the cave will find some new passage, thus making the survey quickly out-of-date.

On March 6th 1971 John Parker returned to push on beyond his previous limit in Wookey 21 using improved lighting. Three other divers, Tony Giles, Alan Mills and Aubrey Newport, visited the 9:2 extensions on a photographic trip while Martin Webster accompanied John to the

20 sump pool which was now the forward diving base. From 20, John followed his old line for 300 feet to its limit where he had previously become disorientated. Tying on a new line reel, he entered a very large passage and, despite having a beam gun, he could not see the walls or roof of the sump. Progress was made by following the ripples on the sandy floor which marked the main current. At his farthest point from the 20 sump pool he had run out 480 feet of line and was ascending a steep ramp with the way on wide open. He estimated his depth to be about 40 feet. Then, unfortunately, he reached his safety margin for air and decided to return to base. The report of this highly significant dive is tucked away at the end of a lengthy account by Oliver Lloyd in CDG *Newsletter* No. 19 which describes the problems of using back mounted sets in Wookey Hole; a point which had earlier been amply demonstrated by James Cobbett. The total dive from the end of the show cave to John's new limit was just over 1,100 feet, much of this being deeper than 50 feet, and, together with Mike Wooding's dive of 1,100 feet in Keld Head on 30th May 1970 represented the longest cave dive in Britain[24]. It was certainly the deepest.

On 20th March and 3rd April 1971, two unsuccessful dives were made; one to try and radiolocate the far end of Wookey, and one to recover Paul Esser's survey slates. But the lack of success of these dives was more than compensated for on 17th April when the redoubtable John Parker discovered Wookey 22. At this time John whimsically talked of having to give up diving because he could not afford the line which he was consuming at an ever increasing rate at Wookey Hole and at other sites in South Wales and Derbyshire. On this occasion, however, he was fortunate in that very little extra diving line was needed. John was assisted by Bob Churcher who waited in 20. Prepared for a long dive, John carried four 40 cubic foot bottles to 20 and used two of these new bottles to dive onwards as far as possible. This was the first time that such an amount of equipment had been used on a British cave dive. At the end of his previous limit, 480 feet from Wookey 20, John picked up his line reel and continued on up the steep ramp. After barely 20 to 30 feet he surfaced and emerged in Wookey 22. The ramp which he had been ascending underwater continued at the same steep angle above water which made his exit from the sump and de-kitting extremely difficult. Despite this, John made a rapid exploration of the major passages of 22, estimated to be about 600 feet of dry cave with a large lake at the far end. The new cave was developed in Dolomitic Conglomerate once more and mainly along the strike. The significance of the Dolomitic Conglomerate was largely lost on the sporting cave divers but of great interest to Mendip 'geolophiles'.

Fig. 13.12
Wookey 22 downstream sump pool.
Photo by Clive Westlake

[24] Mike Wooding died of a heart attack whilst cave diving in Meregill Skit, Yorkshire 10 August 2006, aged 63.

Fig. 13.13
Sketch survey of Wookey 22 by John Parker

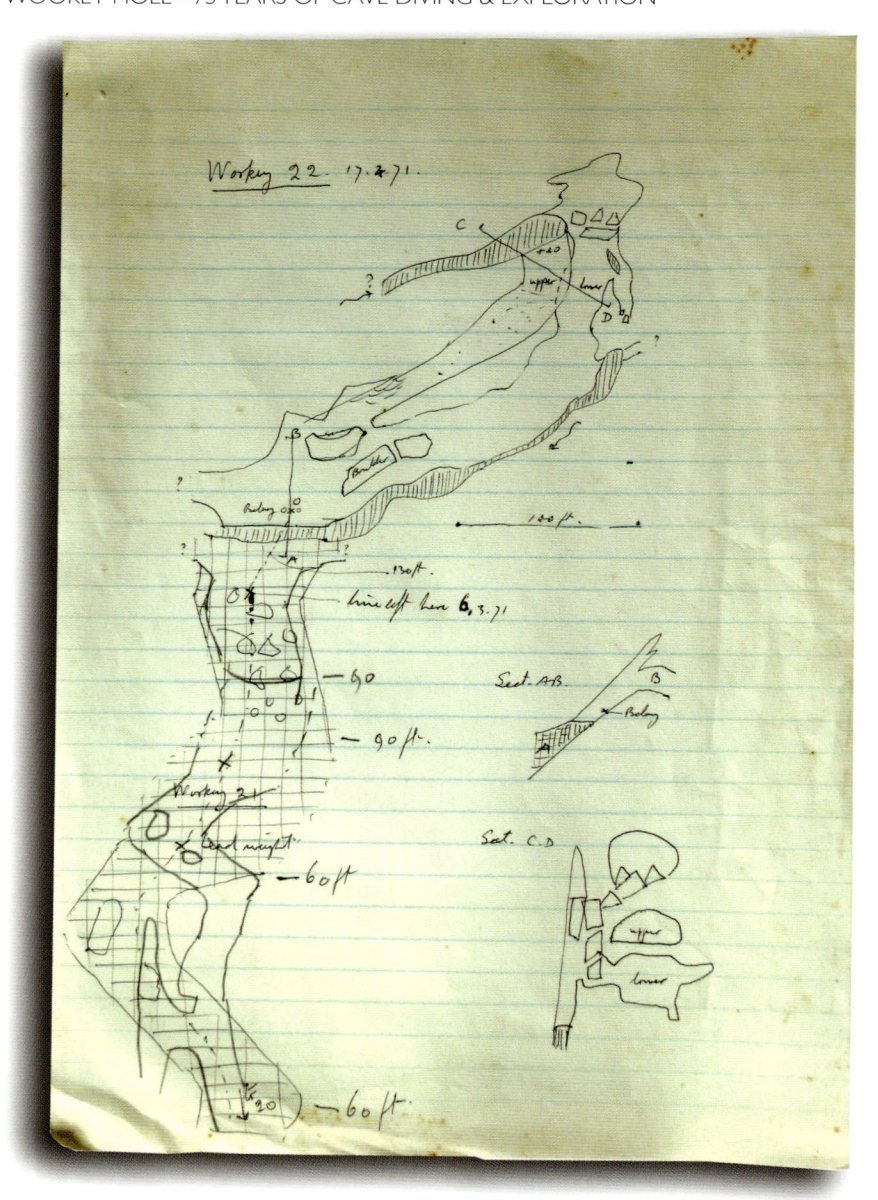

One of the main things John remembers about this dive was that he found a Cadbury's chocolate wrapper in the deeper part of the sump which certainly indicated further open passage and links with the swallet caves at Priddy. Using a crude capillary depth gauge loaned by Martin Webster, John recorded a maximum depth of 90 feet in Wookey 21. This depth has been questioned by other divers who have since reported a maximum depth of 60 feet. The reason for this apparent discrepancy is not clear. Perhaps while laying the line John explored deeper parts of the sump into which the fixed line does not now pass.

Two weeks later on 1st May 1971 John was joined by Tim Reynolds and Brian Woodward on the second trip to 22. Oliver Lloyd and Andrew Brooks were also diving between 3 and 20. The three divers going to 22 all carried four bottles which enabled them to use two full bottles from 20 onwards. This may have amounted to overkill but gave an added sense of security. The size of the underwater passage in the deeper parts of 21 was very large indeed; for much of the sump the divers swam three abreast enjoying the excellent visibility. In 22 the lake at the end of the new passage was dived by John to a depth of about 70 feet where the passage became blocked by mud and silt with no obvious way on. The entry into this lake with diving gear was quite difficult because of a vertical drop of about 4 feet into the water, and the return would not have been possible without some assistance from above by someone on shore. The fact that the lake did not go was disappointing as no other obvious continuation of the streamway had been seen above or below water. Two large avens were noticed but these could not be climbed. The divers also examined a large inclined rift at the northern end of the new passage. This was followed for an estimated length of 200 feet until it gradually narrowed down and enthusiasm waned despite the fact that a definite end had not been reached. The divers decided to call it a day and returned to base in 3. This proved to be a great mistake as twelve years later on 23rd January 1983 Rob Harper and Trevor Hughes pushed along this passage and entered a low crawl which proved to be a high-level bypass to the lake which had stopped John Parker's progress. This discovery, known as Cam Valley Crawl, is described in a later chapter. The lake itself was successfully dived in 1976 by the combined efforts of Colin Edmunds, Oliver Statham and Geoff Yeadon as will be seen later in this chapter.

The dive to 22 in May 1971 marked the end of progress upstream for some time although a number of dives were made to examine the underwater passage in 21 in an attempt to find other possible ways on. The first of these dives took place on 20th November 1971. Tim Reynolds and John Parker thought they had found a way on but this 'new passage' simply led back to 22. Two months later Tim again had a look at the underwater passage in 21, but could only confirm its massive dimensions and that it did not contain any obvious route upstream.

The lack of a decent survey of Wookey 20 was something that needed doing, if only to keep Willie Stanton, Oliver Lloyd and everyone else happy. On 4th March 1972 Aldwyn Cooper and Julian Walford tried to remedy this omission of the earlier divers. However, their efforts were thwarted by problems with a flooded clinometer. Later, on 4th November 1972, they were joined by Adrian Wilkins and, together, they surveyed the major part of Wookey 20. But unfortunately, their efforts were not rewarded as the survey never appeared in the CDG *Newsletter*; an inexplicable omission after all the pleas for such information from the editor, Oliver Lloyd.

On the same day that Wookey 20 was surveyed, John Parker and Brian Woodward climbed back over the 9:1 sump pool. John led up a narrow ascending ledge on the right of the stream and was followed by Brian. A substantial bridge above the 9:1 sump pool was reached and from this it was possible to look down about 20 feet into a water filled rift. On the opposite side of the rift a small eyehole could be seen but the lack of any means of getting back prevented a descent into this chamber. The rift which had been 'found' proved to be the Eighth Chamber.

Two weeks later Phil Collett joined John and Brian with the aim of getting into the eyehole which had been seen on the opposite wall of the Eighth Chamber. An extra rope and some Jumars were carried through to Nine to make this possible. A ladder would have been more sensible but this seemed to be a good opportunity to try a bit of prusiking as this was starting to gain popularity with a small group of Single Rope Technique, or SRT, aficionados. The rope was tied to a natural belay on the bridge and John descended to the water in 8, swam across to the other side and climbed up a large flake to the eyehole. The rest then followed. The eyehole was not very large but was rapidly enlarged by the removal of some loose blocks. From there it was possible to look down into a parallel water-filled chamber. As none of the three divers had yet surfaced in the Seventh Chamber from the sump below, they did not recognise it for what it was. The rope was then put through the eyehole into the Seventh Chamber. All three divers plus one or two large rocks then descended into the water in 7. The water at this point was quite deep but, by swimming to the downstream end of the chamber, it was possible to get out of the water onto a large block. Above this, a high rift ascended. This was climbed for about 20 feet but lack of protection prevented further progress.

On the return from 7 the climb out of the water provided some amusing moments. Phil had never used Jumars before and what information he had read about SRT did not cover treading water while kitting-up. After a great deal of cursing Phil eventually attached himself to the rope. It was then realised that one of the large rocks which had fallen out of the eyehole had trapped the rope underwater and Phil had to disengage himself in order to try and free it. Without any diving weights he could not swim down to where the rope was trapped. By hauling on the rope he eventually succeeded in reaching the offending boulder which he moved. Needless to say, once the rope was free Phil shot to the surface like a cork out of a bottle, much to the amusement of John and Brian. In order to get back from the eyehole across the Eighth Chamber to the bridge above the 9:1 sump pool, a Tyrolean traverse was rigged up just for the hell of it. All in all this was a very enjoyable trip.

Phil followed up these high-level discoveries on 30th December 1972 when he and James Cobbett carefully explored the chambers from 3 to 9 from the stream level. Phil managed to do some climbing in 5 but did not reach the top of the Fifth Chamber. This reappraisal of the cave from 3 to 9 was producing some interesting findings. The earlier divers in their more cumbersome equipment had not been able to do much climbing and it was now becoming apparent that an old high-level route from 3 to 9 was a distinct possibility. Maybe Herbert Balch had been right after all.

The period 20th January to 8th February 1973 saw an intense period of activity involving John Parker, Aldwyn Cooper, Colin Priddle and Adrian Wilkins.

The main aim of these dives was not exploration but filming. HTV were making a film about Wookey Hole Caves which included a section on cave diving. This documentary featured the late Wynford Vaughan-Thomas and included interviews with John and Aldwyn. After one or two mishaps with equipment, some good underwater shots were obtained despite the bad filming conditions; after all, this wasn't Jacques Cousteau in the Red Sea. Although no new exploration was carried out on these dives it did mean that access to the cave was easier for this period of time. Normally, divers were only allowed in the cave in the evenings after all the tourists had gone home and it was not uncommon to return to the surface in the early hours of the following morning. Little had changed since 1935 in this respect.

On 3rd February 1973, John Parker took Colin Edmunds and Peter Moody on a training dive from 3 to 9 with the aim of looking at the high-level passages in 9:2. On this dive John discovered two large rifts above water. These rifts were found off Coase's Loop and comprised part of an outer loop going from the 9:1 sump pool and eventually meeting the main route from 9 to 20 upstream of Coase's Loop. This new passage was called Coase's Loop Extension. Five days later the filming party took some footage in this new passage but bad visibility meant that the quality of the film was not too good. During this period Martyn Farr, who was later to play an important role in the upstream extensions of Wookey Hole, had his first dive in the cave.

After the filming was finished, John Parker returned on 28th April to climb up the rift in the Seventh Chamber. Diving through to 9, he and Sandy Navrady climbed back into 7 and continued on from where John had stopped previously about 20 feet above the water surface. About 80 feet above the water they entered a high-level passage. This was followed by a traverse over the top of the Sixth Chamber, eventually reaching the Fifth Chamber which Phil Collett had started to climb earlier from below. But John and Sandy were stopped at the top as they did not have a rope to descend to water level. So, it was now possible to traverse from 9 to 5 via this high-level route and a free-diving route from 3 to 9 was now a distinct possibility.

It was at this time that Mrs. Hodgkinson, the owner of the cave, decided to sell up and retire to the Channel Islands, thus ending the special relationship she had with the Cave Diving Group and Oliver Lloyd who had made all the arrangements for diving at Wookey Hole since the mid-sixties. Everyone was concerned that the CDG should still be able to dive at this excellent site and it was with considerable relief that we found the new management to be very interested in the activities of the group. Access to the cave was assured and even became easier, thanks to the behind the scenes activities of Oliver Lloyd and Jim Hanwell. The Great Cave and old Paper Mill downstream had been bought by Madame Tussauds and the new manager, Graham Jackson, had a genuine interest in developing the cave in a sympathetic way. As with many tourist caves, particularly those in France, cavers on site became an accepted part of a visit to Wookey Hole Caves.

The possibility of extending the show cave to take in the Seventh, Eighth and Ninth chambers was discussed, and plans were put into operation to achieve this aim. As this project would involve drilling an artificial tunnel linking 3 to 9, and then a new exit from 9 to the valley and Mill, it was essential to have a precise fix on the Ninth Chamber and intervening passages. As the top end of 9:2 was very close to the surface, it was decided to open this up to provide a dry way into 9, thus enabling its exact location to be verified.

On 5th August 1973, Martin Webster, Aldwyn Cooper and Sandy Navrady took Brian Prewer's radio location transmitter to the end of the high-level 9:2 extension. On the surface, Willie Stanton, Graham Jackson, Luke Devenish, Tom Davies and his son had the receiving coil. Once transmission started a good strong signal was picked up 175 feet east of the expected location at a depth of only 20 feet. Over the period 13-15th August 1973, John Parker, Jim Durston and Peter Marshall surveyed the passage from 9 back to 5. On these trips, which were paid for by the cave management, they also took the opportunity of installing ladders and ropes on the climbs.

Before the top entrance to 9:2 was opened up, as described in the following chapter, two free-diving trips were made from 3 to 9 using the high-level route beyond the Fifth Chamber that had been pioneered by John Parker. Colin Priddle had the privilege of being the first person to complete this trip aided by John. In order to do this Colin free-dived from 3 to 4; there was a small air space but it was easier to dive the 6 feet long duck. The dive from 4 to 5 then followed. This sump is about 18 feet long and up to 6 feet deep. Once in 5 Colin climbed out of the water and waited for John who had already dived through to 9 with breathing apparatus. After climbing back via the high-level route, John eventually appeared at the top of the Fifth Chamber. He lowered a line to Colin and a ladder was hauled up, thus enabling Colin to join him. They then proceeded to 9 much to the amazement of four divers from the National Coal Board Rescue Team who were having some sump diving training under the watchful eye of Oliver Lloyd and Colin Edmunds. Four weeks later on 8th December 1973 this trip was repeated by John and Colin together with Clive Westlake and Chris Smart. On this occasion John took the opportunity to climb yet another aven above the bridge joining 9:1 and 8. This was ascended for about 100 feet and connections were made with the higher levels of the Seventh and Eighth chambers, but no major new discovery resulted.

Following this hectic three year period of discovery in the cave, which was largely the result of John Parker's efforts, there was a hiatus in activity. New divers, Colin Edmunds, Richard Stevenson and Bob Churcher were coming along but they needed time to gain experience. There was also the feeling that if John could not find the way on upstream it was not going to be easy. And matters were not helped when the experienced cave diver Roger Solari was lost without trace in the downstream Sump 4 of the giant Agen Allwedd system in South Wales, on 15th June 1974.

The mining of the new tunnel from 3 to 9 was carried out during 1974-75. It formed a great loop to a new entrance in the valley and was graded to allow access to the foot of the high-level route found by John Parker. The mining company Foraky hired local men recently made redundant following the closure of the Kilmersdon Colliery in the exhausted coalfield lying to the north of Mendip. Thus, their tunnel became known as the Kilmersdon Tunnel to cavers; a nice touch inspired by Oliver Lloyd. This tunnel had the advantage that any new pushing dives could now start from the Ninth Chamber so avoiding the 200 foot long dive from 3 to 9.

On 7th December 1974, John Parker went to 22 for the third time and took some photographs both above and below water. There had been two large floods since his last visit there; one in 1972 and one in February 1974. These had left flood marks 50 feet above the normal water level! Bob Churcher then visited 22 in March 1975 and thought that the main flow of water came out of the boulder pile just before 22. On his return he had the unfortunate experience of having complete light failure and made the return trip from 21 using tactile stimuli only; something he had no desire to repeat. Apart from these dives very little was done in the far reaches of the cave until Colin Edmunds returned to 22 in December 1975. It was four and a half years since it had been first discovered.

Colin had started diving under the guidance of John Parker in the early 1970s and from 1974

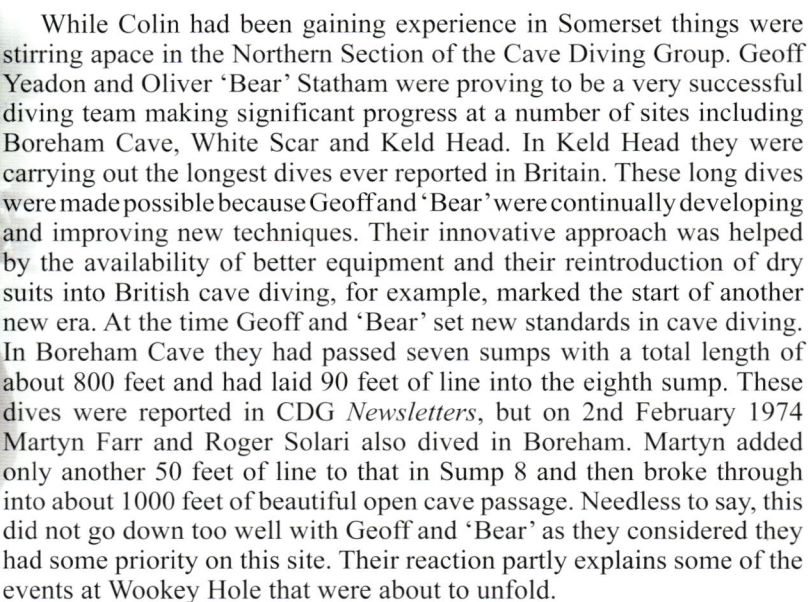

Fig. 13.14
Oliver 'Bear' Statham.
Photo by Lindsay Dodd

Fig. 13.15
Geoff Yeadon in Wookey Hole car park.
Photo by Martyn Farr

to 1975 he teamed up with Richard Stevenson. During this period they explored the underwater passage in the region of 9 to 15 very thoroughly but without making any significant discoveries. In the latter half of 1975, Colin borrowed some large cylinders from Martyn Farr, who was on a caving tour of the USA, and he started to look at 21 and 22 trying to find the way on upstream. On 6th December 1975 he visited 22 and two weeks later searched the left hand wall in 21 going upstream. He eventually emerged in a large rift which now bears his name. This rift is some 50 feet long, 15 feet wide and about 50 feet high. But, due to the steep sides, Colin was unable to get out of the water. A survey of this chamber did not appear in the CDG *Newsletter* and its position in relation to 22 was not clear; consequently it was not until 1984 that this chamber was visited again and Colin's findings were confirmed at last.

On 17th January 1976 both Colin Edmunds and Richard Stevenson made a systematic upstream investigation; Richard searched the right hand wall of 21 going upstream while Colin went to 22. In 22 Colin re-examined all the obvious passages and in addition entered a bedding plane crawl for about 60 feet to a point where it appeared to end in a sump. The exact position of this crawl was not made clear in the CDG reports. One site Colin did not and could not examine was the lake in 22 for John Parker had already dived this in May 1971 and reported that it was silted up at a depth of about 70 feet. And in order to enter the lake remember some assistance was needed as there was an awkward climb down. As Colin was alone he could not risk getting in the water only to find he could not get back out. Consequently, although he had come to the conclusion that the lake needed re-examining, if only to confirm John's findings, Colin had to leave it until he had someone to give him a hand. Richard Stevenson who was diving with Colin over this period often had problems clearing his ears in 15 and because of this he might have encountered further difficulties getting to 22. As a result, Colin had to wait until Martyn Farr returned from his trip to the USA before he could put his plan into operation.

While Colin had been gaining experience in Somerset things were stirring apace in the Northern Section of the Cave Diving Group. Geoff Yeadon and Oliver 'Bear' Statham were proving to be a very successful diving team making significant progress at a number of sites including Boreham Cave, White Scar and Keld Head. In Keld Head they were carrying out the longest dives ever reported in Britain. These long dives were made possible because Geoff and 'Bear' were continually developing and improving new techniques. Their innovative approach was helped by the availability of better equipment and their reintroduction of dry suits into British cave diving, for example, marked the start of another new era. At the time Geoff and 'Bear' set new standards in cave diving. In Boreham Cave they had passed seven sumps with a total length of about 800 feet and had laid 90 feet of line into the eighth sump. These dives were reported in CDG *Newsletters*, but on 2nd February 1974 Martyn Farr and Roger Solari also dived in Boreham. Martyn added only another 50 feet of line to that in Sump 8 and then broke through into about 1000 feet of beautiful open cave passage. Needless to say, this did not go down too well with Geoff and 'Bear' as they considered they had some priority on this site. Their reaction partly explains some of the events at Wookey Hole that were about to unfold.

Geoff had started his diving in Somerset while at Art College in Corsham near Bath and had done some training in the clear waters at Wookey Hole. Tales of these dives were told to 'Bear' and this, coupled with the fact that they wanted to get some experience of doing some 'up and down' diving, meant that Wookey Hole was high on their list of priorities. The main reason they had not been down to Wookey as a team was the usual cash-flow problem that most cavers often seem to have. However, when Oliver Lloyd invited them to 'come and have a go at Wookey' they accepted his invitation. In view of their impoverished state they planned to drive down in Geoff's aged Morris Minor with all the gear at their disposal and spend a week at Wookey, thus maximizing their transport to diving cost ratio. The trip was organised well in advance for the week starting on 21st February 1976. At this time they were under the impression that Wookey 22 had not been visited since Bob Churcher's dive on 22nd March 1975. Colin Edmunds' and Richard Stevenson's recent dives to 21 and 22 had not yet appeared in the latest CDG *Newsletter* and they did not get updated on the latest progress until they arrived at Oliver Lloyd's home in Bristol on the evening of 20th February.

After climbing over the pile of diving bottles in the hall they entered the lounge to find him playing the piano; he did not stop playing. This gave Geoff and 'Bear' some time to examine some of the more interesting cobwebs and artifacts in the lounge. Some time later Oliver stopped playing the 'unbearable music', according to Geoff, and greeted his guests. Over a glass of wine and some foreign cheese, which Geoff thought was definitely mature, the story of the recent dives at Wookey was recounted and they became acutely aware of the fact that a confrontation was inevitable in the circumstances that Oliver had arranged. Earlier on the same evening members of the Somerset Section had also been informed by Oliver whilst training at the University Swimming Baths that Geoff and 'Bear' were arriving and intended to dive the following day. This news, needless to say, added an air of urgency to the plans Colin and Martyn had made.

The following morning at the crack of dawn under the watchful eye of Dan Hasell who was acting as Controller, Colin and Martyn went straight to 22. Once there Colin, entered the lake with assistance from Martyn and descended into the depths. At the bottom, about 60 feet down, he followed a fan of silt to a low arch at the northern end of the lake. To his surprise he could see a passage continuing over a muddy floor. This was followed and it soon started to ascend. Colin ran out all of the 300 feet of line on his reel plus some more unexpectedly! Someone else had loaded the reel and had not tied the line securely to it; consequently, he was now in a potentially dangerous situation, having overshot and lost contact with the line. Some quick thinking on his part enabled him to retrieve the end of the line in rapidly deteriorating visibility and attach it to the empty reel. At this point he had ascended from the bottom of the sump at 60 feet and was now at a depth of about 30 feet. But the lack of more line called a halt to his progress and he had to return in zero visibility to Martyn who was waiting in 22.

The ease with which progress had been made was a surprise and an excellent example of the fact that sumps do change and that the cave diver should always play the role of Doubting Thomas. Since John Parker's dive in the 22 lake in 1971 there had been two large floods. Also, when the Kilmersdon Tunnel was being drilled to 9, the sluice at the entrance to the cave had been opened thus lowering the water level and increasing the flow of water from the cave. It is likely that the mud and silt which had stopped John's progress had been washed out during one of these floods or as a consequence of the drilling operations. This left the way open for the first person willing to dive the sump that John Parker of all people had said didn't go. Colin had been slowly building up his experience and confidence and it looked as though his efforts were about to be rewarded. On returning to Nine, Dan Hasell was informed of their progress and they then left the cave.

Whilst this had been going on Geoff and 'Bear' had left Oliver Lloyd, having survived the night, and went out to Mendip. On arriving at the Wookey Hole car park they were surprised to find Colin and Martyn had already entered the cave and they were starting to get irritated by the competitive situation which had arisen owing to Oliver Lloyd's organisation of the dives: whether knowingly or not, we will never know for sure.

While walking up the path to the cave, Geoff and 'Bear' met Colin and Martyn who were just leaving. A short, uneasy conversation ensued and details of the dive in the lake were exchanged. Geoff and 'Bear' then carried on, while Colin and Martyn whom they had just roguishly renamed 'Taffy Turnip' because of his flushed face in his hot wetsuit, returned to the car park. The Northern divers' main object on this dive was to familiarise themselves with the up and down nature of the sumps between 9 and 22 following a sketch map supplied by Richard Stevenson. On this dive they wore dry suits with which they had gained experience in Keld Head and carried three bottles each. They reached 22 with no problem and had a look around the dry series. At this point they made two decisions: one was to forget about dry suits on their next dive because they did not feel they were needed; and the second was that, on their next dive, they would have a go at the lake regardless of what anyone else said. They had been planning for this trip for about two months and felt justified in making such a decision despite the progress Colin and Martyn had just made. After all, they were totally unaware of Colin and Martyn's intentions until the previous day. And, no doubt, memories of events at Boreham Cave also influenced their decision!

While all of this was going on Derek Tringham, who was having a training dive, climbed up an aven in 9:2, this proved to be a 50 feet oxbow, but its exact position is not clear as a survey was never published. Sadly, Derek was later lost in a sump in Northern Spain whilst diving in the Cueva de Vegalonga, Asturias, on 28th August 1976.

In the Hunters' Lodge Inn at Priddy that evening there was much talk about a joint dive, but the die had been cast. Geoff and 'Bear''s original holiday plans would go ahead. On Sunday 22nd February they returned to Oliver Lloyd's and filled their bottles. On Monday 23rd they arrived at Wookey Hole to carry on where Colin had finished off. Both divers used two 40 cubic foot bottles, one 80 cubic foot bottle and two valves and wore two wet suits plus polar underwear. In 22, the 80 cubic foot bottles were left behind, a valve was swapped and they carried on using two 40s each. 'Bear' was in the lead while Geoff followed making a survey as he went. At the end of Colin's line at a depth of 30 feet 'Bear' attached his 600-foot line reel. The sump could have closed down, but it did not. After reeling out only 60 feet of line, the surface was broken and the long sump passed.

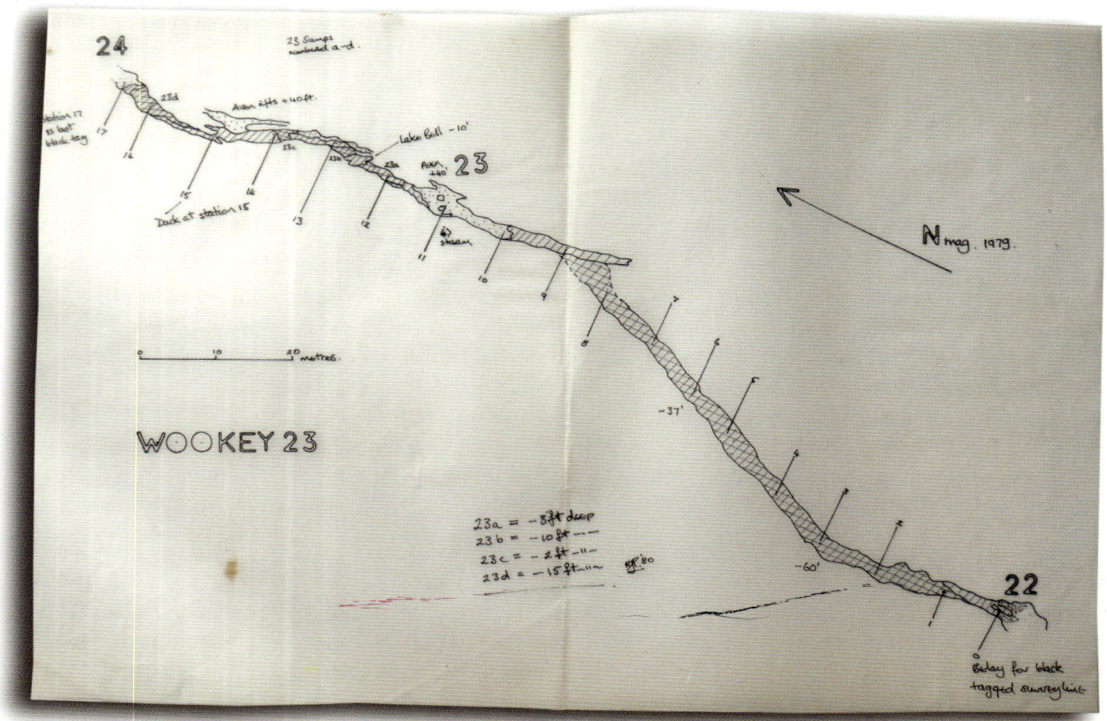

Fig. 13.16 Survey of Wookey 23 by Rob Palmer, 1980

They entered a passage 40 feet high and up to 10 feet wide with a sandy floor. But, after 150 feet they reached another sump which did not look too promising. This was obviously not the main passage they had been hoping for as there was no active stream, only signs of water flow in flood conditions. They then returned to a rift above the sump they had just passed. This rift was climbed by Geoff until he was stopped by a tricky move about 50 feet up. This happened to be the same rift that Rob Harper and Trevor Hughes descended from the Cam Valley Crawl much later on 23rd January 1983; a dry high-level connection between 22 and 23 that had not been pushed, remember, at the begining of May 1971. But back to Geoff Yeadon and Oliver Statham on 23rd February 1976.

With no other option left they returned to the next uninviting sump. After digging away at a mud bank, 'Bear' was able to enter the water, Geoff was to follow after 5 minutes. 'Bear' soon surfaced after only 15 feet in another muddy pool in a rift and was able to make voice contact back to Geoff who then followed through in zero visibility. Three more short muddy sumps were passed by 'Bear', each one looking less promising with no sign of flowing water. The last of these sumps surfaced in another muddy chamber with the only way on being up a steep, slippery mud bank. 'Bear' was trying to thrash his way up this using his diving knife to cut steps up the mud but had not managed to make much progress when Geoff arrived. Using combined tactics they both eventually reached a muddy ledge. 'Bear' started to de-kit but Geoff went on ahead looking for the next sump. As the passage increased in size so did his excitement; Geoff was soon in a large passage 20 feet wide and 50 feet high. He screamed back to 'Bear': 'De-kit, de-kit!' This triggered a mad rush and together they ran off into the unknown casting aside masks, bottles, fins, etc., letting them fall willy-nilly to the floor. The adrenaline was flowing at an ever-increasing rate being given an enormous boost by the distant roar of water. A final climb over boulders presented them with a priceless sight: the River Axe pouring down a handsome passage 40 feet high and 5 feet wide. Sadly, Oliver Statham, who lived life to the full, committed suicide in 1979 and so is unable to contribute his own recollections. Geoff Yeadon sums up their feelings on making the biggest breakthrough to date in the history of cave diving at Wookey Hole:

It was a damn good trip. I thoroughly enjoyed it... it was such a surprise to see that stream passage when we were fully expecting more sumps... and awkward sumps as well. What followed was the finest bit of stream cave I'd seen on Mendip. We were elated!

The Priddy to Wookey Hole Master Cave seemed to be at their feet. They both entered the water moving upstream against a strong current, until they were unable to make progress, because of the sheer volume of water, and a traverse over the stream had to be made. The water here was crystal clear and relatively warm compared to that of the Dales caves and reminded Geoff of some of the clean-washed streamways he had been exploring in the Matienzo region of Northern Spain the previous year. After about 200 feet, the passage enlarged dramatically into a large chamber with a high-level passage off to their left. But the lure of the water was too great to resist. Following the stream they passed three other high-level openings on their left, eventually reaching a large lake 200 feet long and 25 feet wide. They swam across this, but ominously they noticed that water was swelling up from under the left hand wall. At the far end of the lake they clambered up a mud bank and 'Bear' then half climbed, half fell down into a small continuation of the lake, but this closed down. Returning across the lake it was obvious that this was the downstream end of the next sump and so Wookey 24 ended at this point. The high-level passage they had noticed on their left on the way in also rejoined the stream at this lake.

Feeling they had done enough for one day they started back. Geoff made a survey using a wrist compass and pacing to measure the distances, jotting his data down on his diving slates. About 2000 feet of new passage was surveyed between the last sump they had passed and the lake. This included the parallel high-level route which they were able to explore back from the lake to a point where they could look down onto the stream. The climb down to stream level from this point was too difficult without a rope; therefore they had to retrace their steps. At its downstream end, at a sharp bend, the active stream terminated in a sump which was not dived.

After retrieving their diving gear, which had been scattered along the passage in the excitement of the breakthrough, they set off back towards the entrance. But before leaving 24, Geoff could not resist using the mud bank to express his feelings. 'It was easy' read his short message to those who would follow later.

They were tempted to call their new find Whitescar but resisted. Instead they called the sharp bend where the stream entered the downstream sump 'Sting Corner' after the popular film 'The Sting' which described the activities of two con men who, of course, bore no resemblance at all to Geoff and 'Bear'!

On leaving the cave, they contacted the cave management to tell them of the new find and then headed back to Oliver Lloyd's in Bristol where they intended to get their bottles filled so as to continue with their exploration later on in the week. Oliver was not in when they arrived so they left their survey together with a barbed ditty written by 'Bear':

Just when Farr was feeling queasy,

That was when we found it easy.

When Colin Edmunds found it gruesome,

It was easy for a twosome.

Stevenson he had catarrh

He couldn't even get that far.

'Bear' and Geoff

As they were about to leave, Oliver returned to hear the good news. On being told about the events of the last few days and mindful of the fact that he had made the whole scenario possible, his first comment was:

Poor Colin, poor Colin.

Colin Edmunds had been ringing the cave management to find out what had been happening and this coupled with Oliver's regrets made Geoff and 'Bear' feel they had probably stirred things up enough. 'Bear' immediately called Colin on the phone to tell him what they had found. Needless to say, Colin, who was the unwitting victim of these events was very upset as he felt that he should have been allowed to push the extension he had so very nearly found. His understandable reaction to the day's events made Geoff and 'Bear' decide that they had done enough and that they had best head back up North the following day. That night they went across to Geoff's old Art College at Corsham where they met Neil Adcock and celebrated in the finest tradition by getting well-oiled on the local beverage.

So ended what was a memorable weekend adventure for Geoff and 'Bear'. Their raid on Mendip which they had been planning for some time was highly successful, if somewhat controversial. Colin Edmunds, who, in particular, had been on the verge of making the discovery himself puts his own recollections of 1976 as follows:

I had started diving under the guidance of John Parker and Jeff Phillips. This was some time after John's discovery of 20, 21 and 22 and, although I hadn't dived at Wookey with John on more than a few occasions, we talked about the prospects for extension. John had dived into the 22 lake and formed the conclusion that it didn't go and, forgetting one of his favourite maxims, I had accepted this. When I did some diving in Wookey I became puzzled by the fact that 22 remained unvisited for six years; so, when Martyn Farr went abroad for a few months leaving me in charge of some 84 cubic foot cylinders, I decided to use the opportunity to have a thorough search in 21. With the greater margin of air now available I was able to check out both walls in 21 and inspect the boulders at the bottom of 22 for currents or a practical way on. But there were no finds other than one large air bell off the left hand side of the boulder slope and this appeared to offer no potential.

I spent the next few dives similarly investigating 22 as thoroughly as I could; yet, with the exception of one bedding crawl later investigated by Martyn Farr, no immediate or apparent way on in the dry section was found. This left the lake in 22 as the only prospect. I was now dependent on what support I could muster because the access to the lake was a vertical climb of some four feet directly into what I knew from John Parker to be 60 feet of water. Richard Stevenson came to my aid but had ear clearing problems at depth and often failed to reach 15 because of them. For my part, I was unwilling to attempt entering the 22 lake alone in case I could not climb back out. Shortly after I reached this stage, Martyn Farr returned, claimed his bottles and offered to support me as well. We fixed up several dates but water conditions frustrated any dives for a period of several weeks.

We decided to try again on one of the regular Wookey meets on 21st February 1976 and were perturbed to find that Geoff Yeadon and Oliver Statham from Yorkshire had requested that they be allowed to dive on the same day. Their stated intention was to go straight for the lake in 22. I felt that after the work I had put in on the project and in view of the fact that I had not been contacted by either, there was no reason for Martyn Farr and me to alter our plans and we correspondingly went ahead. Diving the lake immediately revealed a fan of silt up which the main current obviously flowed, and following this gave access to an ascending passage. Before diving that morning I had allowed someone to load my line reel, a mistake for which I was to pay dearly. As I swam the passage I became aware of a change of sensation from the hand carrying the reel, and when I looked was horrified to see it empty and the end of the line vanishing in a rising cloud of silt several feet behind me. Frantically grabbing at the loose end I succeeded in wrapping it onto the reel which I stuck into the mud before beating a blind retreat.

On exiting the cave we met the Northern team going in. We had not met before that moment because they had stayed somewhere on Mendip whilst we had travelled direct to Wookey Hole from Wales. They congratulated us on the success of the dive and we suggested a joint expedition to follow it up, the way on now being wide open. It was agreed that this would be one day during the following week since Martyn and I had previously committed ourselves to installing diving kit in Otter Hole, near Chepstow, on Sunday 22nd February, ready for a dive by Martyn in the near future. We would need our bottles to be fully charged for another dive in Wookey. We stayed that night in the Forest of Dean.

Fig. 13.17
Annie Lavender and Chris Milne looking downstream in Wookey 24.
Photo by Clive Westlake

On returning home on Sunday night I received a telephone call telling me that Oliver and Geoff intended to dive again in Wookey the following day and push beyond my discarded reel. I was rather disappointed. On Monday 23rd February Oliver Statham phoned me at work, from the show cave at Wookey Hole after the dive and described the new discoveries in 24. He and Geoff were cooperating with the management over the publicity arrangements necessary with a commercial cave. He went on to explain that, as they had both taken holidays specifically for the dive and did not have a lot of time available, they had, after discussion between themselves, decided to dive that day so that they would be free to follow up other projects and they did not know where to contact us. Oliver apologised for having to do it without us. I accepted the apology and reasons, especially since we had been committed to our own arrangements over the weekend and would not have broken them. I did not, and do not, feel any bitterness over the events of that weekend, since I could just as reasonably be criticised for not putting our dive back to the Saturday afternoon. After all, Oliver and Geoff would have travelled all the way from Yorkshire on the Friday and, spending the evening on Mendip, were most likely to dive the following afternoon. We had preferred the morning. It was the stories afterwards that highlighted the competitive aspects of the discoveries and these did some injustice to both parties involved. On the day, we had missed out and they had succeeded.

Fired up by what had happened on 23rd February, Colin Edmunds, Martyn Farr and Richard Stevenson were back down the cave on the following Friday, 27th February. Richard again had frustrating ear clearing problems and did not get beyond 20 but Colin and Martyn continued to 24. Once in the new stuff they proceeded upstream taking an oxbow on their left so as to avoid fighting the current in the narrow section of the streamway. At the lake which had stopped Geoff and 'Bear', Martyn made an exploratory dive going down to a depth of about 80 feet where the sump bottomed out. He returned to the surface and on his next dive he ascended up the dip on the upstream side of the sump eventually surfacing after a 300 feet dive. He was in a dark lake with mud-covered walls and this oppressive pool was later called 'The Lake of Gloom'. Ahead he was able to swim up to a barrier which he could see over into an adjoining pool about 30 feet in diameter, but he could not get out of the water and over this barrier. Martyn then returned to Colin who was waiting in 24. Together they made their exit after examining the more easily accessible passages in 24. At least they had been able to regain the lead in what had, for a few hectic days, become a race. Wookey 25 was theirs.

Cave diving is a high risk sport which can do without the added pressure of competition. After the controversy surrounding these dives at Wookey and the earlier events at Boreham Cave a new unwritten code of practice emerged whereby divers now respect the rights of the original explorers of a sump. The rights to sumps do change hands but only after consultation with the original divers and the purchase of an adequate amount of beer, the quantity of which is dependent on the potential of the sump (if that can ever be assessed), and the depth of the aspirant's pocket!

A month later on 27th March 1976 Martyn and Colin took some cylinders into the cave to be used on their next pushing dive which was scheduled for 10th April. The amount of air required to reach the end of the cave and start pushing was now getting to the limit of what one person could reasonably carry and a new stage in the exploration of Wookey Hole was about to start. Not only were there problems with air supplies but the repeated diving down to depths of 60 to 80 feet on the way to 25 meant that decompression procedures had to be seriously considered for the first time in British cave diving. Also, the cost of the equipment required to get to the furthest reaches of the cave was becoming a major problem. Because of this, and the commitment required on behalf of the new generation of pushing divers, finance and sponsorship started to become an additional factor which dictated the activities at Wookey Hole and some other sites.

The problems of getting into and out of the sump pool in 22 needed to be overcome and a start on this problem was made on 2nd March 1976 when Richard Stevenson and Martin Bishop took some wooden billets through to 22 for the purpose of making a climbing platform. Later on an iron ladder was installed to make things easier.

On 10th April Colin Edmunds and Martyn Farr returned with the aim of examining the terminal upstream sump in 25. Martyn lost his diving knife on the way in, and at the lake in 22 he also managed to lose a fin! Despite this handicap he and Colin proceeded to the end of 24. They transported one set of diving gear along the passage in 24 which allowed Martyn to dive through to 25 alone. Once in 25 he was still unable to reach the upstream sump pool which was tantalizingly close and so could not push on. He did manage to get out of the water on a small muddy ledge but it became obvious that some assistance would probably be needed if any further progress was to be made. On his return to 24 Martyn got badly caught up in the line through the sump and without his diving knife he was in a tricky situation. For someone who could not clear his mask when he took his pool tests, Martyn had come a long way. Keeping calm, he took off his reserve cylinder and valve and proceeded to sort out the problem with the entangled line. Once he had freed himself from the line he retrieved his spare set and returned to Colin who had been waiting in 24. The return trip to Nine was slow because Colin had to carry most of the spare equipment through the sumps while Martyn limped along with one fin.

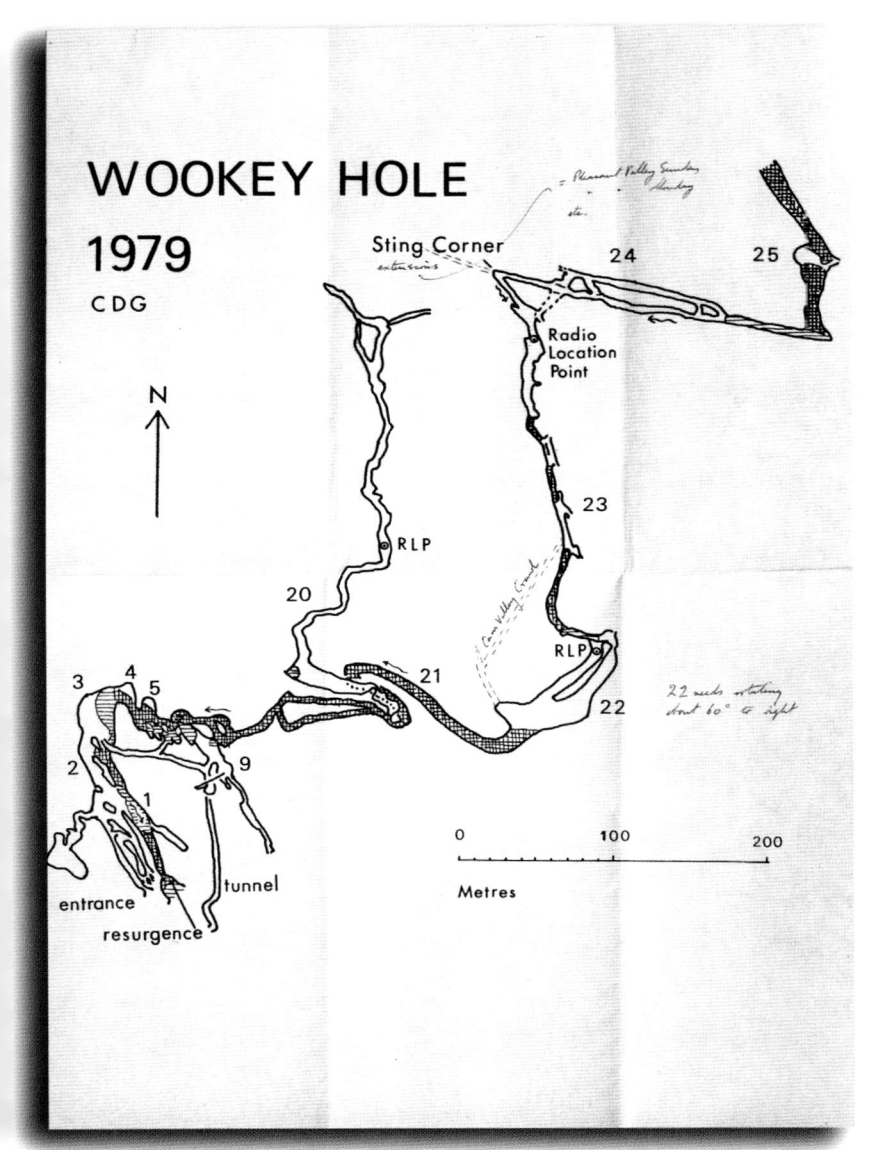

Fig. 13.18
Wookey Hole, 1979.
Annotated with later additions by Oliver Lloyd

Martyn's next attempt on the 14th July was more successful. On this occasion he was supported by Pete Lord and Dave Morris. Once again Martyn dived alone from 24 using two full 45 cubic foot bottles. In 25 he was not going to be turned back again by the climb over into the upstream sump pool. Using brute strength he managed to heave himself over the barrier and enter the pool which he had seen on his previous dives. Diving in excellent visibility he descended vertically and reached a ledge at 50 feet. Beyond the visibility was still excellent and the way on was to his left along the strike in a westerly direction. Martyn followed this passage but found himself being forced down deeper and deeper in a passage inclined at 60 degrees with a large void on his left with no bottom in sight. After running out just over 250 feet of line he reached a depth of 100 feet and could see the way on continuing but was unable to proceed safely with his present equipment and air supply. Reluctantly he left his line reel and returned to the surface in 25 having achieved a new British cave diving depth record.

Further upstream progress in Wookey Hole Caves would require a much more sophisticated approach as the depths involved were greater than anyone in British cave diving had experienced. It was time to reappraise techniques again.

Fig. 13.19
Annie Lavender and Chris Milne looking upstream in Wookey 24.
Photo by Clive Westlake

Chapter Fourteen

Dry to Wookey Nine
by Brian Prewer

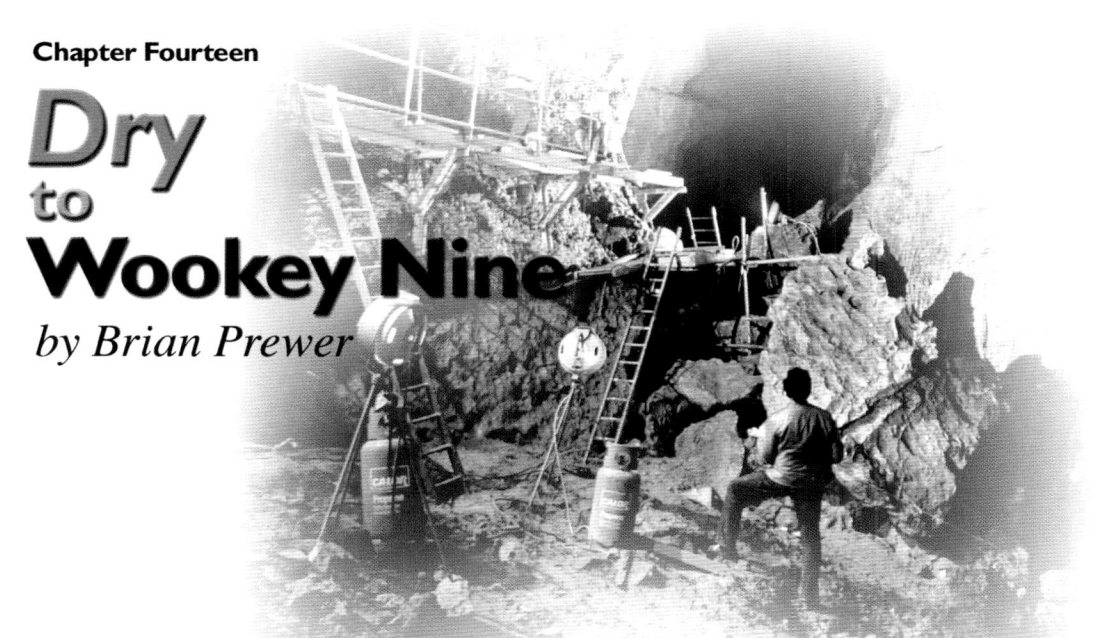

With the increase in the number of tourists visiting Wookey Hole Caves during the 1970s it became clear to the new cave management that a circular route through the cave would help to alleviate the congestion caused by the existing contra-flow system.

Madame Tussauds, the new owners, had considered the possibility of blasting a tunnel from the Ninth Chamber to the valley side opposite the natural entrance. The main stumbling block to this scheme was that no survey existed giving an accurate position for 9. The divers' surveys were not thought to be reliable enough to permit a costly tunnel to be driven with the required degree of accuracy.

Fig. 14.1
Brian Prewer demonstrates how his radio location receiver was deployed in 1973 to find the highest point above 9:2 first reached by John Parker and Brian Woodward in 1971. Open cave lay just beneath the hillside.
Photo by Willie Stanton

During the exploration of the upper dry passages above 9 by John Parker and other divers, a steeply ascending, well decorated, passage had been followed for several hundred feet ending at a boulder choke thought to be close to the surface. Tree roots noticed near to the boulder choke were assumed to belong to a clump of trees in a likely location on the steep hillside above the cave. Ideally, a dry route to Chamber 9 was needed to allow a high grade survey which would establish an accurate position for this chamber. Once this had been done and tied in with the other chambers via the old entrance, it would be possible to draw plans essential for blasting the tunnel to 9. Related information about the rocks and their structures could be mapped to aid the design and ensure that blasting did not disturb or destroy important features in the cave such as priceless stalagmite formations. From the Ninth Chamber a series of mined tunnels would then link it with 8 and 7 and finally to the main show cave in the Third Chamber. The key to the success of the whole enterprise lay in the precise location of the boulder choke at the end of the upper dry passages relative to the surface. Apart from the trees, no other significant surface feature had been identified. Clearly, the only feasible way was to employ so-called radio location equipment to pinpoint the choke position on the surface above.

'Radio location' is strictly a misnomer for a technique which uses the magnetic induction principle to transmit and receive a magnetic field between two distant points. Magnetic fields, unlike the field created by radio waves, are virtually unattenuated by rock and, therefore, by the use of a nulling procedure it is possible to establish sets of intersecting lines of coordinates. At the intersection point, the surface detector is situated vertically above the underground transmitter. From the pattern of the magnetic field it is possible to determine the depth of the transmitter below the surface.

Fig. 14.2
A transmitting coil is set up in the cave below in August 1973.
Photo by Brian Prewer

Radio location equipment was available and had already been used successfully at several other cave sites to check the accuracy of existing surveys as mentioned in the previous chapter. The Wookey Hole location task would require the transmitter to be carried into 9 by divers, and from there it would be taken to the boulder choke and set up to transmit a signal to the surface. It might be thought that a major snag in using this type of electronic equipment would be in the need for a completely watertight, pressure resistant, container to house the transmitter; the normal ammo box would probably collapse due to the water pressure encountered at depth on the journey from the Third to the Ninth Chamber. The answer here seemed simple but drastic; flood the container with water by deliberately drilling holes in it! This electronic equipment operates at low voltages and, therefore, might not suffer in the short term from being immersed in water.

On the day, I was unable to be present and left full instructions with Willie Stanton and his helpers on how to operate the equipment. It required a fair degree of synchronization between the surface and underground parties because the transmitter had a limited battery life. From a distance, I had several anxious moments during which I imagined the possible comments on the reliability of this 'new fangled high technology stuff', and the parentage of the owner of the gear. And who had thought of allowing it to flood en route? Well, in fact it worked; total flooding of the container housing the transmitter did not affect its performance and in due course the equipment, complete with holes, was carried by divers through to the boulder choke and set up. Great care was needed at this stage to ensure that the plane of the transmitting aerial coil was horizontal as this would influence the accuracy of the surface location overhead. But it all took longer than had been expected.

Back on the hillside, the surface party had already switched on at the pre-arranged time and was somewhat startled by the reception of, not the expected 'bleep' tones from underground, but the old BBC Light Programme (now renamed Radio 2). But no one was unduly worried, it seems, and had complete faith in the equipment. After half an hour the BBC was replaced by an unexpectedly loud 'bleep'. This caused considerable jubilation amongst the surface team.

Carefully yet rapidly two or three intersecting lines were set up, determined by the nulling of the 'bleep' with the receiver aerial coil, and the point on the surface directly above the transmitter coil was soon located. Then, by taking a second set of measurements, its depth beneath the surface receiver was determined.

The surface point was somewhat of a disappointment, no clump of trees, no bushes, no nothing, just a plain sloping field. As for the depth measurement, this gave a mere 10 feet! I felt that radio location techniques were now going to come in for some close scrutiny from any sceptics wary of results that showed no surface feature and suggested that the boulder choke was only 10 feet below ground level. It had been a very loud 'bleep' though! Later, the divers carried the rather damp transmitter back to the surface for rapid drying.

Fig. 14.3
A mechanical excavator makes a quick breakthrough, August 1973.
Photo by Willie Stanton

The next stage of the operations would be to employ a massive Hy-Mac excavator to dig down at the location point and see if any obvious features could be found to indicate the passage below. It was an unsuspecting Hy-Mac driver who in fact proved conclusively that radio location techniques do work, and rather more accurately than had been expected! The location point was 'spot-on'. The depth, however, turned out to be somewhat less than 10 feet for the Hy-Mac digging machine broke into the passage barely 5 to 6 feet below the surface. Why the error? Obvious, when you stop and think about it: the height of the receiver coil above the ground surface and the depth of the transmitter coil below the passage roof had not been subtracted from the measured value of the 10 feet! After the digger moved away onlookers observed that no more than a thin layer of soil and a few meagre slabs of rock had been covering the end of the passage. Yet no surface feature whatsoever had been visible as might have been expected if the passage had formed recently. We had unearthed an ancient high-level outlet for the River Axe on the hillside. The pundits were pleased.

Fig. 14.4
The dry route to Nine-two is surveyed by Willie Stanton and Brian Prewer, August 1973. Photo by Jim Hanwell

A dry route to 9 was now open. Eager hands made the entrance safe and it was not long before a party was ready to make the first descent ever into the Ninth Chamber without the use of diving gear. On Sunday 5th August 1973, a small party assembled intent on reaching 9. They reached the lip at the top of the Ninth Chamber but failed to locate John Parker's bypass. On the following day, after consulting John Parker's notes, they returned and whilst some of the party started the survey, Terry Tooth and Brian Prewer found the free-climbable descent to chamber 9, although on this first trip the lower section was laddered. We were the first non-divers to set foot beyond the Fifth Chamber; a full 38 years after the original hard hat divers had started the underwater exploration of the Great Cave.

Now that the Ninth Chamber could be reached without diving, an accurate survey could be completed to enable its position to be established and the artificial tunnel to be driven in from the valley side. This Kilmersdon Tunnel is now the route by which tourists leave the cave to proceed down the valley to the old Paper Mill and its attractions.

Fig. 14.5
A new tunnel is blasted open from the Ravine to link with Nine-two, by former Somerset coal miners from Kilmersdon, 1974.
Photo from Wookey Hole Museum

Radio location techniques for survey checking had now been proven and, during the course of the next few years, several other parts of the Wookey Hole survey were checked: for example, the location of Sting Corner in Wookey 24; a repeat of Mel Davies' location in Wookey 20, and the location of Wookey 22. The most significant part of this work was to show that the upstream direction of Wookey 22 was southeast and not northeast as had been previously thought from bearings taken underwater!

Fig. 14.6
The Kilmersdon Tunnel is shored and reinforced with concrete, 1974.
Photo from Wookey Hole Museum

Following the location of 9, the attempt to relocate Wookey 20 was made in 1979. Previous information gained from the relics of Mel Davies' radio location cairn suggested that the point on the surface was in a field just to the west of Green Lane. In fact, the point was located right beneath Green Lane, a few feet away from Mel's original point; an encouraging result. The lower end of Wookey 20 by the water surface was next to be checked and the site was located at a depth of 270 feet below the surface. Next Wookey 22 and 24 received attention. Wookey 24 proved to be slightly to the east of Green Lane, close to the 600 feet contour line and at a depth of 350 feet.

It was at Wookey 22, however, that the survey and the located point showed the greatest discrepancy. After a lengthy discussion about the twists and turns taken by the divers, it was supposed that the selected point to be located in 22 was situated close to Smokham Wood on the west of the Green Lane up the hillside onto Mendip. When the divers switched on the underground transmitter only a faint 'bleep' was detected. Two hastily measured coordinates were set up and the intersection estimated at several hundred yards to the east! The surface party picked up the equipment and ran to a point on the east side of the Lane hoping that the transmitter batteries would last long enough to get an accurate fix. New coordinates revealed the corrected position as being roughly on the 500 foot contour and well to the east of Green Lane by several hundred yards. The error caused some dismay amongst the surveyors and suggested that in fact Wookey 22 was further east than had previously been thought. Comparison between the 1979 and 1985 surveys in the next chapter shows how the error was eventually corrected.

Fig. 14.7
The tunnel from Nine-two back to the Third Chamber needs no shoring, 1974.
Photo from Wookey Hole Museum

Before the final survey was drawn up radio location techniques were to be put through yet another stringent test. Could the unexpected point in Wookey 22 be repeated in view of the importance of this particular site to the accuracy of the survey? A repeat performance was staged with many important onlookers anxiously awaiting the outcome. The point in Wookey 22 was relocated at precisely the same point as before. It was just too good to be true, but there it was, right on top of the stones that had previously been buried and well concealed in the ground at that point. The onlookers departed bemused by the marvels of electronics but satisfied and sure of its accuracy.

Radio location techniques have been used with considerable success in Wookey Hole providing much needed 'bench marks' to map the rest of the system in relation to the surface. The location of the upper dry series in the Ninth Chamber, therefore, was of considerable importance to the future expansion of Wookey Hole Caves and Mill as an asset to the important tourist industry of the whole area. The surveyors have gained important information that has allowed a high grade survey to be made despite the fact that much of the Wookey Hole system is underwater. And what of the equipment itself? Well, it still functions in spite of being flooded with water on many occasions since.

Fig. 14.8
The final dry link-up into the Third Chamber is made. Radstock miner Jack Gay is greeted by Head Cave Guide Alf Stapleton, 1974. The completed tunnel was opened to visitors on 27 March 1975.
Photo from Wookey Hole Museum

Plate V. — Map of THE WOOKEY HOLE RAVINE. By J. H. Savory, 1913.

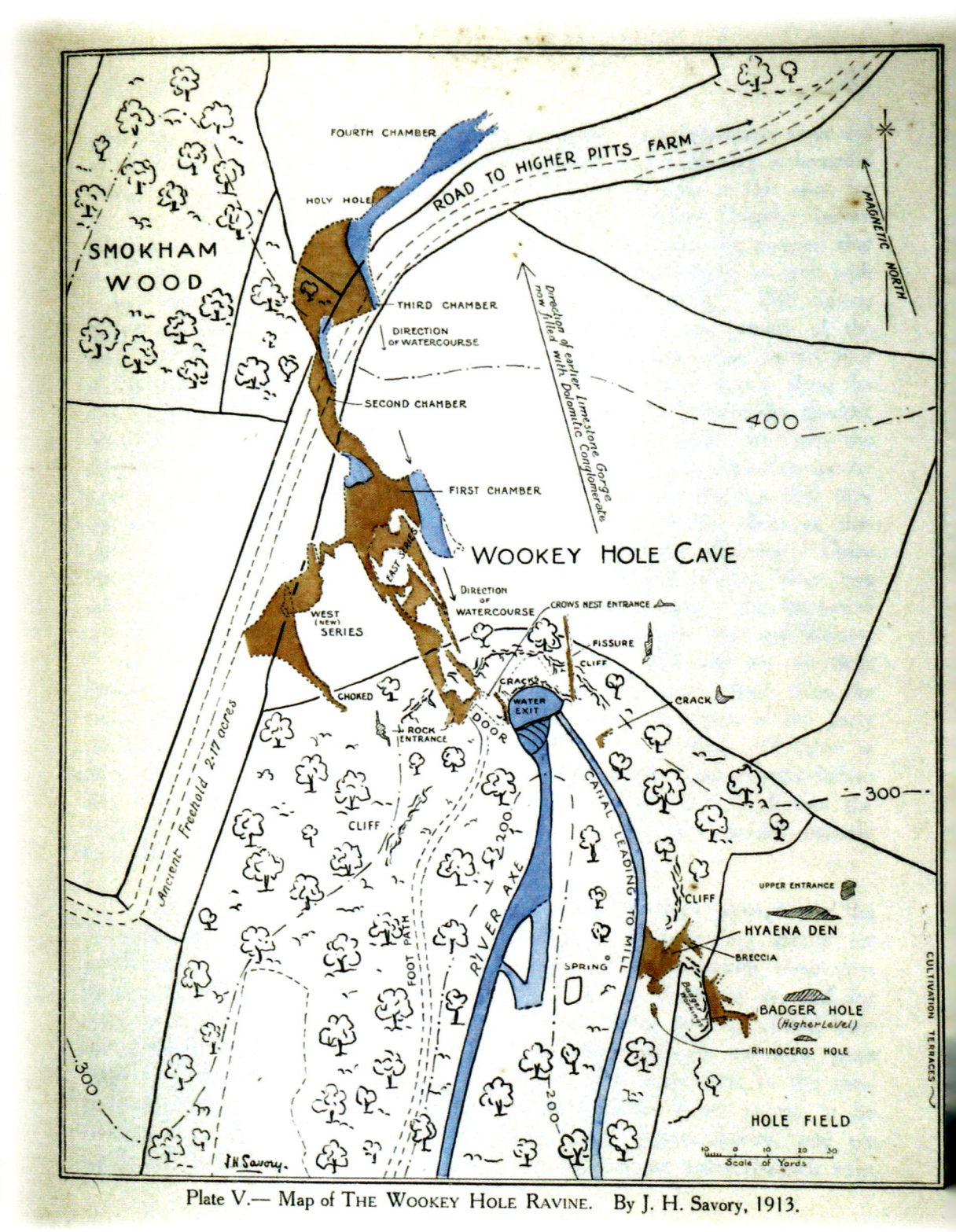

Fig. 15.1 Harry Savory's classic coloured 'Map of the Wookey Hole Ravine', 1913

Chapter Fifteen
The Great Cave Surveyed
by Willie Stanton

*Fig. 15.2
Willie Stanton, who made the definitive survey of Swildon's Hole 1950-53.
Photo by Oliver Lloyd*

Early visitors to Wookey Hole, from William of Worcester (c.1470) onwards, described the Great Cave with varying degrees of accuracy. In 1757 appeared the first published sketch of 'the famous tripple grotto, call'd Ochie-Hole or Wockey, but more commonly Oxey-Hole', illustrating an account of the cave by 'Botanista Theophilus'. The three chambers are shown in a stylised pictorial fashion similar to that used by William Simes in his map of Wells (1735), which to be both effective and accurate demands great skill. Chambers One and Two are brightly lit from above, as shown by the visitors' shadows, but Three is dark and mysterious as befits the 'first rise or source of the Axe' (see Fig. 1.1).

This account and sketch are reproduced in *Wookey Hole, its Caves and Cave Dwellers* (1914, pages 236-246) by Herbert Balch, who criticises the representation of a steep descent made by the river between Three and the resurgence. The fall does seem exaggerated, although there must have been a drop of several metres before the resurgence weir was built about 1852.

In *Cave Hunting* (1874) Professor William Boyd Dawkins presents a diagrammatic section of the Wookey Hole ravine, the Great Cave and a hypothetical continuation linking it to a swallow-hole at Priddy. The latter was of course St. Cuthbert's Swallet, and the connection had been proved 14 years earlier during the Hodgkinson vs. Ennor lawsuit. In the diagram the three chambers of the cave are barely recognisable, and the Priddy link has turned out to be along very different lines, but in fairness to the Professor the diagram was largely drawn to illustrate his idea that limestone gorges are formed by the unroofing of caves (see Fig. 1.2). He elaborates this theory on pages 54-56 of *Cave Hunting*, giving a section of Malham Cove in the Yorkshire Dales almost identical to that of Wookey Hole.

James Parker, who had made the plans of the Hyaena Den used (without acknowledgement) by Boyd Dawkins in *Cave Hunting*, also did some surveying in the Great Cave about 1860, but his map was never finished. The fragments that survive, covering parts of the entrance passages, seem perfectly accurate.

An industrious but largely unsung cave surveyor of the early 20th century was Balch's friend Reginald Troup. He made the first plan and section of the Great Cave, which was drawn up by Balch and published (at so small a scale that much of the writing is illegible) in the book he wrote jointly with Ernest Baker in 1907: *Netherworld of Mendip*. The north point is magnetic; a sign of inexperience. The overall shape of the cave is clearly set out, but there are some differences of detail compared with later surveys, in particular the outlines of chambers One and Three. The direction taken by Chambers Four and Five is entirely wrong, probably because the raft party that had explored them in 1901, on a rare occasion when the Paper Mill allowed the water level to be lowered, had not carried a compass. Nevertheless Balch had been able to formulate his theory that deeply submerged waterways feeding Glencot Spring beyond the village branched off the River Axe in the vicinity of Five, and he showed such a waterway in the section.

Troup was also responsible for the first published survey of Swildon's Hole as far as Sump 1, beautifully drawn up by Harry Savory, which appeared in the Mendip Nature Research Committee's *Report* for 1921. This is evident from Balch's accompanying account ('The necessary survey has been largely the work of Mr. R.D.R. Troup.' and '...the survey party, under Mr. Troup, were following behind...'). Unfortunately, when Balch republished the survey in his book *Mendip, its Swallet Caves and Rock Shelters* (1937) he omitted Troup's name from the surveying party; thus Troup's pioneering work as a Mendip cave surveyor has been little known.

Plate III.—Wookey Hole Cavern. Plan and Section by H. E. Balch & R. D. R. Troup, 1912.

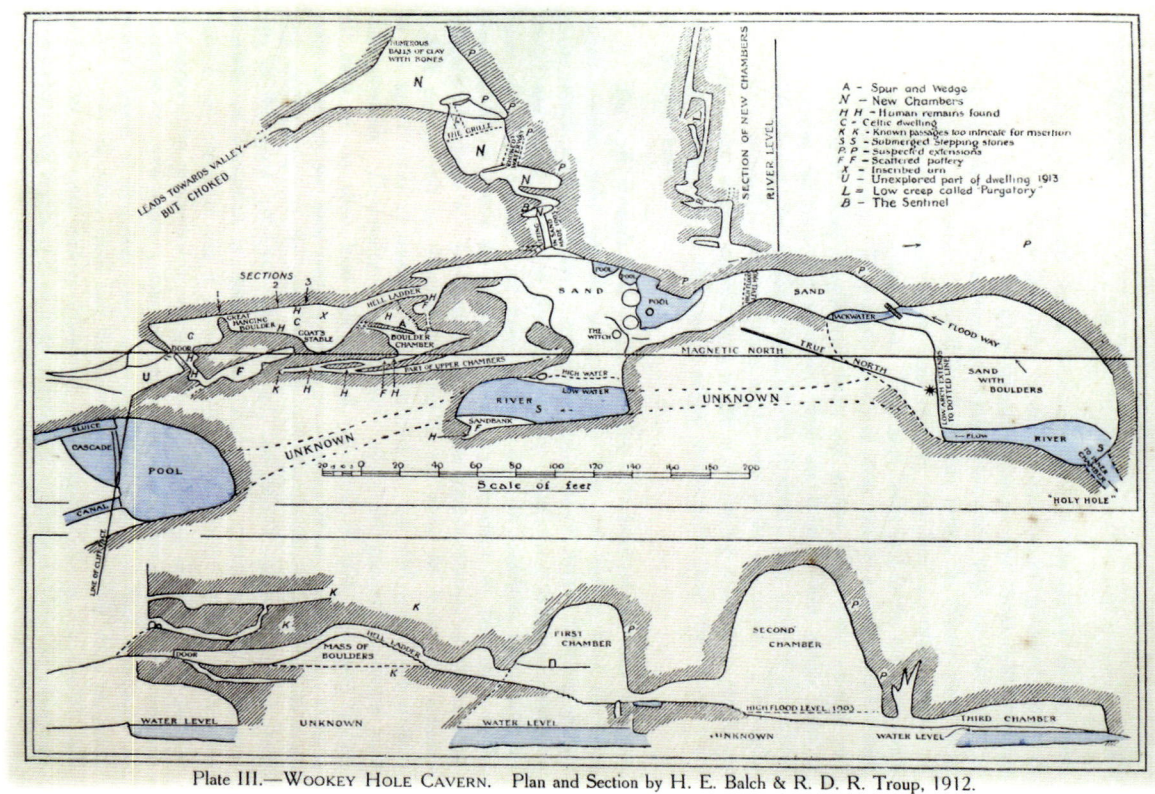

Fig. 15.3 Reginald Troup's survey of the Great Cave as redrawn by Herbert Balch, 1912

Troup's Wookey Hole survey had not been superseded in 1912, when it was redrawn by Balch for inclusion in his monograph *Wookey Hole, its Caves and Cave Dwellers* (1914). This time True North appears, the scale is larger, and some details of the complex passages above the entrance gallery are included. The writing, carefully hand-printed by Balch, is entirely legible, but it varies in size with occasional slightly comical effects.

On page 33 of *Wookey Hole, its Caves and Cave Dwellers* is a map showing the relationship of the Great Cave to the ground overhead and to the Hyaena Den and other minor caves of the ravine. Drawn by Harry Savory with his customary expertise, it is based on Troup's survey, with the addition of chambers Four and Five (not thus numbered) leading in a direction different from that shown in *Netherworld of Mendip*. In Balch's monograph this map is hand-coloured: dry caves brown and water (surface and underground) blue.

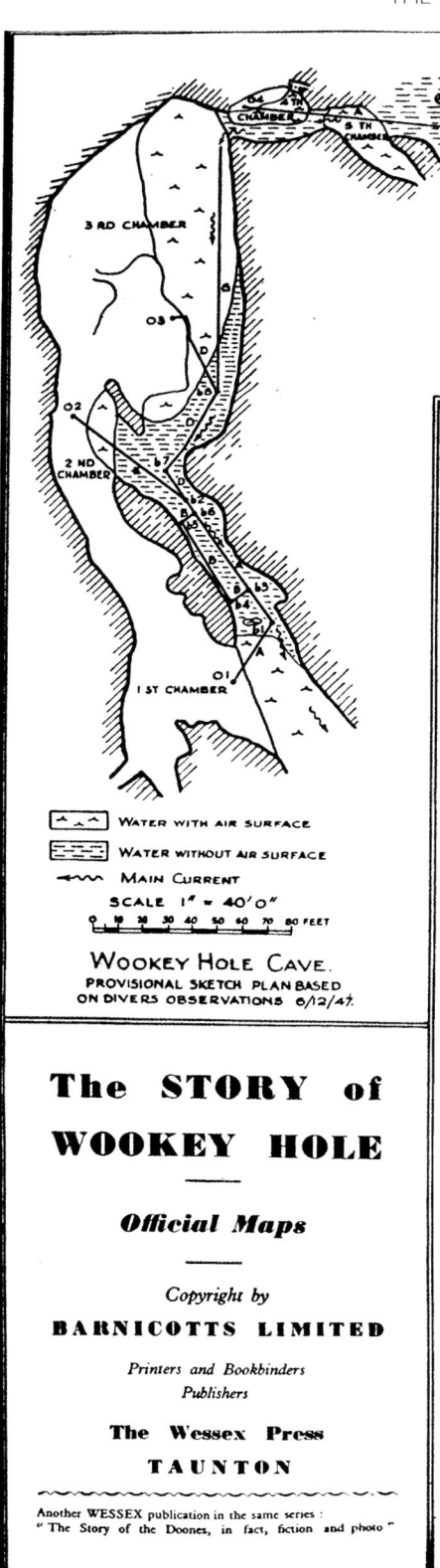

*Fig. 15.4
Thorneycroft's souvenir
'sketch plan', 1948*

The first book of Balch's Mendip trilogy, *The Great Cave of Wookey Hole*, appeared in 1929. Strangely, it did not include a survey of the cave. Balch was conscious that Troup's map was incomplete, for in 1939 he applauded 'the completion of a full survey of Wookey Hole and its upper galleries' by Jack Duck's party. This map, the best that has ever been made of the old cave, included the main upper passages, Charon's Chamber, and chambers Four and Five, but omitted, of course, Six and Seven which were 'accessible only by diving'. In fact that statement (which appears on the plan) was incorrect, and Duck's plan and section both show the aven in the roof of Five that, 35 years later, was descended by John Parker when he pioneered the non-divers' high-level route linking Five, Six, Seven, Eight and Nine.

Duck's original survey was framed and exhibited in the Gift Shop at Wookey Hole Caves. It was reproduced at a small scale in the Mendip Nature Research Committee's *Report* for 1938, and it has formed the basis of all later maps of the old cave (see Fig. 1.16).

After the Second World War the souvenir booklets available to visitors to the Great Cave became more elaborate, and most included a cave survey. The map in *The Story of Wookey Hole* (1948) by L.B. Thorneycroft, a professional writer, must have puzzled many people, for it illustrated the divers' route from One to Seven, not the tourist cave. The anomaly was corrected in the next *Story of Wookey Hole* (1953) by Ted Mason, who used a slightly simplified version of Duck's survey. A note on the plan informs the visitor that divers have now reached Chamber Thirteen.

In fact, divers using the new aqualungs were now making important discoveries as described in earlier chapters. In 1966 Dave Savage entered Chamber Eighteen, and two years later he produced the first compilation survey of all the known cave beyond Three (Wessex Cave Club *Journal*, Vol. 10, page 191). As he pointed out, surveying underwater was not very accurate, but two things were clear: divers had almost doubled the length of the Great Cave since 1935, and the new techniques would permit further rapid advances if open cave could be found.

John Parker soon located the open cave, and a compilation survey of the whole known system was drawn by Jim Hanwell in 1970 (Wessex Cave Club *Journal*, Vol. 11, page 33).

By now the divers' discoveries including the new 'chambers' Nineteen and Twenty, amounting to 1000 metres of passages, were beginning to dwarf the old cave with its passage length of only 520 metres. The divers' survey of Twenty was little more than a sketch, but Hanwell was able to plot it with some confidence thanks to the new technique of radio location, which related a known point in Twenty to the ground surface vertically overhead. Apart from Norman Brooks' experiments in Longwood Swallet in 1955, this was the first known use of radio location in drawing up cave surveys on Mendip.

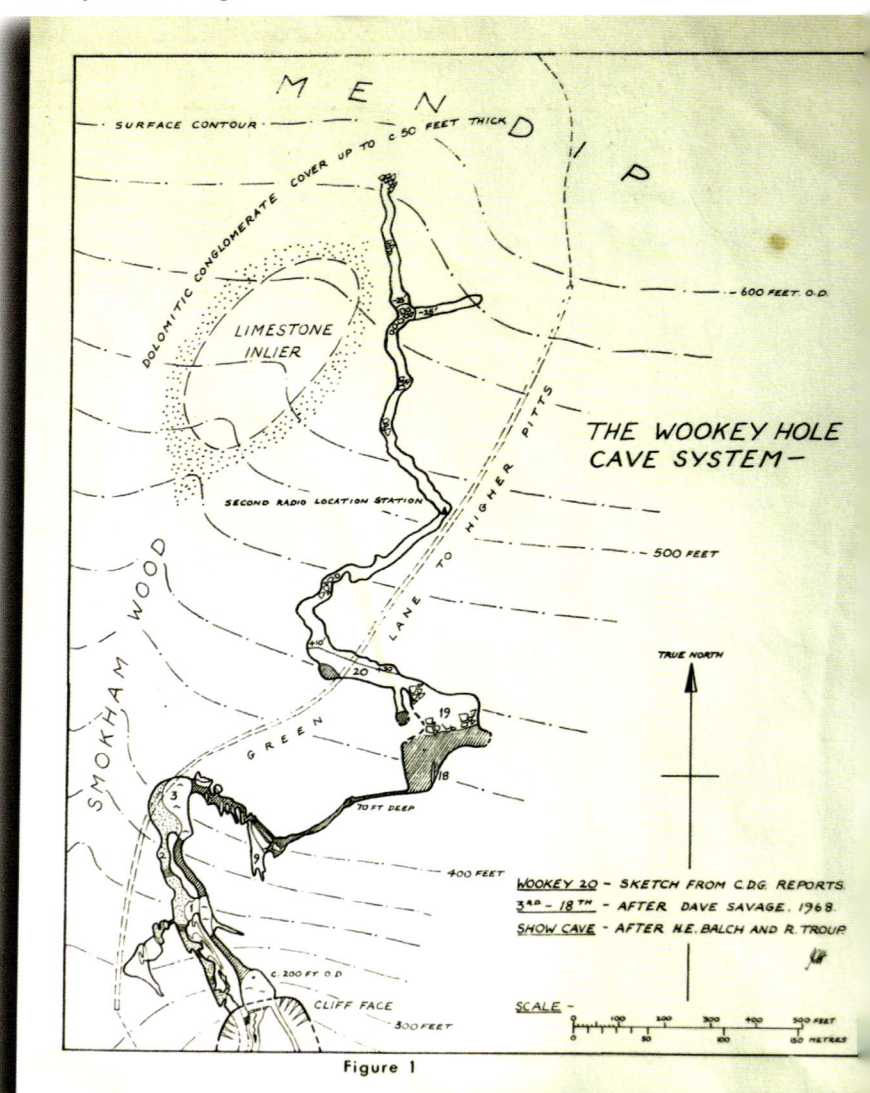

Fig. 15.5 Jim Hanwell's sketch survey showing the 'old cave' and new discoveries of 'Wookey Twenty' in relation to surface features, 1970

In an accompanying article Hanwell presented an informative block diagram illustrating the topography and geology of the ground between Wookey Hole and Priddy, including sections of the Great Cave and its feeder system St. Cuthbert's Swallet. Important natural features of the Mendip plateau, including Ebbor Gorge, Sandpit Hole and Twin T's, were portrayed in relation to the underground cave systems.

Olive Hodgkinson, owner of the Great Cave, wrote her own lavishly illustrated souvenir booklet *Wookey Hole Caves* in 1970. Jim Hanwell's compilation survey, neatly redrawn by himself with a few last-minute additions including the dry passages above Nine, adorned the inner front and back covers. Visitors to the show cave must have been impressed by the vast extent of the chambers beyond their reach.

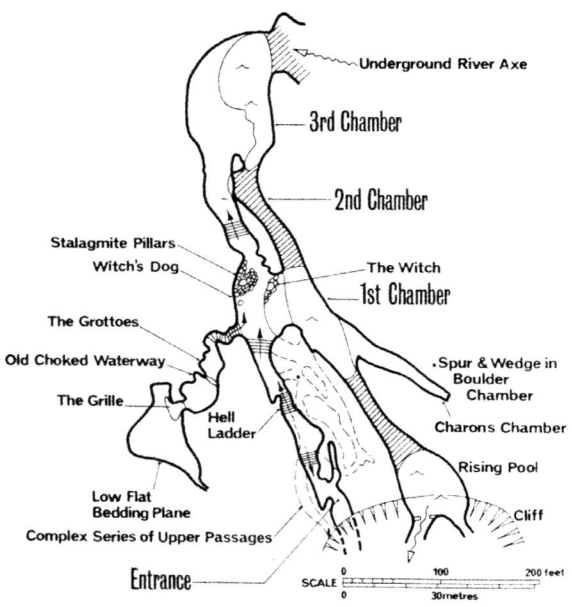

Fig. 15.6
Jim Hanwell's sketch plan of the old 'Show Cave' used by Olive Hodgkinson to illustrate her own souvenir booklet, 1970

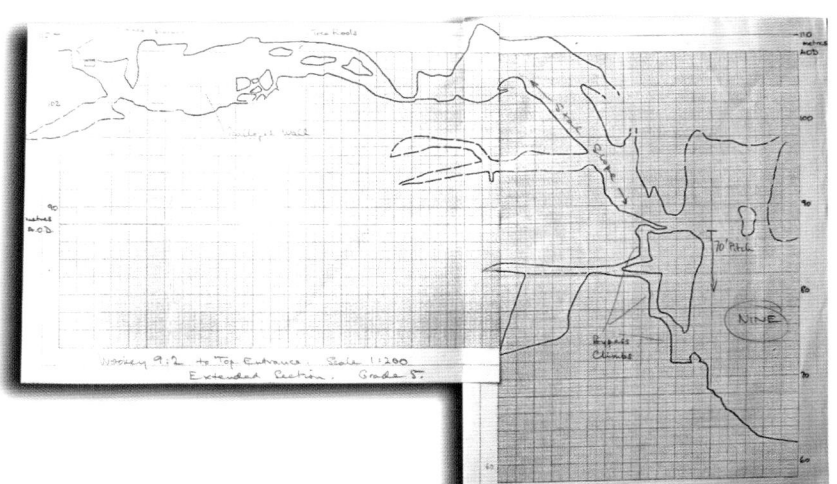

Fig. 15.7
Willie Stanton's preliminary drawing of his Grade 5 survey of the dry passages above Nine-two, 1973 (see Fig. 14.4)

John Parker had dived to Twenty-two by 1973 when the Great Cave, which had been owned by the Hodgkinson family for more than a century, passed out of their hands. It was bought, together with the Paper Mill and much adjacent land, by the London firm of Madame Tussaud's. The new owners at once prepared to drive a tunnel that would take visitors beyond the old cave to chambers Seven, Eight and Nine, and then out to the resurgence and Mill. Planning the tunnel required accurately resurveying parts of the old cave and all the dry passages from Six to Nine. As described in the previous chapter, a new entrance to Nine was radio located and dug open, enabling Willie Stanton to make the necessary surveys.

The tunnel was driven in 1974 and opened to the public in 1975. The new souvenir booklet *Wookey Hole* (1975) included a plan and section of the visitors' route through the show cave and Paper Mill. The Great Cave was now only a part of the Wookey Hole entertainment complex.

Stanton and Robert Crowther drew a compilation section (1975) of the Great Cave in relation to Swildon's Hole and the Mendip plateau that was exhibited in the Paper Mill. The 'unexplored region' between the extremities of the caves so inspired a French visitor to Wookey Hole that he and his family went to Priddy, found and descended Swildon's, got lost, and caused a minor rescue operation. Recent observations by the divers have shown that the relationship of the Dolomitic Conglomerate to the Carboniferous Limestone is more complicated than is shown on the section.

Fig. 15.8 Willie Stanton and Robert Crowther's compilation section of the known discoveries in Wookey Hole Cave, 1975

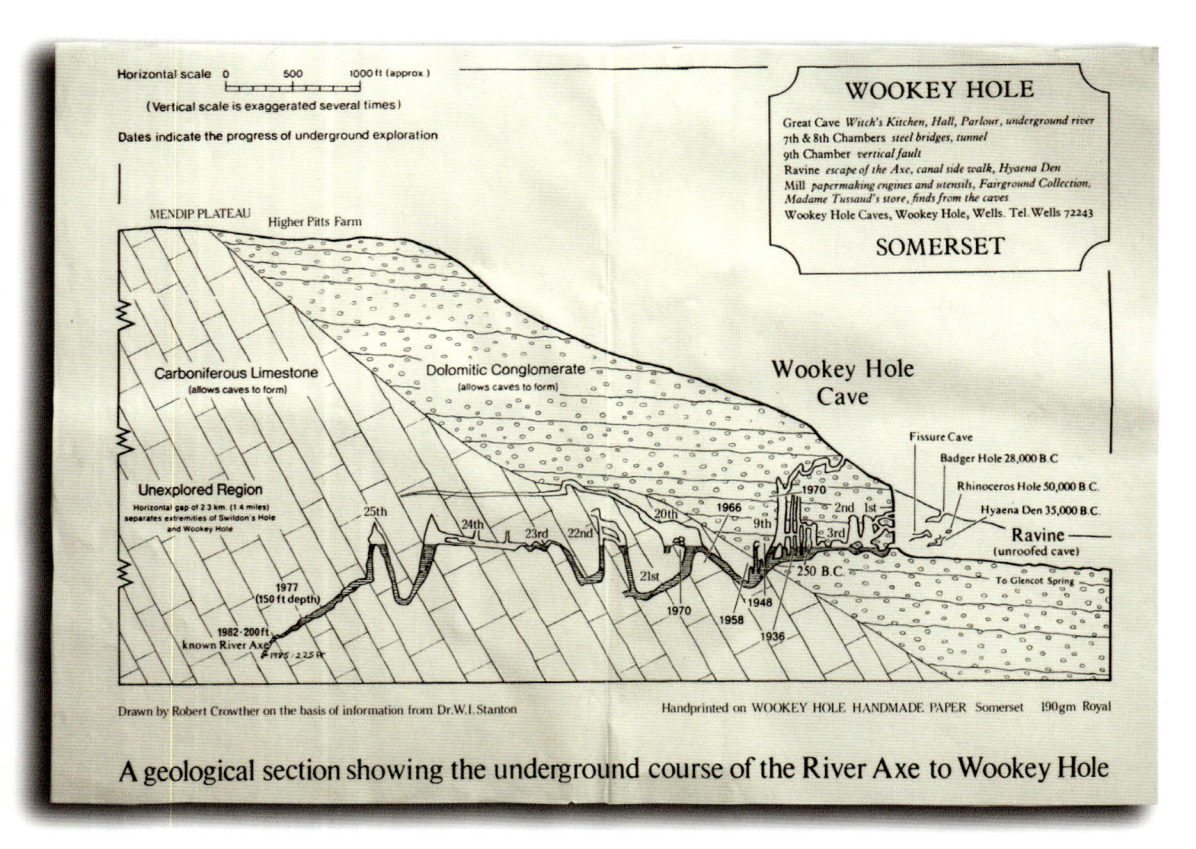

A geological section showing the underground course of the River Axe to Wookey Hole

Meanwhile the divers were pushing ahead. Colin Edmunds found the way to Twenty-three in 1976, and the exploration of Twenty-four and Twenty-five followed at once. Nearly 1000 metres (3280 feet) of passages were added to the known cave in only seven days. The rough surveys made by the divers were included in a compilation survey in the guidebook *Mendip Underground* (1977) by Dave Irwin and Tony Knibbs. The given scale appeared to deny the Great Cave its proud position as Mendip's third longest cave.

A new museum was opened in the Paper Mill in 1980. One of the exhibits was a large scale coloured plan of the whole system, compiled and drawn by Stanton and Crowther. Its accuracy was much improved. High grade surveys of Twenty and Twenty-four had been made in 1978 and 1980 by Geoff Yeadon and Dave Yeandle. These extensive dry caverns were linked by long underwater reaches that, in spite of valiant efforts by Rob Harper and other divers, defied accurate surveying, but the problem was largely overcome by the use of Brian Prewer's radio location equipment. Fixed points were established above the Lake in

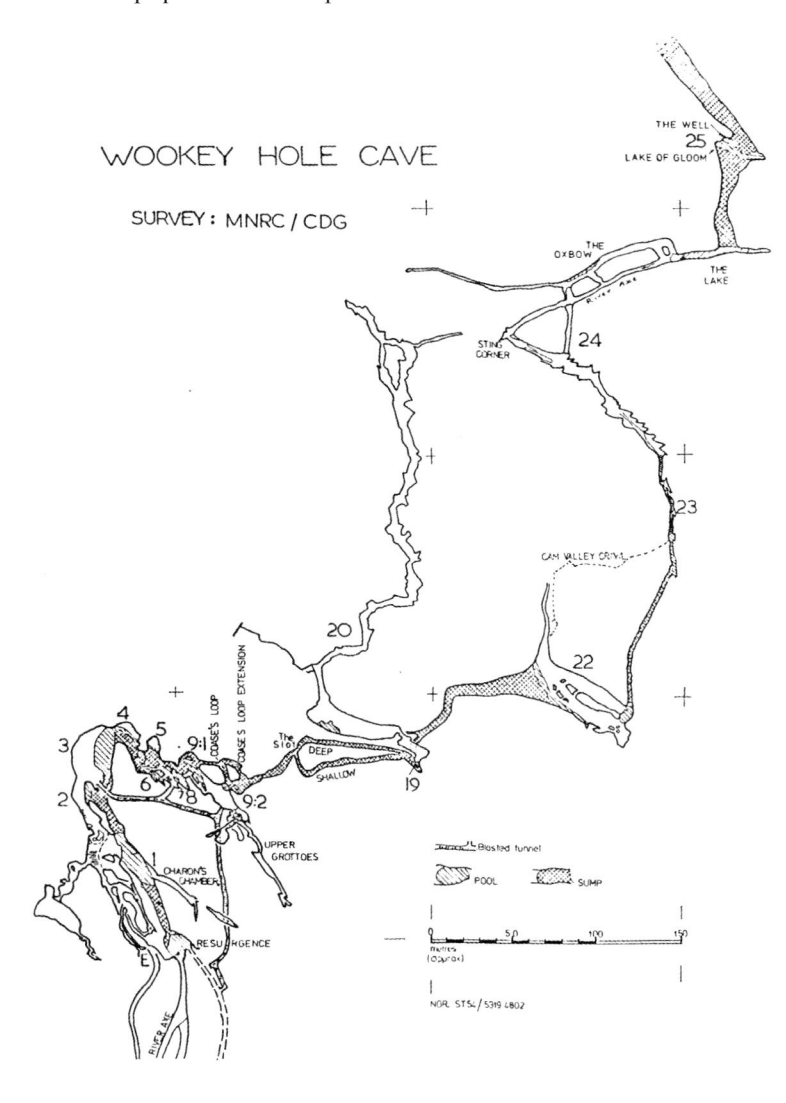

Fig. 15.9
Dave Irwin's and Tony Knibbs' illustration of Wookey Hole Cave in *Mendip Underground* (1977)

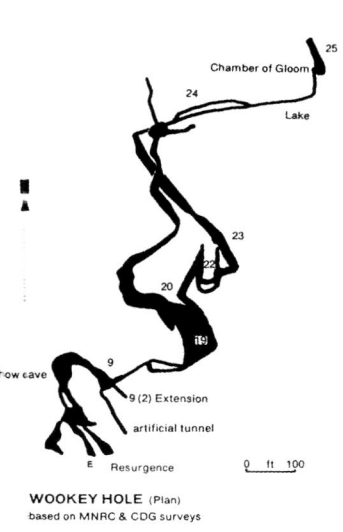

Fig. 15.10
Survey of Wookey Hole Cave published by Wessex Cave Club in 1985

Twenty, the centre of Twenty-two, and a chamber near Sting Corner in Twenty-four. Sting Corner was found to be no less than 140 metres (460 feet) east of its 'surveyed' position. As Brian described in the previous chapter, the precision of the radio location method was verified when the Twenty-two fixed point was located a second time, 2 years after the first, at exactly the same position in the field above.

Since 1976 the divers have added some small but interesting extensions to Twenty, Twenty-two and Twenty-four, bringing the total length of the Great Cave to more than 3,600 metres (11,800 feet)[25]. The extensions figure on a small-scale copy of the museum's plan that appeared in the Wessex Cave Club *Journal*, Vol. 18, page 44 (1985). The only currently available section of the whole system is a rather diagrammatic adaptation of Stanton and Crowther's 1975 section that is printed on hand made paper and sold to visitors in the Paper Mill.

Fig. 15.11
The large scale plan of Wookey Hole Caves prepared by Stanton and Crowther for the new museum in the Paper Mill, 1980

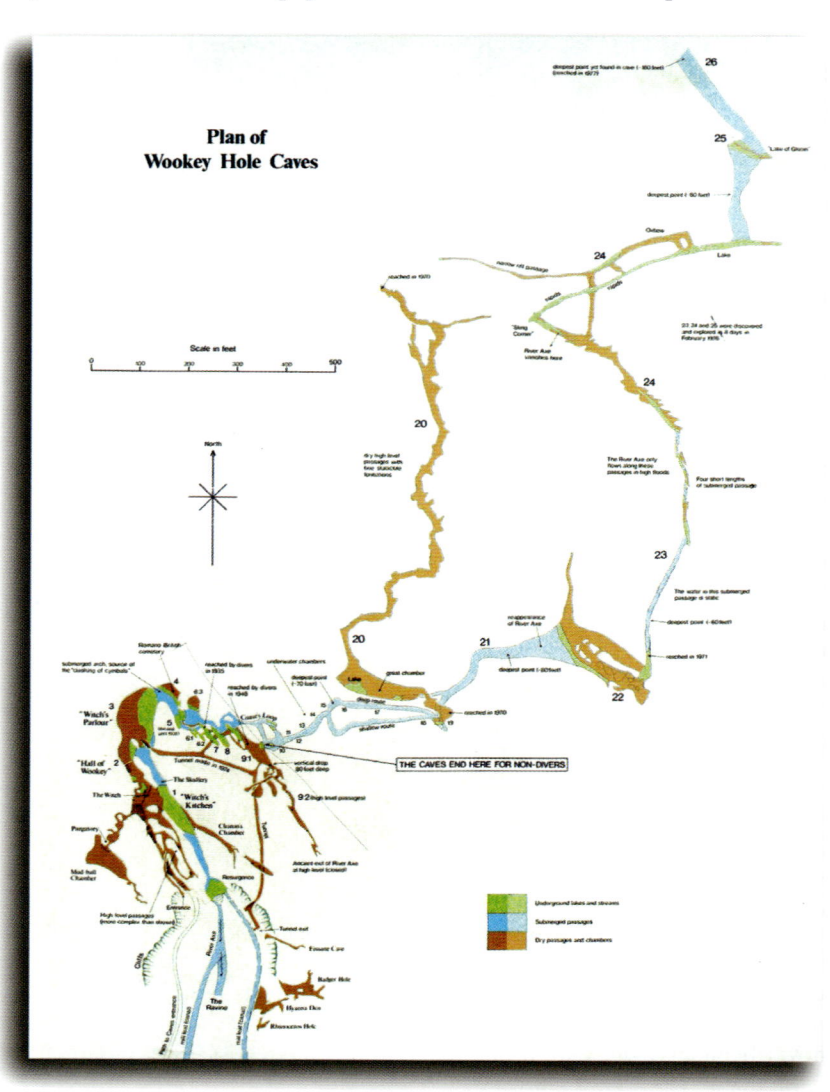

[25] The present extent of the cave is now over 4 km (13,000 ft).

Chapter Sixteen

Photography and other Finds

by Peter Glanvill

Photo by Peter Glanvill

I have from an early age had a fascination for caves and can remember browsing through the copy of *Pennine Underground* my grandfather kept at his retirement home in Arnside, Cumbria. There were no other cavers in the family, my father having leant initially towards diving in the mid 1950s. He trained with Captain Trevor Hampton in Dartmouth (much later I learnt Trevor had taught Steve Wynne-Roberts and Luke Devenish as well) and I was lucky to enough to be allowed to do the training at the tender age of 14. I have dived every year since but a year or so after completing my dive training my interest in caves blossomed. After peering into White Spot Cave in Cheddar Gorge on a school natural history trip I persuaded my father to take us caving. Armed with a copy of *The Caves of Mendip* we decided to find an easy cave to start on and so arrived in Burrington Combe one Sunday afternoon to explore Goatchurch Cavern. We were hooked and slowly built up our experience in the Mendip cave systems over the next few years, eventually joining a caving club – the Cerberus Spelæological Society (the secretary Jack Hill was an ex-diver known to members of my father's diving club). Cave diving seemed to be an arcane activity fraught with danger and requiring extensive training so it wasn't until I had settled into General Practice in the late '70s that I decided to take it up seriously. Prior to that I had had one shallow dip in a flooded Devon cave and supported my father on his brave and epic (because of its constricted nature) attempt to dive the sump in St. Dunstan's Well Cave.

I began my cave diving training under the supervision of Oliver Lloyd in that cradle of cave diving, Wookey Hole, and quickly realised I felt comfortable both under water and under the rock. An encounter with Martyn Farr on an expedition to Iran in 1977 gave me the opportunity to dive in Welsh caves but practical reasons (a restrictive on-call rota and a wife at home) led me to concentrate my activities on Wookey Hole. Over the years I have had to fit my diving and caving activities around work and family commitments which explains the gaps between some of my exploratory ventures. This juggling of work/family/interests has never been easy but I am proud to say I now have two adult daughters who, uncoerced, share the very same interests and a wife who has stuck with me for the past 30 years despite her contact with the tragic side of cave diving.

Fig. 16.1
Peter Glanvill in Wookey Hole car park, 1980.
Photo by Ian Caldwell

In December 1978 I made my second cave dive which consisted of a short trip towards the Third Chamber from Wookey Nine; the opposite direction to the usual way of going upstream first. After more than fifteen years of caving and diving I had managed to fuse the two activities. However, it was another year before I was able to get going again when I returned to Wookey in March 1980. It was that year that my diving activities took off. By the year's end I had visited Wookey Twenty and Twenty-two, dived in Ogof Ffynnon Ddu, Dan-yr-Ogof, Pridhamsleigh Cavern, White Lady Cave and to Swildon's Twelve. I began to think about specific projects within my scope. Photography took my fancy. I had developed a reasonable competency in cave photography and felt I might have a contribution to make.

Photographs of Wookey beyond the Ninth Chamber were quite rare in the 1980s. One of the main deterrents to photography there was that camera containers need to be well waterproofed to pass the deep sumps. Underwater cameras were expensive and the associated equipment bulky. Some nice black and white pictures of Wookey Twenty had been taken by the original explorers, but they were few and far between and beyond here photography was a rare event. The first person to do any serious and presentable photographic work in the far reaches of the cave was Martyn Farr who had managed to obtain pictures both above and below water using a Nikonos underwater camera and bulb flashes with slave units.

Fig. 16.2
Bob Shepherd in
the Wookey Resurgence,
May 1985.
Photo by Peter Glanvill

Martyn helped me with my first pictures of Wookey in May 1980 when he kindly accompanied me to Wookey 20. At that time leaving the water in 20 was regarded by the majority of divers as a tedious diversion possibly because de- and re-kitting in the cramped conditions are so fiddly. One emerges after the long swim from Chamber Nine in a flooded chamber only a few metres long and wide. De-kitting is accomplished by standing on a submerged boulder and placing kit on a narrow ledge in boulders before one climbs out and upward between the boulders and roof. I was most impressed with what I saw of the huge Chamber 20

Fig. 16.3
Jim Durston, Chamber 20.
Photo by Peter Glanvill

with its fluted rock walls, sandy floor and clear blue lake. I took a few poor colour slides and impatiently planned my return. My next photographic trip was to Wookey 22, which contains an impressive inclined rift sloping down to a sump pool in which one emerges from 21. This section of the cave is always notable to me for its conglomerate beds including a striking pillar which seems to support a large proportion of the roof. The sump pool leading to Wookey 23, which fluctuates in depth by several feet, lies in a large, lofty and potentially photogenic rift. Tony Boycott gave me a hand but, even with fast 400 ASA colour slide film in the camera, the electronic flash and slave were not up to the size of the place. A visit to Twenty again early in 1981 with Jim Durston yielded some interesting pictures including one of Jim doing a bit of original exploration by squeezing into a rift we had dug at the far end. More work was done in 20 a few months later and the photographic results of this trip yielded two pictures for the museum at Wookey Hole including one of the extraordinary fluting on the walls of the big chamber. By then I had started using Magicubes for extra light – for those living in the 21st century these were mechanically ignited flashbulbs which were mounted on cheap snapshot cameras. Magicube firers made from aluminium blocks with a prong in the base allowed them to be triggered manually. On this occasion I was assisted by Chris Milne, Jim Durston and Keith Potter.

Fig. 16.4
Keith Potter in Wookey 20,
4 April 1981.
Photo by Peter Glanvill

Later in the year, however, tragedy was to strike. On the afternoon of Saturday 14th November 1981, Keith Potter inexplicably drowned when diving to Wookey 20. Over thirty-two years had elapsed since Gordon Ingram-Marriott's death in the cave. This time, Wookey Hole Caves claimed an experienced and enthusiastic young caver. Once again, exactly what went wrong remains a mystery. Keith came from nearby Wedmore, had been in the Scouts and became a very active cave explorer whilst attending the Kings of Wessex School, Cheddar. He was a talented scholar and went up to Exeter College, Oxford, to read Medicine. At the time he was 22 years old.

Fig. 16.5
Memorial plaque to Keith Potter at the bottom of Pozo del Xitu in the Picos de Europa, Spain. Photo from Oxford University Caving Club

Martyn Farr, Ray Stead and Keith Potter planned a dive to Wookey 24. I had been due to make the dive but after stupidly lacerating my hand 10 days previously and sporting a collection of stitches, found my role limited to base support and the supervision of trainee divers Rob Parker and Julian Walker. Having already been once before to 20, Keith was given the benefit of the clear water and chose the Deep Route from the Ninth Chamber base. Martyn followed along the Shallow Route after a short while and found Keith stationary but some 12 feet below the surface of the sump pool entering Wookey 20. Somehow he had lost his gag. Martyn rapidly landed him and started resuscitation straightaway. Ray soon arrived to help, and about a couple of hours were spent trying to revive Keith – a grim and fruitless task in a location such as this. Eventually Ray returned to raise the alarm and Martyn later followed having managed to single-handedly manoeuvre Keith's body through the flooded passages between Wookey 20 and the Ninth Chamber. The following day Tony Boycott and I retrieved Keith's equipment from the 20th Chamber, finding it to be functioning normally, a fact confirmed by later detailed inspection, making the mystery of his death even greater.

At the Inquest later, Mr. Fenton Rutter, the East Somerset Coroner recorded a verdict of accidental death. In giving his conclusions, the Coroner noted that explorers throughout history had taken risks, and that the world would be a poorer place without them. Keith's parents graciously requested that any donations should be given to the Mendip Rescue Organization; a particularly apt gesture, since Keith himself whilst even a schoolboy had expressed a keen interest in rescue work, and had been invited to attend MRO meetings at Priddy by Jim Hanwell. Subsequently, MRO was able to up-date its sump rescue equipment by replacing the pioneer Normalair apparatus with a new Kirby Morgan Bandmask system; one which largely overcomes the problem of inadvertently flooding the mask by maintaining a positive pressure of air inside. It would not have saved Keith Potter in Wookey Hole Caves that day but is now available if needs be, of course. Naturally we hope that it is only ever used in training, for the recovery of an injured cave diver through a short sump is difficult enough let alone from the far reaches of Wookey Hole Caves.

My first dive to Wookey 24 had been in April 1981 with Tony Boycott and Jim Durston in fairly low water conditions. At that time it was still rarely visited and even more rarely photographed. One reason for this is that although the diving is not tricky, a considerable amount of bottle carrying is required between the landing point in Wookey 22 and the static sump pool that leads to Wookey 23. My first experience of the dive from 22 to 23 was marvellous as I had inveigled myself to the front of the team. In a static sump the water can attain an awesome clarity which unfortunately is lost after the passage of the first diver. Most of the sump consists of a large bore tube from the floor of which mud banks rise up the walls. The mud lies all over the sump and emerging from the pool in low water is a tricky affair. Soon we were all quickly slithering our way down the high muddy rift of 23 to the grotty series of sumps which lead to 24. Although not long these sumps are low and awkward to get in and out of, particularly with tackle, and the last one is especially difficult.

Fig. 16.6
Dany Bradshaw tries on the Kirby Morgan Bandmask, 1985. Photo from MRO photo archive

The first sight that greeted us was a Union Jack (thank Dave Morris for that) and a bottle of wine put in exclusive storage by Martyn Farr. After the de-kit, 24 begins as an impressive dry passage leading to a large chamber. A feature here and throughout 24 is a black 'tidemark' on the walls about 12 feet up. The rumble of the streamway led us along a traverse to a natural weir where the stream disappears down a rift, not to been seen again until 22. From here on is a photographer's paradise. Plenty of superb and varied streamway features exist: natural weirs, a canal, rapids, cascades, a very unusual and bypassable rift containing deep fast flowing water, and a big deep terminal lake. We rushed through the cave to the lake before returning in leisurely fashion through the high-level oxbow which lies 25 feet or so above the streamway. Several photographs were taken but I was still not satisfied; I needed a tripod as well as a more waterproof container for my photographic gear. The trusty ex-W.D. ammo box I

Fig. 16.7
Chamber 20. Photo by Peter Glanvill

Fig. 16.8 Jim Durston approaching the slot between Wookey 9 and 20. Photo by Peter Glanvill

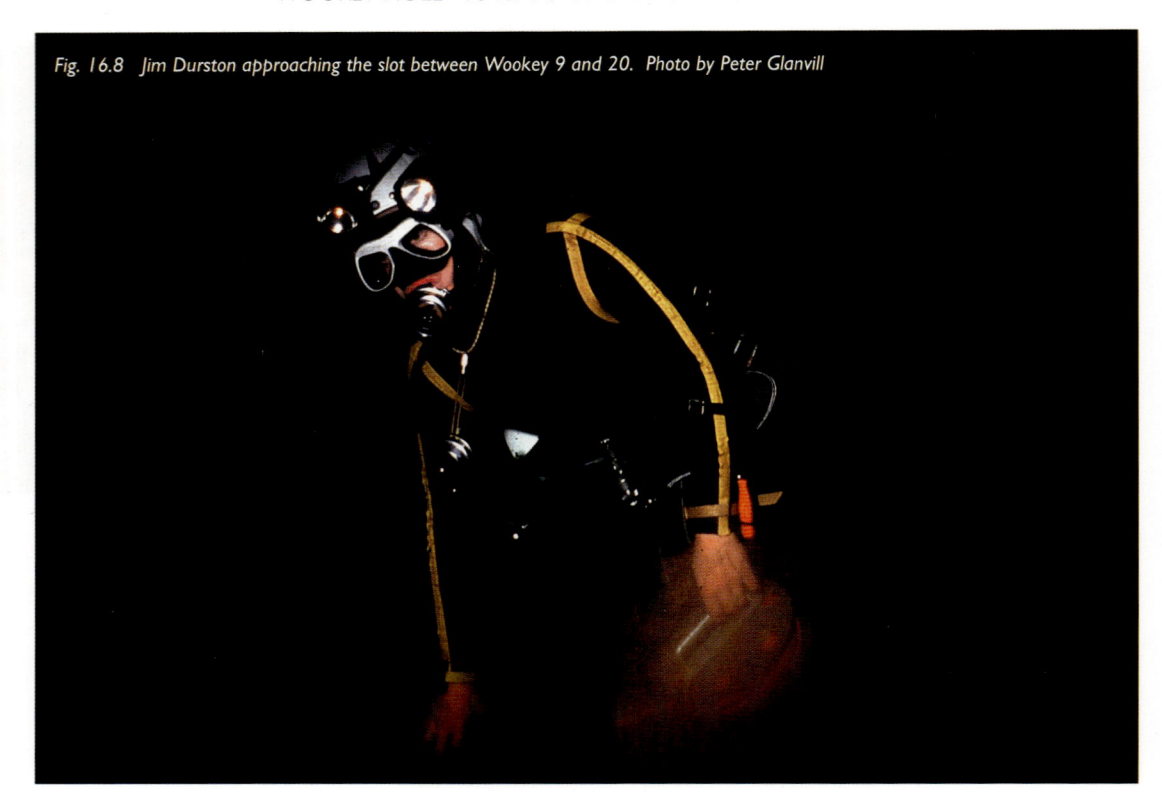

used collapsed in spectacular fashion on the return trip having negotiated one deep sump too many. Subsequent photographic trips were made carrying camera and flash in ex-Army rocket tubes.

Tony and I returned less than a month later. It was a memorable occasion. Not only was the show cave crowded as it was a Bank Holiday but also heavy rain had swollen the underground river and rendered the visibility extremely poor. Tony dived first and after a few minutes I followed him. The struggle up to 20 was considerable, particularly with an extra bottle and kit bag. The current is noticeable here at the best of times and, in the light of my Aquaflashes (pre-LED diving torches), particles swirled dizzily past me as I almost crawled up the steep bedding towards 20. Tony and I later agreed that if either of us had turned back we would not have been very surprised. At 22 the second sump pool was 10 feet above its normal level and the water extremely murky. We found 23 to be virtually a canal, except for a patch of high ground near the far end, and a large shower bath entered it towards 24. At 24 itself, a thin stream was beginning to flow out of the sump and water in the big chamber was backing up on the floor from the stream sink.

We gained the start of the streamway. It was an awe-inspiring sight; the Axe was thundering down into the rift, something like 3 feet deeper than our previous visit. Tony attempted to enter the water and make some headway. We knew a haul rope lay here which ran up to the cascades beyond Sting Corner; unfortunately it was invisible, lying under a raging torrent. The original explorers had not bargained for quite such desperate conditions and had belayed the near end of the line in the stream bed. Tony's attempt failed but, being larger and heavier, I had a go. By sliding along the left hand wall where the current was weaker, I was able to get to a point where I could make out the haul line, and a lunge across the streamway enabled me to scoop it out of the water and make it accessible to Tony. At the canal we could only make progress by hauling ourselves along the line. Its upstream end, where there are rapids, was a seething mass of white water. We clambered onto a precarious boulder where Tony declined further progress which involved a wide straddle over the torrent, here compressed into a rift only 3 feet or so wide. I made my way up into the oxbow where I found the descent to the terminal lake a complete impossibility. We had to scrap our plans to dive a static pool at the far end of the lake. Returning to Tony, I decided to photograph the torrent. Holding the camera and flash gun whilst perched on a tiny boulder, I told Tony to jump in and float downstream whilst I took the picture at a judged moment. The resultant photograph may not be artistically perfect but it does capture the violence of Wookey 24 in high water. The return from that trip was almost marred when I dropped my diving mask into the stream. Thankfully, Tony managed to rescue it so I didn't have to discover what it's like to dive about a thousand feet of sump without a mask!

In late October 1982 Martyn had begun his assault on Wookey 25 (described in chapter 17) and a porterage trip to 24 gave me an opportunity to do some photography in 22. Ray Stead kindly assisted me in my photography which, by then, involved carrying a tripod! I dived early from 9 with the rest of the party seeming to be a long way behind. As each member surfaced in 22 they seemed most resentful of my attempts to photograph them. It was not until I opened *The Observer* newspaper the next day that I discovered that they had spent some time in the 9 sump pool posing for another photographer. I suppose it was a bit much asking someone to pose for pictures at both ends of a long dive! Ray and I had a very successful photographic session which covered most of the photogenic parts of 22 as well as enabling me to check on the location of a long-lost passage only visited once 11 years previously. I also explored a tight little oxbow near the landing point in 22.

*Fig. 16.9
Tony Boycott battling high water conditions in Wookey 24.
Photo by Peter Glanvill*

A month later Rob Parker, Martyn Farr and I returned to 24 to pick up the remaining equipment left from the assault on 25. Water levels were again high and I had the satisfaction of being able to tell the other two at 22 how conditions were going to be in the 24 streamway! The 22 to 23 sump was a murky haze, whilst 23 was again a canal. In 24, a tiny stream was flowing out of these sumps towards Sting Corner. Martyn had brought his camera this time so there was a bit of competition between us over where the model (Rob) should stand. The other two were suitably impressed by the sight of 24 in flood. We set off upstream to the underground camp site in the high-level oxbow where the diving team had stayed overnight during the attempt on 25. Martyn rushed about with camera and flash until, with all the photographic work done, we started to lug the kit out. It comprised some large and heavy 105 cubic foot diving bottles and several waterproof containers which were long cylindrical tubes used by the military to store and carry rockets. Martyn advised extreme caution with the bottles when carrying them near the streamway; they were not buoyant and a fall into the river wearing one would have been fatal. We also decided that an accidental descent of the weir at the end of the streamway would have doom-laden consequences.

The carry was going well and we had moved everything to a precarious perch at Sting Corner when I set off down the streamway, haul line in hand and kitbag on my back. As mentioned earlier the downstream belay of the haul line was a boulder in the stream bed and this proved to be my undoing. The technique for leaving the water at the weir was to anticipate the point of departure and step off, rather like an escalator. Unfortunately, I found my hands following the line to its belay in the streamway whilst I straddled the weir. The sensation of sheer unalloyed terror that hit me then was the worst I have ever had. The River Axe relentlessly backed up behind me until it was surging around my back and shoulders. I knew with awful certainty that I was going to go down the rift, and when I did that I would drown. Inch by inch I was pushed over towards the roaring hole until I was gone. A few seconds of black confusion ensued as I was spun over underwater. The survival instinct made me struggle upwards and I broke surface in the rift sans spectacles with my kit bag floating beside me, but alive and unharmed. Martyn and Rob appeared anxiously at the top of the rift. 'All right Pete?' they shouted. I called back to say that I was, and struggled out of the rift. My survival was due to the mill pond effect in the rift; the water was deep enough for the current here to be much weaker which allowed me to surface before being swept into the sump 18 feet away.

Somewhat chastened and rather blind, for I am extremely short-sighted, I helped the other two continue the carry through the dry section to the next sump. With my diving mask on I could see again (prescription lenses) and Martyn set about distributing the gear. I was given a kitbag containing several very buoyant rocket tubes and I realised I was going to have fun as soon as I entered the sump. A wrestling match with the hyper-buoyant kitbag ensued; it was only defeated by my walking through the sump upside down whilst pulling hard on the bag. Several breathless struggles later plus a tangle with the line in zero visibility and I was in the last airbell before 23. I checked my pressure gauge and was alarmed to find I had used virtually all of my air.

Rob came past and I handed my bag to him extremely thankfully before leaving the sump. This incident resulted in a redistribution of kit before the 23-22 sump which is much longer and deeper; I did not fancy a struggle through it with depleted air reserves. The dive went well although halfway through the bag became mysteriously easier to handle. After this the exit from 22 was a delight with the strong current virtually flushing us out of the cave. Once outside, in our tired state, we backed the car all the way up to the entrance rather than carry the extra kit down to the car park in the darkness.

As can be imagined after such a trip I needed a drink. However, I had not brought my spare spectacles! This explains my entry to the Hunters' Lodge Inn that night wearing my diving mask – there were hoots of derision from the usual crowd! Needless to say I had to drive the thirty miles home wearing the self-same mask and praying I was not stopped by the Police. Can you imagine anyone believing me if I had said I had lost my spectacles when falling down a hole in Wookey 24?

There is a postscript. Martyn rang me a week or two later to ask if I had carried out his 'Thinsulate' suit. I denied any knowledge of its whereabouts and it was not until Geoff Crossley was swimming through the 22-23 sump some weeks later that the mystery was solved. There bouncing about on the roof of the sump was a black cylinder; the missing rocket tube containing the suit. It had broken out of my bag during the dive and escaped in the poor visibility.

Apart from my photographic endeavours my contribution to exploration in Wookey Hole Caves has been to help survey and search for ways on that might have been overlooked. After all, John Parker had found upper dry routes bypassing the earlier sumps. It is equally possible that there are others in the further reaches of the system. They could be the key to the way on.

Most of the publicity featuring exploration in Wookey in the last few years has focused on the attacks on the terminal sumps. But this is not the only part of the cave where progress has been made. At the time I started diving regularly, upstream pushes had almost ground to a halt. Then Rob Harper and Trevor Hughes started to get interested in the climbs in 22 and 24. Tony Boycott and I noted an interesting site at the far end of 20 where a rift led off only to be choked by mud and stal. A return visit with Jim Durston and some digging tools a few weeks later and we were in business. Half-an-hour's digging revealed a descending squeeze which was bravely forced by Jim to a short continuation and the head of a narrow vertical drop which would require 'chemical persuasion' to pass; in other words, we needed explosives. Closer inspection of the stal blockage higher up the rift revealed a slight draught and the presence of several small but interesting holes. The next trip was a thin man's team consisting of Jim, Ray Stead, Tony Boycott, and myself as the obligatory fat man. Whilst Tony had a look at our little find, Ray and myself wandered off into some nearby boulders, finding them extremely unstable until we were called back when Tony got stuck. Having extricated him, Jim, the eternal punster, complained it should have been Ray so that he could have been Steadfast! We showed him the holes in the stal and he wriggled through but found them all to close down. And that is the way things stand there, although this passage lies very close to a similar rift in 24.

Fig. 16.10
John Parker's extension off Chamber 20.
Photo by Peter Glanvill

Fig. 16.11
Simon Brooks by the big boulders in the lake chamber of Wookey 20.
Photo by Peter Glanvill

Fig. 16.12
Ray Stead at the downstream end of Chamber 22.
Photo by Peter Glanvill

Fig. 16.13
Simon Brooks in Attila the Hun's Sardine Cannery, 1983.
Photo by Peter Glanvill

The next area to attract my attention was 22 where, at that time, the Harper-Hughes team was climbing a high aven in the hopes it might link with Hallowe'en Rift, a surface dig on the hillside above. In an eleven-year old diving report, I found a reference to the original visitors to 22 having explored a long rift passage which they got fed up with and left. It appeared to be entered at the opposite end of the pool in 22 where one normally exited, yet was marked on the sketch surveys as leaving it in the roof. On my first trip to 22 I looked at the top of the rift but, with a dim light and lack of enthusiasm for a muddy traverse into what looked like a dark, bouldery corner which closed down, I ignored it. It was only when, with five minutes to spare on my photographic trip in October 1982, that I got right to the far end of the sump pool and realised there really was a passage there. Unfortunately, nobody else seemed very interested at the time so I decided to visit the site again myself as soon as possible. I wrote to Trevor Hughes suggesting we took a look, but he couldn't make the date. A few months later I was in the Hunters'. He greeted me with a big grin and the comment, 'You're going to be really sick!' He and Rob had decided after a climbing trip to take a look at this passage and, in doing so, had found Cam Valley Crawl, the dry route to 23. Trevor deserves a lot of credit for finding the Link Crawl and accomplishing the climb down into 23 which I would have definitely left well alone!

Jim Durston and I did the hard part of the 22-23 link when we volunteered to survey it. I earned Jim's undying enmity when I decided to take my camera and tripod. Mixing photography and surveying in low grotty crawls does not work too well. We managed to just about do the survey but the pictures were mediocre. Ray Stead returned with Jim at a later date to tidy up the loose ends. Several cavers have since had good reason to curse Cam Valley Crawl for it is not an enjoyable alternative to diving a sump! Incidentally, when Clive Westlake and myself (assisted by Simon Brooks and Rob Harper) surveyed 22 properly in April 1984 I was galled to find that, if I had looked rather more carefully two years earlier, I could have entered the crawl then as a route exists from the top of the inclined rift in 22. This cuts out the tricky scramble directly up from the sump pool. Incidentally whilst surveying 22 we also found an interesting and previously unremarked oxbow which acts as a high water overflow from Wookey 22 into the static sump.

Having been thwarted in 22, I became a little more secretive. I wondered if the Twelfth and Thirteenth Chambers, scene of Bob Davies's escapade in 1955, warranted closer inspection. A few people had tried before me but took Davies's route and encountered problems getting out of the sump into the Thirteenth Chamber. However, Rob Harper reported a possibly promising aven. In October 1983 I laid a line in the direction of the Twelfth Chamber and discovered it to be an easy 80 to 100 foot shallow dive from the Ninth Chamber. De-kitting was tricky so was left for another occasion. I later learnt that nobody else had ever bothered to approach from what turned out to be a much easier end.

Meanwhile, Rob Harper had been busy in Wookey 20 finding Attila the Hun's Sardine Cannery. I persuaded Simon Brooks to come and help me survey and photograph what was a wet suit shredding bedding crawl.

In February 1984 it was back to the Twelfth and Thirteenth Chambers with a small party. After photographing all the passage, which was very cramped, we examined Rob's aven. In doing so we discovered a loose rock problem. A squeeze at the bottom of the aven seemed to promise great things and I returned two weeks later with Clive Owen and a lump hammer. Unfortunately, the squeeze led only to a small chamber and a climb linked with the aven.

I still had one more site I wished to examine. There was a three line reference to a find made by Colin Edmunds in December 1975 in an old CDG *Newsletter*. He claimed to have entered a rift unlike 22, just before 22 is actually reached. The rift was never revisited because, only two months later, he and Martyn Farr found the way on towards Wookey 24. In August 1984, conditions were perfect in Wookey following a long dry spell. After a successful underwater photo session with Jim Durston and Clive Westlake, I set off into 21 with Clive and armed with a line reel. At the point where I thought Edmunds Chamber (as I had already dubbed it) led off, I tied on the reel and set off into the darkness. Ahead, a wide rift led on and up until I quickly broke surface in unknown territory. Edmunds Chamber had been relocated after a lapse of 19 years. The rift was large – at least 50 feet high, 15 feet wide and 70 feet long. There was no dry land, the rock walls sinking uncompromisingly into the depths. At one end a huge boulder rested across the rift. After a quick look round, the pressing problem was to find a belay for the line reel and this was eventually done before I returned triumphantly to a slightly concerned Clive. Several return visits were made over the next few years to climb the chamber walls. This climbing venture was complicated by the lack of solid or level floor necessitating the attachment of every piece of bolting or climbing equipment to the climber or belayer.

After Brian Johnson and I had started the climb Rob Harper brought his expertise to the fray. To reach the roof we had to leave the water via a sloping ledge that Brian dubbed the Ramp of Masada and then traverse across the chamber on another angled ledge to the far end. Rob spent some time doing this, dislodging huge slabs from the rotten rock walls which fell into the sump pool with deafening crashes. He finished the traverse but was convinced that the passage was no more than an alcove and unworthy of further attention.

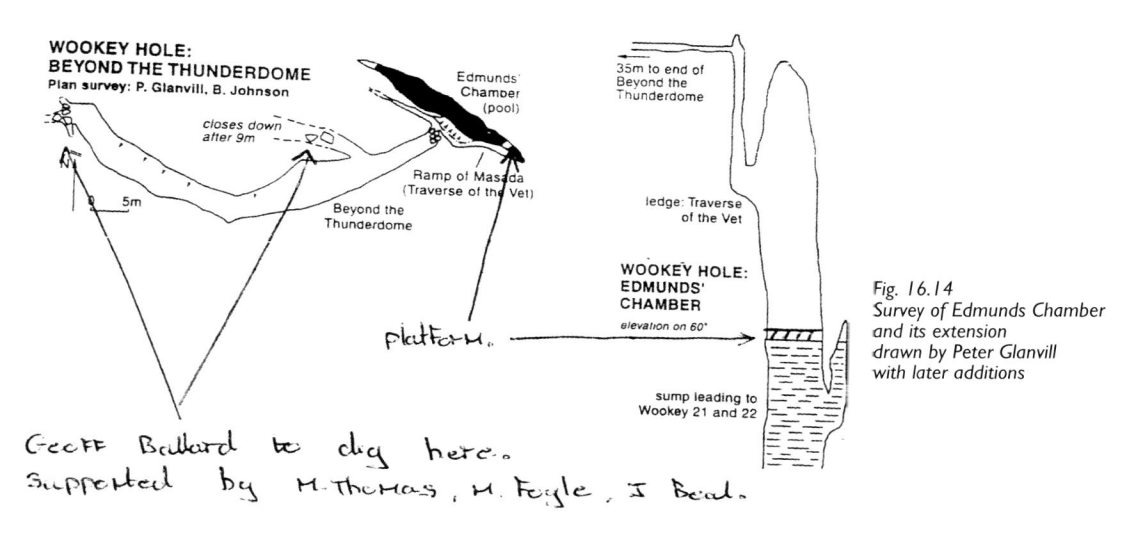

Fig. 16.14
Survey of Edmunds Chamber and its extension drawn by Peter Glanvill with later additions

However Brian and I returned 15 months later in March 1986 determined to finish the climb. The trip was dogged with problems not least of which was Brian losing his fins! This time I was lead climber and, after placing my first belay bolt at the far end of the Traverse of the Vet (as I had dubbed it), I contemplated the climb in front of me. Although not a technically difficult move there was a considerable sense of exposure before I stepped out and into a chimney formed by a flake of conglomerate. A few awkward thrutches and I had reached the roof. There in front of me was open passage. Pausing only to place a final bolt for the return I scuttled off along scalloped elliptical bedding that, after 50 metres, ended in a potentially diggable choke. One final trip a couple of years later was made to photograph and survey the site before it was again left to the darkness. Its name Beyond the Thunderdome alludes to a popular contemporary film and the sound that rocks made falling into the sump pool. Beyond the Thunderdome may be worth further attention although digging here would certainly be in a remote location. However there is a draught and it runs directly towards Chamber 20.

Finally in 1986, scouting through old reports I found another site that hadn't been surveyed – a series of passages off Wookey 20. They had been visited only once by Paul Whybro and Brian Woodward and were reached by a rope climb not far from the lake. Brian Johnson and I decided to photograph and survey the series. Unfortunately the tight bedding squeeze into the extension defeated Brian and I managed it only by removing my wet suit (not an experience I would repeat). The series of passages beyond were in Dolomitic Conglomerate, draughted slightly and had several dig sites. To my knowledge only three individuals have been here and I dubbed the passages the 2 W's Extension after the original explorers. This series beckons to visitors wishing to extend Wookey but who don't want a series of long dives.

My cave diving activities were curtailed by my wife's anxiety about the trips and my trips into Wookey had ceased by the early 1990s. I am eagerly waiting for a new generation of digging divers to extend the cave further and, maybe, find that dry link back to Chamber 9!

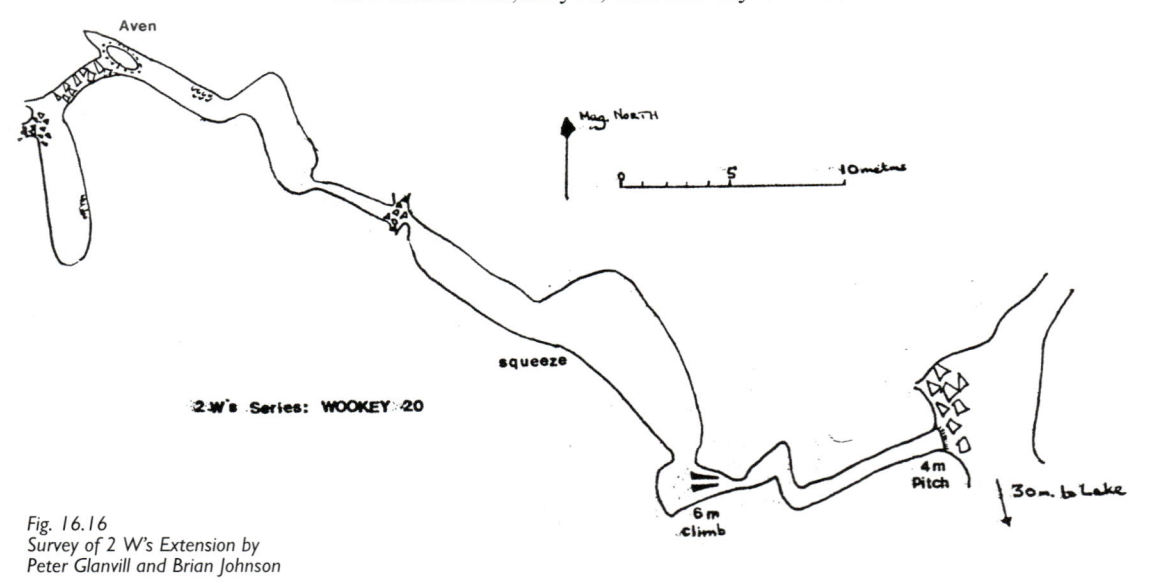

Fig. 16.15
Sting Corner, Wookey 24.
Photo by Peter Glanvill

Fig. 16.16
Survey of 2 W's Extension by Peter Glanvill and Brian Johnson

Chapter Seventeen

Farthest by Farr

by Martyn Farr

Fig. 17.1
Martyn Farr 1977.
Photo from Martyn Farr collection

It had been a hard slog and we were both hot and weary. Chamber 24 was tremendous: battling against a strong flow of the river, traversing along narrow rifts, swimming, climbing, and finally a relatively easy amble along an abandoned oxbow to this, the next sump.

Colin Edmunds was seemingly deflated, possibly not so much by the hardships of the past three hours leading to the head of Chamber 24 but more because his moment of triumph, his discoveries, had been unceremoniously snuffed out. Four days ago he should have broken through from Chamber 22 and Chamber 24 would deservedly have been his. But all that had now to be forgotten. By contrast my enthusiasm was fired by the thought that the next chamber, Chamber 25 somewhere beyond this deep blue lake, could well be an even bigger breakthrough, the one that would ultimately connect with the caves of Priddy and establish one of the finest through trips of all time. The half mile of passage discovered by Oliver Statham and Geoff Yeadon was rightfully ours but, rivalries aside, I viewed it as a stepping stone; a key breakthrough leading to even better things beyond.

From a letter left by our northern colleagues we fully expected a big sump, but probably one that could be passed relatively easily. So, for the last hour we had carried kit through the winding rifts and rapids of Chamber 24, kit sufficient for just one diver. The honour of the exploration ahead was now mine.

Set at the end of a long spacious rift, the sump pool confronting us was a deep tropical blue. It was an inviting prospect. Leaving Colin sitting astride a large boulder I slid into the water and pulled my way against the current. Within a few metres a steep descent loomed from under the left hand wall. By 10 metres depth positive buoyancy had been corrected and I swam comfortably down a sizeable tunnel floored with sand and coarse silt.

With the aid of Colin's beam gun torch and the tremendous visibility the sides of the passage were easily checked; there were no side leads. At about 15 metres depth, I stopped to clear my ears which were just becoming painful. In the process the hand held beam gun, which was positively buoyant and attached to my wrist by a short lanyard, became entangled in the diving line. This was the first occasion I'd used such a big powerful light to supplement my single NiFe cell and of an instant all the advantages had been lost. It took several minutes to sort out the twists and wraps; very frustrating and annoying.

By the time I was moving forward again precious air reserves had been depleted and, on reaching the elbow of the sump at 23 metres depth, one third of the air in my first bottle had been used. I pushed the reel into the silt and returned to 24 to brief Colin and change over valves and cylinders. This done, I quickly returned to the deep point, picked up the reel and started on the ascent.

The passage was now a muddy bedding plane of indeterminate width inclined at a very steep angle. Expired air rushing up the slope had the effect of destroying the visibility well ahead of me, and it was a rather unnerving experience negotiating the odd slab or detached boulder in a premature cloud of mud.

Up, up, up; my head began to swim with that heady feeling experienced on surfacing. Then, suddenly, a silvery surface appeared to one side and in an instant I was there, floating in Chamber 25. A spontaneous assessment spelt no spacious chamber, no open streamway and generated little elation. Disappointingly Chamber 25 was a muddy, desolate airbell and immediately christened The Lake of Gloom. There appeared to be no dry ground and, therefore, nowhere to land. The only obvious way on was over a sort of low wall which reached about one metre above water level. As the floor was steeply angled my hands were only just able to reach the top of this and with effort I was able to hoist myself up and peer into a similar pool next door, The Well. Short of climbing over this obstacle there was no easy access to the adjoining water, which on reflection was clearly the way on. The lake that I had surfaced into was now heavily muddied whereas the water in The Well was clear and deep. The cave wasn't about to yield all just yet.

The line was secured to a submerged boulder and the exit commenced. In a silt-out I wandered too far to the left into a constricted area. Worrying moments were experienced until the elbow was regained. Beyond it was straightforward.

Before leaving Chamber 24 we checked out several of the more promising side passages. It was soon confirmed that the likelihood of any further dry extensions was slim. There were climbs and several tortuous sections of rift to push but seemingly 24 had already given its best.

Over the next couple of months all easily accessible dry leads were exhausted and diving resumed. It was quickly found that it was impossible to dive under the wall in Chamber 25 but that a muddy alcove at one end of the pool gave at least a comfortable patch of muddy ground above water level.

In July, supported by Dave Morris and Peter Lord, the first dive was made into The Well. Once again this was a solo operation beyond Chamber 24. A short section of passage was followed at 10 metres depth and suddenly I found myself following a sloping route down a narrow, steeply inclined rift. Forced ever deeper by the descending roof I reached 30 metres depth and realised that it was wholly unwise to continue. By now there was no indication whatsoever of the floor and the passage ahead gave no indication of levelling off. To continue would have been foolhardy, especially as I had no watch or decompression meter. Such a deep dive was completely unexpected and it was immediately clear that future dives would be very serious affairs.

Further progress beyond Chamber 25 required considerable preparation and a major assault was planned for the summer of 1977. The operation was structured upon the lines of an expedition and, while sponsorship was being sought and plans were being made, considerable work went on in the cave itself. Ropes and ladders were installed on short climbs and a swim, and equipment was stockpiled in the further reaches. Planning the actual push dive was itself a lengthy process especially as the potential hazards, decompression and narcosis, surpassed any so far encountered in Britain. In this respect invaluable assistance was given

Fig. 17.2
Dave Morris and Brian Hague in Chamber 22.
Photo by Martyn Farr

Fig. 17.3
The approach to dive base via the Oxbow, Chamber 24.
Photo by Martyn Farr

by Frank Salt. Frank had been closely involved with a very technical deep diving operation at Sinoia Cave in Zimbabwe and he finally came up with the idea of using oxygen/nitrogen mixtures on open circuit. For the dive as far as Chamber 25, cylinders containing 50% oxygen/50% nitrogen were to be used, thereby lessening the effect of narcosis. This mixture, however, could only be used safely to a depth of 30 metres. Below this limit a 30% oxygen/70% nitrogen mixture could be adopted, and would be safe down to 45 metres depth. At this point the narcosis and nitrogen absorption levels would be the same as when breathing compressed air at 40 metres.

Further, the use of pure oxygen, again on open circuit, was obviously beneficial for decompression. Apparently at shallow depth, i.e. 6 and 3 metres, oxygen could reduce the diver's stop time by a half. One always had to remember that to incur a bend at an isolated spot such as the further reaches of Wookey Hole could well prove disastrous. There was absolutely no question of being speedily evacuated to a recompression chamber and, as this was the first occasion in Britain when decompression was a major consideration in the dive plan, it was extremely important to do it right.

There was also the question of depth training, acquiring confidence and familiarity with the equipment that would be used on the final push. Mastery had to be achieved over the constant volume dry suit which had only very recently been introduced to cave diving. Not only was the neoprene suit expensive but it was also bulky and hot to wear; not easy to move in through a British cave. It also had to be treated with great care as a leak would prove extremely awkward and uncomfortable, especially on a long decompression stop in chilly water. There were, therefore, many facets to the Wookey problems necessitating a lot of time and effort before the dive.

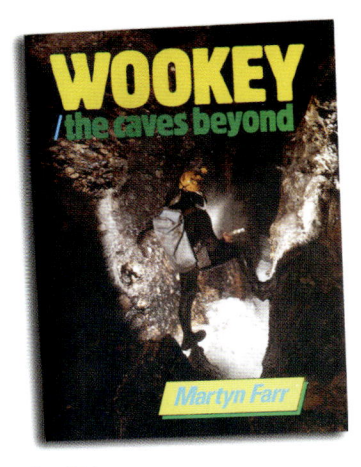

Fig. 17.4
Looking downstream in Chamber 24.
Photo and 'Wookey/the caves beyond' book by Martyn Farr

17.5
Brian Woodward, Richard Stevenson, George Bee, Colin Edmunds, Martyn Farr and Dave Morris, Wookey Hole 1977. Photo from Martyn Farr collection

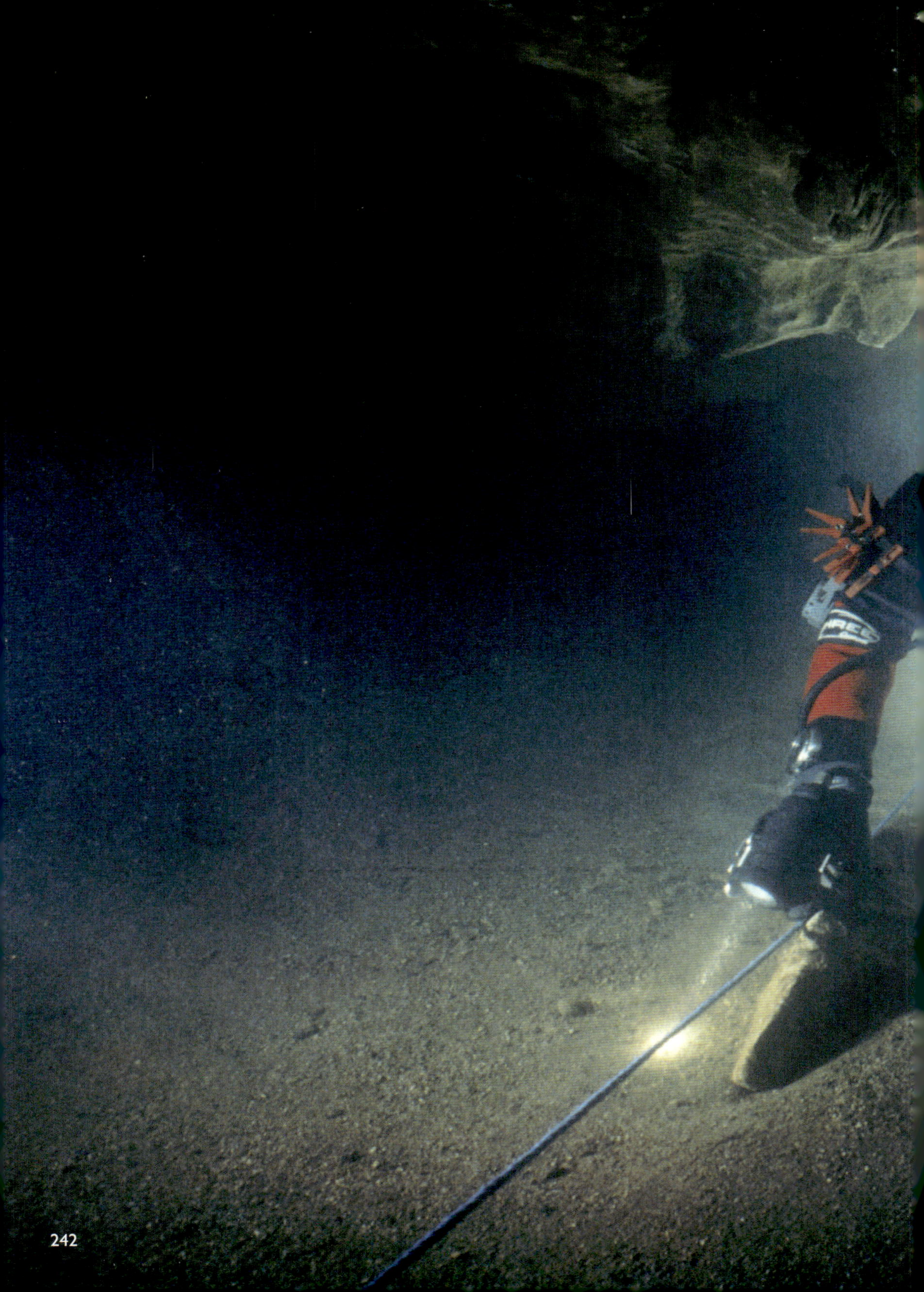

Helen Rider, Chamber 15. Photo by Martyn Farr

The 18th June 1977 was a tense day. Well over a year had been devoted primarily to this one project. It was such a relief to dive away from the Ninth Chamber, leaving behind the melée of noise and the glare of the powerful TV lights; great just to get on with the job.

There was plenty of support from people such as Paul Atkinson, George Bee, Colin Edmunds, Brian Woodward and Dave Morris. For once, fortunately, I had little to carry along the strenuous sections en route to the head of Chamber 24. Here the final preparations commenced. Dave and I organised ourselves ready for the penultimate dive. By now the sump to Chamber 25 was familiar to us both and, despite the fact that several items had to be taken through, the passage of this section also went uneventfully.

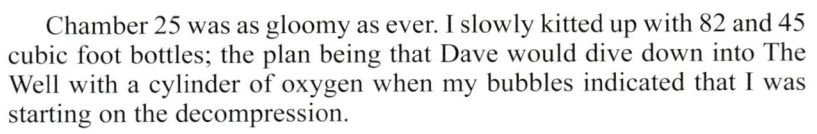

Chamber 25 was as gloomy as ever. I slowly kitted up with 82 and 45 cubic foot bottles; the plan being that Dave would dive down into The Well with a cylinder of oxygen when my bubbles indicated that I was starting on the decompression.

A couple of minutes out from Chamber 25 I reached my previous limit. With three lights switched on and fair visibility it was clear that there was no sign of any levelling off; the way on was downward. At 40 metres depth the narrow rift suddenly belled out on the lip of a drop. Cautiously I swam over to land in a bouldery area at the foot. Still the way on was down, but much less steep. About 10 metres later my depth gauge registered 45 metres the safe limit of my mixture. The exploration was over.

I dropped the reel to one side and started back. Suddenly the place felt extremely isolated and all I now wanted was to regain relatively shallow water. But hardly had I started up the sheer ascent, leading to 40 metres depth, than I realised that my breathing was difficult. Visibility was down to less than a metre and I could feel the stress mounting.

I began to breathe very heavily, but even so failed to get a satisfactory lungful of air. If it was my valve at fault I dared not swap onto the reserve supply as I felt that I had no air in my lungs to last the vital seconds necessary for the switchover.

I held the gag in tightly with one hand and went up. Somewhere or other positive buoyancy developed and the next thing I knew I was wedged against the roof at about 30 metres depth. Fortunately, I was still in full control of my physical reactions and immediately vented the suit, removing surplus air and restoring neutral buoyancy. My breathing, however, was still far from satisfactory and I moved on without delay. By the time I had gained 18 metres depth the situation eased. Even so, it was an incredible relief just to hang on to the shot line in The Well.

The minutes passed, calm was restored and eventually Dave appeared in the gathering murk. By now everything was back to normal. The bottle of oxygen was gratefully received and Dave retreated to dry ground.

When I surfaced there was little delay. Dave was cold following the period of inactivity; wearing a wet suit he really needed to get moving. He left almost immediately with just one task still to be undertaken. Prior to my surfacing in Chamber 24 I had another 10 minutes decompression stop to make and Dave's job was to bring the second bottle of oxygen

Fig. 17.6
Martyn Farr checking gear.
Photo from Martyn Farr collection

down to 6 metres depth where I would be waiting.

It all went smoothly and the exit commenced. The majority of the equipment was transported back to Chamber 22, from which point it was retrieved over the following weeks, and the final dive began. One more decompression stop was made before surfacing in the Ninth Chamber and 11 hours after entry the whole operation was over.

Despite the fact that no further section of dry cave was reached the exploration did establish a British cave diving depth record. It was also the first occasion on which decompression was deliberately adopted in a British cave, the first time that a gas mixture was used on aqualung equipment and quite possibly it was one of the most technical undertakings in the history of British caving.

To progress beyond 45 metres depth was a daunting proposition. In 1977 there were few cave divers with experience of deep diving and inevitably time was essential for the psychological acceptance of such extreme techniques. But tragedy struck in November 1981, when the promising young explorer Keith Potter drowned on a routine dive close to Chamber 20 as described in the previous chapter. Once again the inherent risks of the sport were illuminated all too clearly and the incident served as a severe deterrent to any renewed assault on the system.

Following two successful British expeditions to the mysterious 'Blue Holes' of the Bahamas the scene was set for another advance. Long and deep dives had been commonplace in the Bahamas and, having established a world record for submarine penetration into a cave network, the challenge at Wookey Hole was suddenly a feasible prospect.

It was October 1982 when the renewed assault was mounted. The logistics of the operation were even more complex than on my previous dive and the preparations took many weeks. Once again air and oxygen had to be stockpiled deep inside the cave and a camp established at the head of Chamber 24. This was fully equipped with stove, food and sleeping bags as, owing to the severity of the decompression problem, advisors from the Royal Navy recommended that the team remain at that point overnight to allow the nitrogen to be safely expelled, prior to the return to the surface.

On the day four divers were involved: myself making the solo exploration; Rob Palmer and Rob Parker supporting in Chamber 25, and Ray Stead who assisted through the dry section of the cave in Chamber 22. We departed from the glare of the movie cameras and eager reporters at 1 p.m. and quickly realised that a larger support team of divers would have been desirable. We were much too hot in our dry suits and over-burdened with so much equipment.

Fig.17.7
Rob Palmer at the campsite in Chamber 24, 1982.
Photo by Martyn Farr

Fig.17.8
Ray Stead and Martyn Farr, Chamber 9.
Photo from Martyn Farr collection

Six hours after leaving the Ninth Chamber, we reached Chamber 25. Once again the moment of truth was close at hand. What would the cave do beyond the 1977 limit? Would it shortly start to ascend, hopefully to dry cave? Would it level off at the 45 metre mark and continue horizontally? Would it continue to descend? What would we find on 31st October?

Rob Palmer shot some film a short distance into the final sump; then I was on my own. I was equipped with two large 105 cubic foot bottles, enough air to last three and a half hours on the surface but less than half an hour below 45 metres. Fresh line was tied on at the previous 'end' and, in visibility of just under 10 metres, it was evident that the passage here was unusually large, about seven metres square. Ahead the route dipped away at a steep angle with a gradual diminution in passage size. The least likely option had become the reality and, after laying less than 33 metres of line, I suddenly reached the magical depth of 60 metres. This was not the bottom but it was the limiting depth of my decompression tables and a physical barrier. At this point the roof closed to within 0.3 metres of the sand floor and any continuation would necessitate a tricky offset squeeze. Further progress was out of the question; the ascent commenced.

The first two decompression stops, at 20 and 15 metres, were made under deteriorating visibility. Despite minimal movement, the copious quantities of fine silt were disturbed and, thereafter, an hour or so was spent in a complete blackout, unable to see any instruments at all. It was an apprehensive experience cocooned in the dry suit clinging tightly to the diving line, daring not to move up or down. To sink could mean oxygen poisoning and blackout; to rise the possibility of the bends. At any moment one had to be prepared to replace one exhausted supply with another. There could be no substitute for years of training, instilled reactions to deal with any emergency even before it arose. Expecting the worst, at least I was still comfortably warm.

After a total immersion of two and a quarter hours the dive was over and I was able to rejoin my two loyal supporters shivering quietly in Chamber 25. Once again there had been no breakthrough but, even so, there was a sense of satisfaction. The previous depth record which had stood intact since 1977 had been bettered by 15 metres. Equally as important was the fact that future explorers would have that much better information with which to plan their assault.

All that remained now was to prepare for exit to Chamber 24 and the overnight camp. The real problems began as it was found that visibility in the outgoing sump had been reduced to zero. Three attempts had to be made before Rob Palmer was able to locate the way out and it was with great relief, severe depletion of our air reserves and the loss of several pieces of valuable equipment that I arrived back in Chamber 24 well into Sunday morning[26]. A bottle of wine, a warm meal and a comfortable sleeping bag were real bliss.

The next morning we struggled carefully back into our dry suits, now rather wet, and were ready to combat the next set of problems. Despite failing lights, little air and both the Robs' suits flooding we made our way out. As we cruised back into the crowded and floodlit showcave at the Ninth Chamber we were three hours behind schedule. For those at base the 25 hours of uncertainty was at an end and clutching bottles of champagne we could at last enjoy a brief moment of triumph.

Fig.17.9
Martyn Farr, Rob Palmer and Rob Parker in 9:2 at the end of the 1982 push.
Photo from Martyn Farr collection

[26] Rob Palmer disappeared whilst diving in the Red Sea on 5 May 1997, aged 45.

Chapter Eighteen
Up to the Way On
by Rob Harper

*Fig. 18.1
Rob Harper surveying
in Meghalaya.
Photo by Stuart McManus*

With the apparent closing-down of the deep sumps at the upstream end of Wookey Hole it was felt that more emphasis should now be placed on the search for a possible bypass at higher level.

After successfully diving a sump for the first time, there is obviously a great temptation for divers to head straight for the the next one rather than being distracted by smaller side openings to the right or left, and certainly above. So, it was understandable that the main stream itself had taken priority throughout the seventies, and it seemed likely that any side passages both below and above water had received less attention. John Parker's discovery of high-level routes above 9:2 by dint of removing his diving kit and climbing was a splendid example of what might be achieved elsewhere.

When I returned to work in Bristol in 1980 I thought that the time had come for a closer look at some of the less well frequented areas of the cave. Realising my limitations as a diver I chose the above water sections. Before I could get started it was necessary to find someone who was similarly motivated. Fortunately for me, if not for him, the Royal Navy decided that they could get by without Trevor Hughes and he came to live in Wookey Hole village.

On our first trip into 20 we found a short section of passage with only a minimal amount of rock-shifting. It turned out, however, to be no more than an oxbow. Trevor was eager to get onto bigger things so we left 20 temporarily and moved on to 22.

Once in 22 we quickly found ourselves a rather grim little passage with a series of ducks which turned out to be another oxbow found by Pete Glanvill a couple of weeks before. Was this cave going to yield nothing but tight oxbows? Moving through to the area around the upstream sump we thought to have a look at the roof. What roof? Two huge avens soared out of sight. Even Trevor was rendered speechless, which was no great loss in those days as nearly all his conversation underground appeared to be lifted verbatim from Biggles books! My suggestion that these climbs were best left to another generation with fewer scruples and less imagination was treated with withering scorn. Over the next few weeks we climbed the smaller and more easily accessible of the two, 'Genghis's Revenge' which closed down at a height of about 80 to 100 feet. We looked at the other aven and decided to call it 'The

*Fig. 18.2
Trevor Hughes in the Sludge Pit entrance, 1975.
From the Trevor Hughes collection*

*Fig. 18.3
Extracts from Trevor Hughes'
logbook, 1983*

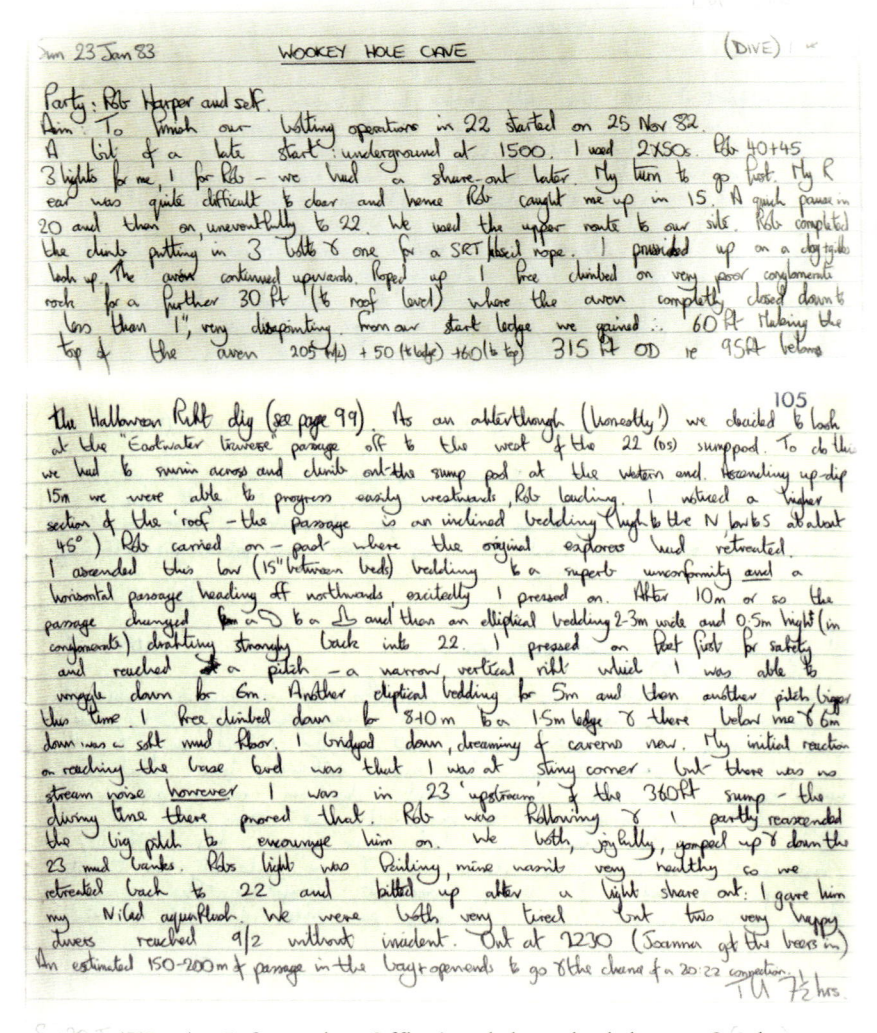

Mongol Hordes Information Office' and then slunk home. On the way out we decided to see if we could find a small passage mentioned many years before during the very first few trips into 22. Parker and others had reported following an 'Eastwater-type' rift until they had got fed up. This was supposed to head off from the far side of the sump pool and sure enough, after a cold swim, it was spotted. Trevor got out of the water at the second attempt; his first attempt ending in a double back flip with twist into the pool again, which adequately demonstrated what some of us have known for years - mud formations make very unreliable handholds! Advancing up the slippery slope with all the consummate grace of a pair of cart horses on skis, we gained a narrow inclined rift very reminiscent of the steep Traverse in Eastwater. Here, Trevor found a high-level squeeze leading into several hundred feet of crawl which eventually debouched in the roof of Wookey 23, creating a bypass to the long, static sump between 22 and 23. This was dubbed the 'Cam Valley Crawl' (after the Cam Valley Morris Men; a group of traditional dancers that had been formed at Hallatrow where the River Cam drains towards the Avon at Bath).

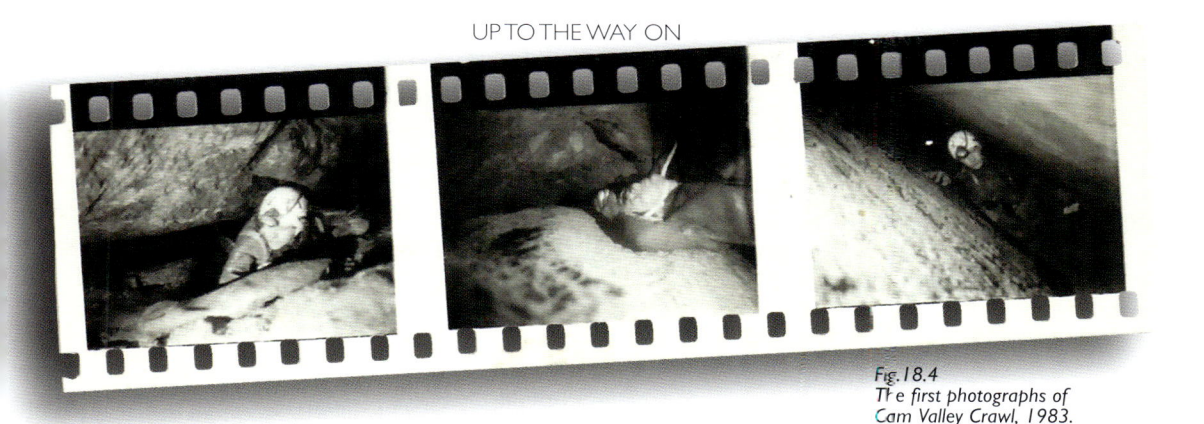

UP TO THE WAY ON

Fig. 18.4 The first photographs of Cam Valley Crawl, 1983. Photos by Peter Glanvill

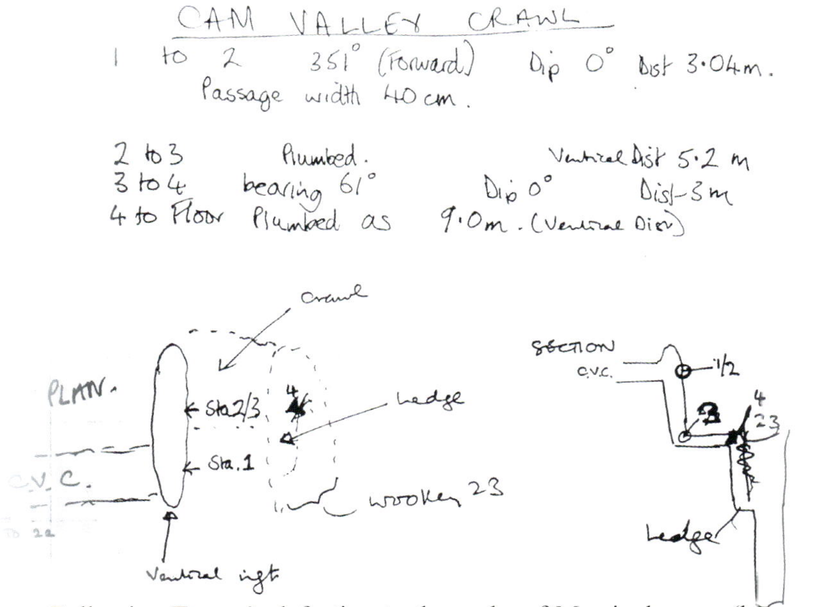

Fig. 18.5 Sketch Survey of Cam Valley Crawl by Peter Glanvill and Jim Durston, 1983

Following Trevor's defection to the ranks of Morris dancers (known to us as the hanky-waving, hey-nonny-noe, folksy brigade), he acquired a regular girlfriend and became less keen on cave diving and more so on the pleasures of the flesh! I managed to tie up a few loose ends either on solo trips or in company with whomsoever could be conned into going with me. During this period I examined the various airbells, starting with those in Coase's Loop Extension, and found that they were in fact the same airspace that is shown separately on the survey as Wookey 10. It also occurred to me that apart from Mike Thompson and Steve Wynne-Roberts early in 1964 the only person who had spent any length of time in Wookey 12 and 13 was Bob Davies in his epic near tragedy and that, under the circumstances, he might be forgiven for not having pushed all the available leads. Finding 13 took two trips. The first I would rather not go into in too much detail about; suffice to say that underwater there is a tight blind rift right next to it! The second trip was successful and after a few tense minutes getting out of the water and dumping kit I spent some time looking around 13. In particular I noted an aven above the sump which I climbed for about 30 feet until it got rather tight. This was subsequently pushed by a set of emaciated dwarves from the UBSS to give a somewhat technical round trip in a vertical plane which went nowhere.

Wookey 20 yielded a series several hundred feet in length following some dedicated climbing and digging by Chris Milne, Paul Whybro and Brian Woodward in particular. One of their passages led to a large perched static sump which unfortunately closed down almost immediately.

In a couple of solo trips I looked at all other possible leads in 20 noting two avens for later inspection. Apart from 150 feet of well decorated passage, 'Attila the Hun's Sardine Cannery' which ended in a too-tight bedding plane, there was nothing that was open. However, at the high point of the series there is a rift heading towards 24 if the survey is to be believed. The Wookey survey does need to be interpreted with extreme caution but it is intriguing that there are two passages in 24 which look as though they might connect. The rift in 20 is unfortunately closed off by stal formations. More of this in a moment. Despite diligent searching nothing new was found in either 22 or 23.

Fig. 18.6
Martin Bishop about to dive B12 Well in Box Mines, 1970. Photo from Bob Scammell

On a trip into 24 with Martin 'Steve' Bishop in the summer of 1983, I thought to have a look at a side passage mentioned to me some years before by Martyn Farr. This led off from the large high-level oxbows and had been roughly surveyed and explored by Bob Cork and Dany Bradshaw and reported as being too tight. A hundred feet or so of thrutching brought us to a pitch of about 75 feet which subsequently proved to be free-climbable although on this occasion I used a rope for assistance. Sure enough the way on at floor level was too tight and obviously flooded in times of high water. Just before leaving I chanced to look up and 20 feet above there was an ascending tube. It was small and tortuous and led to a tight, angled rift and then to a T-junction with an abandoned streamway. To the right at the junction led to a short drop which was not free-climbable, later shown to lead to a blind chamber. To the left was an increasingly tight, but still negotiable rift. Following a spirited rendering of popular songs from our youth, Bishop and I elected to call this 'Pleasant Valley Sunday'[27]. I managed to tempt Trevor out of retirement for a further look at the tight rift and he pushed it to a point where the way on is open if only a boulder jammed in the passage can be removed.

Later in the year I traversed over the top of the 75 foot pitch to enter a short section of passage about 20 feet long. This was particularly exciting since it is a stalled-up rift which is in the correct position to connect with the afore-mentioned passage in 20 and its size and shape is very similar. It was mooted that we call it 'Pleasant Valley Monday' if it goes, since it is right next to 'Pleasant Valley Sunday'. The late OCL did not approve of such frivolity.

Elsewhere in 24 the aven above the terminal sump has been climbed by Paul Whybro and shown to close down. I managed to overcome a reluctance to get totally submerged for long enough to dive the sump at 'Sting Corner' where the River Axe disappears, not to be met again until somewhere in 21. This sump can most definitely be left to a generation with fewer scruples. It is a tight rift with the negotiable sections at varying heights. The only good thing that can be said about it is that, even in the drought of 1984, there was a strong current flowing and visibility remained extremely good.

For these exploits I was accompanied by Simon Brooks who appeared on the Mendip scene with tales of Russet Well and Peak Cavern dives in Derbyshire and so was immediately commandeered. As far as I am concerned Simon is an ideal diving companion; he lets me take most of the glory whilst he does most of the work! Also, in my experience, he has no nerves whatsoever.

Deciding to take full advantage of his enthusiasm we rapidly disposed of the only two avens in 20 that looked hopeful. One was quite an epic involving some complicated aid-climbing and two falls, one of which dislodged a large boulder weighing about half a hundredweight. Simon's lightning reflexes allowed him to nod this away into the corner of a non-existent goal just before it landed on the fallee, myself. Despite copious quantities of blood we were able to carry on. The aven went nowhere!

Eventually, in 1984, the evil moment could be put off no longer. We made a start on the Mongol Hordes Information Office climb. Traversing out of a high-level passage and then free-climbing vertically for 10 feet got us into a small alcove about 40 to 50 feet above the terminal sump. The next move was a step out onto a large semi-detached 3-bedroom block hanging from one wall. Another previously unnoticed aven going straight upwards from the sentry box was climbed as a prevaricating measure but it thwarted us by closing down about 40 feet above. So it was back to the bold step. Spurred on by Simon's firm assurances: 'I suppose it might take your weight'; 'There doesn't appear to be much holding it up', and so on, I scampered across and, after ten minutes of tight-sphinctered speculation and some instant Christianity, put in a couple of bolts to ensure that I would not accompany the block on any downward movement. By this stage my nerves had had enough for one trip.

Desperate for reasons not to go back, we were overjoyed when Pete Glanvill and Clive Westlake re-discovered an airbell off the main underwater route in 21 which had been found by Colin

[27] Martin Bishop died on 3 March 2005, aged 56.

Edmunds during earlier dives. In his honour they named it 'Edmunds Chamber'. A quick inspection showed that there might be a passage at roof-level. On a trip with Pete earlier in 1984, I climbed up to an inviting looking phreatic tube only to find that it was an alcove. To be sure that nothing is missed, someone will have to climb the remaining 15 or 20 feet to the roof which is itself about 60 feet above the water; but, I feel that there is very little chance of an extension[28].

To summarise, the current situation is that most of the cave has been fairly thoroughly investigated above water. In 20 there are two good digging sites: one at the far end of the passage, where a series of interlinked rifts filled with small boulders and sand were dug by Chris Milne before he got married; the other more exciting prospect is the passage heading towards 24. The latter, however, will require an application of Dr. Nobel's 'Amazing Rock Remover' to break up the stal flowstone before it can be dug.

Edmunds Chamber off Wookey 21 and the Information Office climb in 22 are still to be completed. Although these may be blind, there is the enticing prospect of a connection to Hallowe'en Rift, an old outflow cave on the wooded hillside directly above 22. This was found by Trevor Hughes one day while he was out collecting firewood and has since been sporadically dug by himself and various other BEC members.

Not all of the roof level of 23 has been inspected but most of the easier free-climbs have been done and shown to go nowhere. There is one slight possibility; an aven which terminates at the underside of a boulder pile. This might repay prodding with a crowbar by someone with no imagination and a strong death-wish. Come back Trevor, all is forgiven!

The Pleasant Valley passages in 24 must be investigated further as both offer possible connections to 20. For instance, Pleasant Valley Sunday does seem to have a slight draught on occasions, prompting speculation about a possible surface link. In addition, there are several enticing dark holes high up in the walls of the main 24 passage.

I have no personal experience of 25 but believe there is very little possibility of above water extensions. But we will never be sure until someone takes a closer look, of course.

Photo by Peter Glanvill

[28] A view confirmed by Geoff Ballard and others in 2005 after extensive efforts had been put into constructing a landing stage in Edmunds Chamber.

Chapter Nineteen

Another Year Older and Deeper in Depth

based on Rob Parker and Bill Stone

Once again a stalemate situation followed Martyn Farr's dive beyond Chamber 25 in the autumn of 1982. The deepest sumps yet lay ahead in Wookey Hole Caves. In the three years leading to the Fiftieth Anniversary of the very first cave dives back in 1935, most attention had been given to finding high-level bypasses as seen in the previous chapter. As the years roll by, not only do the distances divers have to go to penetrate new passages mount up, but the technical problems grow bigger. Inevitably, more time and money have to be spent on each successive expedition. A special effort was appropriate for 1985 and the time was ripe for those members of the Cave Diving Group more professionally involved in big caving expeditions around the world to have a go at pushing Wookey Hole Caves.

Rob Parker, a 22 year old carpenter from Bristol, led the challenge[29]. He was supported by Dr. Bill Stone, an American engineering scientist and noted 'spelunker' on the USA circuit. Costs would be met through trans-Atlantic sponsorship, national press coverage and television film rights for the project. Rob and Bill first met and made their plans in April 1984 on an expedition to explore Pena Colorada Cave, Mexico. Both had dived deep in the so-called 'Blue Holes' off the Bahamas and arranged to train together with their team of Harlech Television (HTV) camera men in similar 300 feet deep warm water springs in Florida. Their bid at Wookey Hole would take place late in June 1985.

As a major tourist attraction, any publicity just before the main summer holidays is good for business at Wookey Hole Caves. Gone are the old days when cave diving had to be undertaken in the off season and at night to avoid the paying visitors. It is now a bonus for guided parties to see some real live action underground. Commercial interests, insurance cover and the Health and Safety at Work Act of 1974 must be complied with these days. Only those in the lead are allowed to take risks, on paper at least. Detailed and strict conditions had already been agreed between the Cave Management and Cave Diving Group through

Fig. 19.1
Rob Parker.
From the Bill Stone collection

Fig. 19.2
Bill Stone.
From the Bill Stone collection

---
[29] Rob Parker died whilst cave diving in Four Sharks Blue Hole, Bahamas, 17 August 1997, aged 35.

Fig. 19.3 Wookey Resurgence. Photo by Bill Stone

Oliver Lloyd. When Oliver died just before the event, Bob Drake[30] took over this responsibility; not an easy task when only Rob Parker, Julian Walker and Ian Rolland[31] of the team were CDG members with previous dives to their credit in Wookey Hole Caves. Bill Stone and his wife, Pat, had vast experience of caving and diving in North America, however, even if new to British conditions. Their vital HTV camera team, on the other hand, comprising sub-aqua divers Leo Dickinson and Peter Schoones, had to learn whilst on the job. As well as careful organisation the seven of them would need some luck.

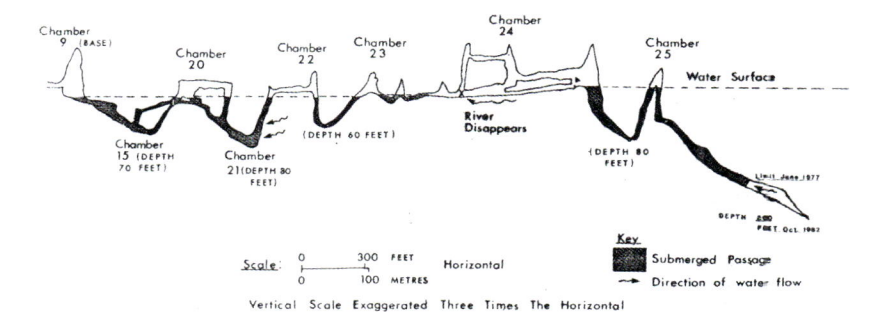

Fig. 19.4 Leo Dickinson. From the Bill Stone collection

The scene was set for the country at large in a well-informed article by journalist David Rose in the weekend edition of *The Guardian* newspaper for 22nd June 1985. Dave, an Oxford University Caving Club member involved in descents of deep caves in the Picos de Europa, Northern Spain, had cause to appreciate the dangers having been at university with the luckless Keith Potter who had drowned on the way to Wookey 20 only four years before. Dave and Richard Gregson were also writing a book on their Spanish finds, and *Beneath the Mountains* published in 1987 is dedicated: 'For Keith, hombre'. His *Guardian* article about Rob Parker's attempt to push further in Wookey Hole Caves, entitled 'The diver with three tickets to staying alive', contained the following extracts. Only the unusual numbering of the far sumps in Wookey Hole has been altered to follow the accepted convention on Mendip of the open passages found *beyond* a dive taking the next number in the sequence:

At about lunchtime on Tuesday a 22-year old carpenter from Bristol called Rob Parker will dive beneath the surface of a black pool in an underground chamber in the far reaches of Wookey Hole Cave in Somerset. Alone, without the slightest prospect of rescue if anything goes wrong, he will be going where no man has gone before.

Parker's dive, if successful, will pass a barrier that many cave divers thought impossible. He will be going deeper underwater than anyone has ever been in a British cave, facing intense pressure, cold, and the limits of existing technology. His potential reward is the discovery of further caverns and the source of the River Axe, the blue stream emerging into the daylight at Wookey's entrance.

The deeper the dive goes, the more difficult and more dangerous his progress becomes. As pressure increases, nitrogen starts dissolving in the blood, creating the hazard of nitrogen narcosis, a drunken, hallucinatory state in which divers have sometimes, believing themselves to be above water, removed their equipment and drowned. More deadly, there is the hazard of 'the bends' the formation of nitrogen bubbles on the return to the surface or to lesser depth. It can induce fatal, agonising seizures.

Fig. 19.5 Survey from 'Wookey Hole/the caves beyond' by Martyn Farr

[30] Bob Drake died from heart failure whilst sea diving off of Brixham, Devon on 1 June 1990, aged 40.

[31] Ian Rolland died whilst cave diving in the Systema San Agustín, Mexico on 27 March 1994, aged 29.

Martyn Farr's 1977 attempt on sump 25 was already the deepest penetration by a British diver. He got down to 160 feet (actually 45 metres): the flooded tunnel continued inexorably down.

Five years later he tried again. This time, he had special decompression tables, calculations which told him how to avoid the bends by spending given times at given depths waiting on his way back up. The tables and his compressed air aqualung imposed a maximum depth of 200 feet on his dive.

He reached it only minutes after leaving chamber 25. The river Axe flowed up past him, but still the only way on was down. Worse, the passage beyond appeared to narrow to a fierce constriction, a slot no more than a foot wide, bad enough in the shallowest of sumps, let alone 200 feet underwater. It is through this slot that Rob Parker, defying many sceptics, intends to go.

The logistics of his expeditions are alarming. Most of his equipment has been spread out in his Clifton Hill living room: diving cylinders, thermal clothing, food, and sleeping bags. Parker and his four man support team are setting up camp for five days in Wookey 24, and they packed much of the gear into water-proof canisters in order to ferry it into the campsite in a series of preliminary dives.

Wookey has already claimed two lives in earlier phases of its exploration, but Parker has faith in his equipment. Some of the diving cylinders, 43 will be needed in all, are lightweight fibre glass models developed originally by the American space programme.

To combat the cold, the water, he said, will be 39 degrees Fahrenheit[32], particularly bitter during the long decompression stops. He will wear thermal underwear, Thinsulate (an artificial fibre twice as warm as down, volume for volume), and a unique, sealed rubber dry suit, made for the dive by Brian Bickell, a specialist marine designer.

Finally, there is his air supply: helium trimix, pioneered by deep oil rig divers, used together with decompression tables computed by Fort Bovisand, a diving firm in Plymouth, on the basis of Parker's own physiology. No existing published tables were any use: they all make the dive completely impossible.

'Essentially,' Parker said, 'I've got three tickets home I can't afford to lose. There's the line I lay through the sump to follow out when the silt gets stirred up and the visibility falls, perhaps to zero.

'There's the need not to get the slightest nick in the dry suit. If the water got in, I wouldn't be able to decompress without dying of hypothermia, and even though we'll be taking a load of drugs down, if I came straight up I'd almost certainly get the bends.

'And of course there's the air supply. If I blew out a high pressure hose, I'd obviously be finished.'

The team will enter the cave on Monday morning, reaching camp for dinner and the night. On the following morning, Parker will set off through sump 24 with two of the others. With no dry ground there, they plan to wait in hammocks fixed to the cave walls while he makes his first exploratory dive.

They may be there a long time. If, as he hopes, Parker finds the upward continuation of sump 25 beyond the slot, he will try to surface: 'I think we may be near the bottom now. A lot of the other Wookey sumps have a squeeze at the bottom.' If necessary, he is ready to go down another 100 feet.

If he does surface, he will first have to tread water decompressing for 90 minutes. He will not, on this first occasion, explore beyond the end of the sump: to do so would risk a fatal nick in the dry suit on the sharp conglomerite for which Wookey is notorious. Diving back, because of residual nitrogen in his blood, he will have to decompress for a further three hours underwater.

Parker feels that it may, however, take a second attempt to reach the far side of the sump. But whenever that happens, it is then that the fun will really begin.

After a further night at the camp, Parker will go back through the sump with Bill Stone. This time they will be carrying clothes to change into, and bivouac gear. 'I have a feeling that there may be a lot of open passage on the other side and it would be a pity to waste it,' Parker said. He plans to stay with Stone in that impossibly remote place for two days and two nights to explore whatever lies beyond and make a survey to add to the map of Wookey Hole.

Bill Stone, on the other hand, was not a believer in adventure, and is firmly on record as 'reserving that term strictly for ill-planned expeditions gone amuck'. His professional expertise and enthusiasm were essential during the team's preparations in Florida, particularly in getting the

---

[32] Water temperatures in Wookey Hole Caves stay a constant 50°F (or 10°C).

technical details absolutely right for the trimix of helium-oxygen-nitrogen proposed. In an article entitled 'In pursuit of the River Axe' published in the USA by the National Speleological Society (NSS News, March 1986), Bill detailed the problems that they had to overcome:

It has been known since the 1930s that one can decrease, or eliminate altogether, the effects of narcosis by replacing nitrogen with helium in the breathing mixture, since helium is an inert gas with no known neurological side effects. However, helium has its own set of problems: due to its exceptional ability to transfer heat, special actions must be taken to avoid excessive body heat loss in a diver. Some of the heat loss can be reduced by filling drysuits with a separate compressed air supply and by replacing only part of the nitrogen in air with helium, leading to what is known as trimix. Narcosis problems are still present, but the effective working depth is substantially increased. The second, more serious, problem is that presently available decompression schedules for helium mixtures call for switching to pure oxygen at -20 m on the ascent. This is a particularly dangerous procedure for sport divers since pure oxygen breathed below -10 m can induce oxygen poisoning. Millitary and commerical divers will almost always have the benefit of performing decompression in a warm, dry pressure chamber, where corrective changes in the breathing gas can be made by an attendant. Cave divers, and particularly sump divers, working in remote locations are thus placed in the position of either risking an oxygen 'hit' or violating the decompression procedures.

The Louis Holtzendorff incident in 1975 is still fresh in the memory for many Florida cave divers. It was in the autumn of that year that Court Smith and Louis, both very competent cave divers, decided to experiment with helium-oxygen to assist in combating narcosis in a spring where depths were known to exceed 100 m. They used an 80% helium-20% oxygen mixture.

Smith and Holtzendorff did a limited duration dive to -86 m and had no difficulty with the breathing mixture. However, they had not been on the oxygen at -20 m for more than a few minutes when Holtzendorff convulsed and spat the regulator out. They had no air cylinders with them on which to switch, and it is doubtful at that stage whether Smith could have made Holtzendorff hold the regulator and breathe. Holtzendorff drowned and helium diving in Florida was abandoned until some eight years later when Dale Sweet did a 110 m dive in Diepolder II. This was passed off as a novelty and the memory of Holtzendorff remained strong. It was better to dive on air and deal with the narcosis. That viewpoint was solidified when less than a week after Sweet had planted an American flag at -110 m in Diepolder, Sheck Exley came in on air, saluted the flag, and laid another 70 m of line to -116 m[33].

The recent helium dives in the Vaucluse by both the French team and by Jochen Hasenmayer, served to rekindle interest in mixed-gas diving. However, there was still the oxygen problem to be dealt with and the use of portable habitats (as used by the French) were not practical for sump diving operations. It was with this background that I met Parker in Florida in April 1985 for two weeks of training in deep diving on air.

During their training in Florida in the deep Diepolder sinks north of Tampa, they discussed the situation with John Zumrick, chief medical officer for the U.S. Navy Experimental Diving Unit. More importantly, he was also a top-rank cave diver. Bill Stone continues his article:

At breakfast one morning John and I got into discussion of narcosis and gas mixtures, as my experiences with 'the wall' in Diepolder had again fired my interest on the subject of using helium for Wookey. 'Even if we could get the gas,' I said, 'we would still have to deal with those 20 m oxygen tables.' There was a strange pause in the conversation. 'You know we've been conducting intermediate helium dives at the unit during the past six months,' Zumrick said. He began tapping the fingers of his left hand on the chair, as if pondering whether or not to say the rest. 'You're not going to believe this, but the helium tables are starting to look more and more like the exceptional exposure air tables.' His hesitancy was understandable, since there were two people at the table who were going to interpret such a statement as a trimix diving license. 'Of course, I would still recommend using oxygen at the six and three metre stops for insurance,' he added, realizing what was going to happen.

In short order I performed some calculations in Parker's notebook which indicated that with the addition of 25% helium, we could create an effective air depth of -45 m while diving at -73 m. John stressed that we would want to maintain 20% oxygen content so that the mix could be breathed from the surface down. The calculations used ideal gas theory (relatively accurate since there were no polar molecules present) and came out as follows: (1) pull a vacuum on a bank of tanks; (2) pump 307 psi of oxygen into the bank; (3) add 1300 psi of helium for a total tank pressure of 1607 psi; (4) top off with air to 5000 psi, the working pressure for our filament wound Acurex tanks. The greatest problem would be finding a place to get the gas and having the mix

---

[33] Sheck Exley died whilst attempting a dive to -300 m in Zacatón, Mexico on 6 May 1994, aged 45.

verified in a gas chromatograph.

Needless to say, they did get hold of the gas and checked that it was the right mixture. Within days they returned to Diepolder, reached an encouraging depth of 73 metres with no ill-effects and reported back to Zumrick:

I told John that Parker and I had been back to Diepolder and had felt clear at -73 m and he said, 'I doubt that.' Then I added, 'on helium.' And his eyes lit up. 'You got the gas!' he said, somewhat shocked. He then interrogated us as to each step we had done, both in the filling of the tanks and checking of the gas mixture, as well as the dive procedure. We then began discussing Wookey and Parker said that he might like to push it as deep as -85 m. John then said, 'Well, if you want to hit that, you probably ought to think about increasing the helium content. As long as you use air to inflate your drysuit, you won't have any problems with heat loss.'

Parker said he would approach Air Products of Britain immediately upon his return to see if they would be willing to supply the gas, since we had estimated needing about 1400 cubic feet of trimix alone for our work beyond Chamber 25. John, an expert in diving physiology, agreed to develop a special set of decompression tables which would take us from Chamber 24 to Chamber 26. The complexity of this dive in Wookey was just now starting to hit home: we would need three Acurex tanks (about 330 cubic feet) of trimix for each diver, a bottle of oxygen for decompression on the far side, and a small bottle of air just to inflate the dry suit for each attempt beyond 25.

[Once again, the terminology has been changed in these extracts for the sake of consistency with usual practice.]

By the time that Bill and Pat Stone flew into Heathrow on 21st June, Rob Parker's home-based team had begun to set up the camp in Chamber 24. He had lined up an impressive array of equipment and drafted in Marco Paganuzzi as an additional diving sherpa. In a matter-of-fact account, Rob described his own strategy, equipment, helpers and safety precautions as follows.

This year marks the 50th anniversary of cave diving in Wookey Hole. Its challenging resurgence sumps have long proved irresistible to explorers, who have pioneered new techniques and equipment in the search for the elusive source of the river Axe. Water sinking at Swildon's Hole and other major caves high on the Mendip hills has proved difficult to follow on its journey through to Wookey. This water reappears in the 25th chamber, 2 km into the hillside. This is at a point beyond 600 m of flooded passage at the head of a sump over 60 m deep.

To explore this point and discover the next section of the cave, the diver would have to overcome many problems. Most of the difficulties are created by the great pressures experienced at depths below 60 m, the most significant of these is nitrogen narcosis, encountered when breathing nitrogen gas at high pressures. Large quantities of breathing gas are also required to sustain any length of exploration at this sort of depth. Complicated decompression schedules are required to avoid 'the bends', compounded by the possibility of surfacing beyond the sump and then returning. Finally, the pressure reduces the insulation provided by compressible neoprene suits, exposing the diver to the cold. In addition to these pressure-related problems, the normal cave diving hazards of bad visibility and small passages when diving alone would still be present.

Having been involved in the 1982 attempt to find Chamber 26 it was realised that an expedition-style approach was required. In order to avoid excessive accumulation of residual nitrogen in the bloodstream a camp was to be set up in the dry Chamber 24 to allow comfortable delays of over 12 hours between dives. Using this camp as a base, three attempts to pass the terminal sump could be made. In the event of a successful passage of the sump to Chamber 26, lightweight exploration and bivouac equipment (supplied by Lyon Equipment) was on hand. In this fashion, it was hoped that another section of the river Axe could be mapped.

Owing to the problems previously described, standard British cave diving equipment was thought unsuitable. Lightweight composite pressure vessels developed by Acurex Airotherm Ltd., provided the only feasible method of carrying large quantities of breathing gas. As these cylinders work at a pressure of 333 bar, specially modified Poseidon demand valves supplied by Underwater Instrumentation were used. A flexible

ANOTHER YEAR OLDER AND DEEPER IN DEPTH

Fig. 19.6 Junction of Deep and Shallow Routes below Chamber 19.   Photo by Martyn Farr

high pressure manifold was developed, linking the hip-mounted cylinders to give full access to gas supplies in the event of a single demand valve failure.

To overcome the narcosis effects encountered using compressed air at these depths, a mixture incorporating 36% helium, 19.5% oxygen and 44.5% nitrogen (Trimix) was used instead. However, this introduced additional problems with decompression schedules and heat loss factors. Special decompression schedules were developed in the United States by Dr. John Zumrick to suit the precise gas mixture and predicted dive profiles, whilst extra thermal insulation was provided by a modified Poseidon Unisuit combined with Thinsulate and Damart underwear.

Extensive training with this equipment was undertaken both in the long, deep Florida springs and the cold British sumps of Yorkshire. Great importance was also placed upon a high level of personal physical fitness.

In order to help Rob Parker accomplish the objectives, a small efficient team was assembled. This comprised Marco Paganuzzi, Julian Walker and Ian Rolland to help sherpa equipment to the camp location. In addition, the latter two acted as support divers during the attempts to pass Sump 25. Dr. Bill Stone was imported from the United States for his knowledge of deep diving techniques and the use of Trimix. Pat Stone, Bill's wife, aided Leo Dickinson in the filming of the project for a television documentary. A total of 6 people were to camp in Chamber 24 for four nights.

Fig. 19.7
Rob Parker filling composite cylinders with Trimix.
Photo by Bill Stone

Due to the serious nature of the dive, a number of safety precautions were also incorporated. After consultation with Dr. Maurice Cross of the Diving Disease Centre at Fort Bovisand, it was decided to surface-feed oxygen to a full facemask for all decompression at -6 m and above. This reduced the chances of both decompression sickness and drowning in the event of oxygen convulsions. This equipment, supplied by Underwater Instrumentation, could also be used with relevant schedules and oxygen supplies to recompress a bends victim.

Both Pat Stone and Julian Walker were trained as paramedics and in the event of a serious incident could control recompression and administer first aid. Ian Rolland was constantly ready to make a fast exit from the cave to alert Dr. Peter Glanvill and Maurice Cross through the Mendip Rescue Organization. Working within limitations imposed by the cave environment these measures reduced the risks involved to an acceptable level.

Bill Stone quickly found the essential differences between Florida's warm clear water and the numbing River Axe, charged with mud after days of rain on Mendip. Even more unaccustomed to such places, cameraman Pete Scoones was soon out of action. After getting shots of Bill's first trip to 22, he slipped on a steep mud bank and injured his ankle. A painful and plucky swim back to safety, then a check-up in Bristol Royal Infirmary, found his ankle fractured in three places! Pat Stone was detailed into the team so that the show could still go on, for HTV wanted to see it through following the big build-up. Such are the pressures on today's cave divers. Getting to the comparative comfort of the camp in 24 was a challenge in itself as Bill Stone describes:

At the end of June there was a three-day stretch of sunny weather and we made our move. We had substantially delayed the push already, since poor visibility would have prohibited effective exploration in Sump 25. Given that this was the first traverse of the cave for Pat, the others had given us an hour's lead time to sort things out. At the head of each sump, I went over the moves with her, as Parker had done for me. Given zero visibility in all the sumps beyond Chamber 22, I had found it extremely useful to know where to switch hands on the guide line to avoid the narrow spots, and when to skip incoming lines. The two of us would then go down a pre-dive check at the head of each sump and be off. I was concerned about Pat going through from 22. Even though it was only 120 m long, there was a very low section at the bottom of the loop at a depth of 20 m. The tunnel was only tall enough in the very centre for a diver laden with equipment to pass. There was a trough worn into the floor there where all the divers had been. To either side it was impassably low, and since it was common practice here (unlike in the States) to leave substantial slack in the guide line, it was easy for one not familiar with the route to find themselves wedged off to one side, believing that the fanatics who had originally explored this place had actually forced themselves through that 20 cm-high crawlway. Since you could not see where the centre was, you had to feel around for the trough, but once there it was clear sailing.

It later struck me that these little troughs were traceable all the way from Chamber 9. They were the sherpa trails to Chamber 24. In the beginning sumps where the visibility was better, you could see them like beaten-down paths on a slope. They were, of course, a testimony to bad silt control technique. When Rob informed me that Wookey was the best cave diving in England, I could see how silt control was not high on the list of priorities. Most of the time you could not see anything anyways, so why not put yourself along the bottom; it was faster than finning.

Chamber 23 and the two short sumps beyond were strikingly different from the rest of the cave. The active stream had long ago been pirated (captured) into a parallel (and as yet unexplored) gallery between Chamber's 24 and 22. Chamber 23 had been left standing, its sumps stagnant. In its wake the Axe had left great loamy silt mounds, a veritable delta of slimy, greasy brown mud. It was now easy to see why the visibility was zero in these sumps, for at either end a diver was forced to claw his way out with a resulting cloud of silt tumbling down the underwater slope. Once stirred, the sediment would remain suspended for several days. Rob and the others, realizing the difficulties that would be encountered by a heavily laden diver trying to climb out of the water in some of these places, had taken the uncommon measure of bringing in four-metre lengths of aluminium ladder.

I was just getting out of my drysuit when Rob appeared in camp. The others were half an hour behind. Ian had brought a trenching tool and with this we set about levelling a sleeping area for the six of us. It was while doing this that I happened to notice that the back end of camp was bordered by a rift which dropped a good 15 m to the River Axe. Ian promptly walked up and relieved himself. 'This is the camp urinal, I reckon,' he said smiling. I was about to question the merits of polluting our water source, but figured they planned to get it at some upstream location. It later proved to be a bad decision, as the smell of fermented urine crept

*Fig. 19.8
Ian Rolland, Rob Parker and
Julian Walker at the campsite
in Chamber 24.
Photo by Bill Stone*

back up that fissure and hung over the camp like a smog for the remainder of our stay. Some of the crew had never camped underground and did not recognize the criticalness of the location of the waste disposal area. I later went off down the passage and dug the latrine, downwind from camp.

That evening at dinner we stepped through the procedures planned for Parker's initial dive from Chamber 25. Pat described the symptoms of oxygen poisoning: tingling or itchy feeling around the nose and face, numbness, nausea, and tunnel vision. According to Maurice Cross you had around three to five minutes of these symptoms prior to the onset of convulsions. It would be up to Parker to diagnose these when he went onto the band mask at -6 m and to give three sharp tugs on the surface feed line to signal those above that trouble was afoot and to switch to air, which would be hooked up on the parallel supply line. By this time everyone was getting tired and began sacking out. Because of the difficulty of underwater equipment transport everything at Camp 1 was at a premium. Foam sleeping pads, being bulky and positively buoyant, were left on the surface. Instead we slept on top of our Thinsulate drysuit garments. Rob had commissioned the fabrication of six one-pound mummy-type synthetic sleeping bags that kept us fairly warm; you could get four of them in a dry tube. The River Axe was softly roaring in the background and drowned out the rustling of people shifting positions. We had a relatively restful night's sleep.

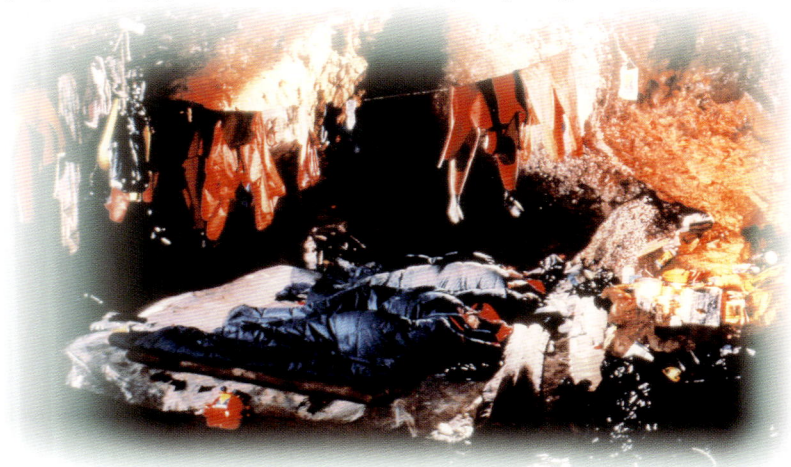

*Fig. 19.9
Camp in 24.
Photo by Bill Stone*

Bill Stone's article in the *NSS News* continues its fascinating blow-by-blow account of Rob Parker's dive beyond Chamber 25, but let the man who got there finish the story of this trip himself:

Before exploration of the terminal sump could begin, diving, camping and filming equipment had to be stockpiled throughout the cave. Rigid ladders were installed on the most dangerous and awkward climbs to aid the transportation of the heavy, fragile loads into the system. Large watertight containers, weighing 20 kg each, were used to protect camping and filming equipment above and below water. More than 50 man loads were eventually required to equip the 5-day camp.

After three weeks of work we were ready to attempt the dive. However, good underwater visibility was essential to aid exploration and as usual the British summer did its best to deter us. It was not until Sunday 30th June that the six divers left Chamber 9.

A six-hour trip saw the team at the camp site in Chamber 24. It took a further 4 hours to establish the camp but by the end of the day the cavers could relax in the relative comfort of their lightweight mountain sleeping bags.

During the course of the following day, the line to Chamber 25 was relaid and oxygen decompression equipment carried through. The breathing system for the deep dive was also assembled and checked to ensure correct functioning after its arduous carry into the cave. Meanwhile Leo and Pat filmed as much of the activity as possible.

The first exploration dive took place on Day 3. After an early start, Rob Parker dived through to Chamber 25 with Julian Walker in support. Once in Chamber 25, it took 30 minutes for Rob to make the final adjustments to his equipment. Bill Stone arrived with a cine camera to record the occasion for Leo.

Carrying 315 cubic feet of Trimix and 40 cubic feet of oxygen in four separate tanks, the diver descended into the cold water. Visibility was good, in excess of 15 m and at a depth of 45 m he dumped his first depleted 105 cubic feet stage cylinder. At -60 m he encountered the constriction which marked the limit of exploration and tied on his line. The gravel floored passage was dipping steeply, forcing the full flow of the river Axe through a gap 2.5 m wide by 0.4 m high.

Fig. 19.10
Rob Parker passes a cylinder up to Pat Stone in Chamber 24.
Photo by Bill Stone

Pushing through this constriction a roomier section of passage was followed ending in a second constriction. Belaying the line to a lead block to secure the return route, the second constriction was attempted. After 4 m of hard won progress at a depth of 67 m the diver conceded to the cave. Having ploughed a channel into the gravel to reach this point, it was now slumping in behind him and threatening to block his exit route. Three metres beyond the diver, the gravel slope appeared to level off, but the passage to this point was impossibly tight. No further progress could be made via this route.

Returning through the now cloudy water, the first decompression stop was reached at a depth of 18 m. The ascent to Chamber 25 took over two hours, this being the penalty to pay for the 25 minute dive to -67 m.

The team did not give up the search for Chamber 26. Being in their unique position with plentiful gas supplies and manpower, a thorough search of the sumps upstream of the camp was made in the hope of finding an alternative route. No such route was found.

On Thursday July 4th, the team surfaced in Chamber 9 to be welcomed by a champagne reception. News from the camp had been brought out the previous night by Rob Palmer after a solo trip to Chamber 24.

Despite the best possible equipment and techniques and a highly competent team, Chamber 26 had not been found. It is felt that at this time no further progress is possible along the route attempted.

The expedition, however, was successful in many ways:-

A new British cave diving depth record was achieved. It is also believed to be the deepest sump explored beyond other sumps in the world. Considering the adverse conditions under which this was accomplished, the achievement is even greater.

The techniques employed in this project have extended the limits to which SCUBA diving has previously been used. This involved the use of open-circuit Trimix and tailor-made decompression schedules combined with the unique fibre-glass cylinders and their specially-adapted demand valves. Help and advice was gratefully received from both the Diving Disease Centre at Fort Bovisand and the U.S. Navy Experimental Diving Unit.

The expedition achieved a high level of teamwork and co-operation in a harsh and unforgiving environment. It combined the diving expertise of Dr. Bill Stone from the U.S., the experience and talent of Leo Dickinson and the enthusiasm and effort of three British cave divers to execute one of the most ambitious cave diving projects ever.

For the record the expedition was filmed for Channel 4 by Leo Dickinson and HTV. This was the first time that a professional camerman had recorded such activities which is a tribute to the dedication of Leo. His film was shown on Channel Four in 1986.

Little new ground may have been gained but a lot was learnt on this trip. Not least, it confirmed the theory that Mendip's deep groundwater movements occur as great vertical loops. The rest of the River Axe under Mendip is for future generations of cave divers and diggers...

Fig. 19.11
Cover of CDG Newsletter No. 77
October 1985

Chapter Twenty

Back to Basics in St. Cuthbert's

by Alan Butcher and Stuart McManus

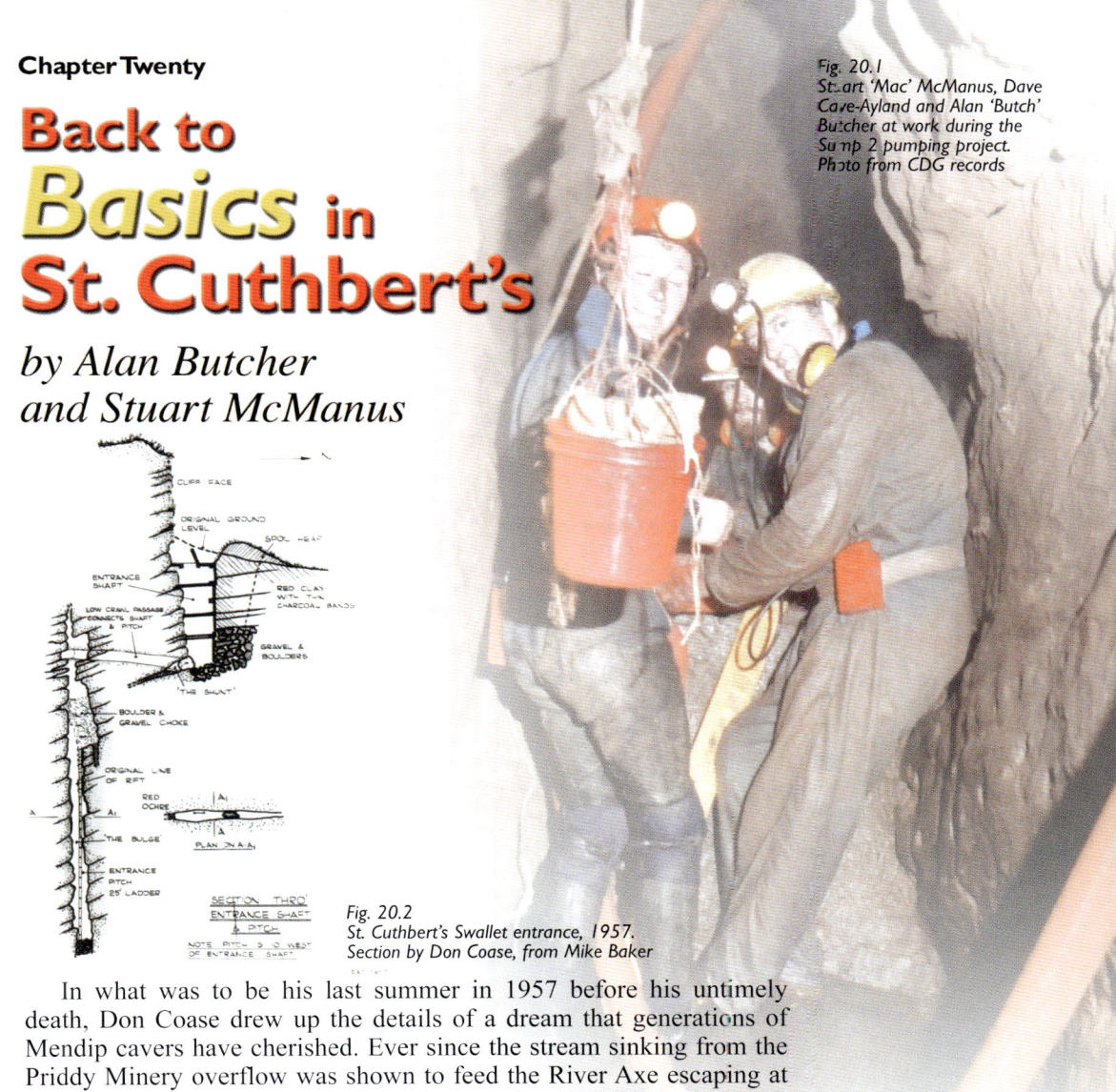

Fig. 20.1
Stuart 'Mac' McManus, Dave Cave-Ayland and Alan 'Butch' Butcher at work during the Sump 2 pumping project. Photo from CDG records

Fig. 20.2
St. Cuthbert's Swallet entrance, 1957. Section by Don Coase, from Mike Baker

In what was to be his last summer in 1957 before his untimely death, Don Coase drew up the details of a dream that generations of Mendip cavers have cherished. Ever since the stream sinking from the Priddy Minery overflow was shown to feed the River Axe escaping at Wookey Hole Caves in 1860, an underground connection had become a challenge. But it was not until 30th August 1953 that St. Cuthbert's Swallet was first entered. By that time divers had already reached the brink of the submerged slope beyond the Eleventh Chamber in Wookey Hole. Don's cross-section of the underground drainage between both systems reveals what members of the Cave Diving Group and Bristol Exploration Club respectively had achieved in the Wookey Hole to St. Cuthbert's route by 1957. Don was a leading personality in exploring

Fig. 20.3
St. Cuthbert's Swallet to Wookey Hole 1957. Section by Don Coase, from Mike Baker

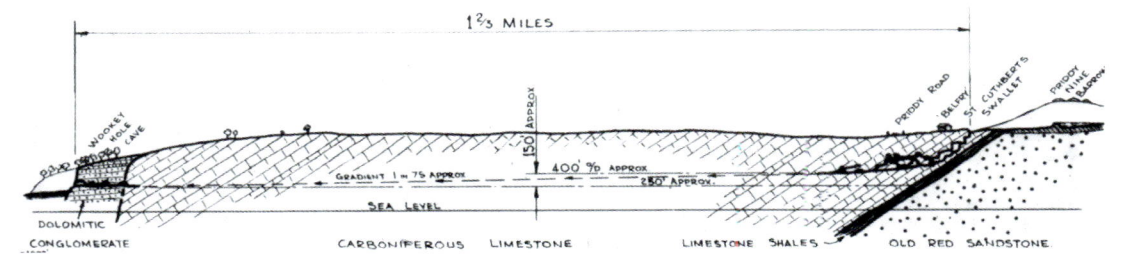

Fig. 20.4
Self portrait of John Cornwell passing The Duck, 1964.

Fig. 20.5
*St. Cuthbert's Swallet entrance pipe and dam, 2009.
Photo by Mark 'Gonzo' Lumley*

both systems, of course. Maybe the splendid but choked downstream passages in St. Cuthbert's would prove to be an easier proposition than the deep upstream dives in Wookey Hole?

On his last major caving trip on June 9th 1957, Don dug open the first section of choked streamway at the end of Gour Rift with John Buxton and turned it into 'The Duck' as it is known today. But they soon reached another unyielding silted-up pool which has since been called Sump 1. Progress would not be easy for St. Cuthbert's diggers and divers either! In fact, digging whilst underwater and breathing from base-fed air lines was first tried here by Mike Thompson, Fred Davies and Steve Wynne-Roberts in the early Sixties. Some progress was made but downstream sumps usually silt up again between trips and hard won gains can be lost during a single flood. The following accounts about the eventual passing of Sump 1 and sieges on Sump 2 are taken up by Alan 'Butch' Butcher and Stuart 'Mac' McManus respectively - two of the most persistent diggers with a strong belief that the way on to Wookey would be found.

Alan Butcher recalls:

After passing The Duck and reaching Sump 1, there were several attempts made to dive on and yet it remained an obstacle for another twelve years. During this period the most notable dives involved a well co-ordinated push over the weekend of 5th-7th February 1967 by three

BEC divers, Mike 'Fish' Jeanmaire, Phil Kingston and Barry Lane. They had joined forces with Roy Bennett and Dave Irwin to organise a massive inter-club digging and diving operation which would involve over sixty members from half a dozen clubs.

It was unfortunate that the weekend chosen was rather wet and the water in the cave was quite high. My memory of the occasion is one of building temporary dams in the approach passage in an attempt to stem the onrush of the stream, a task at which we were so successful that the old duck became a two foot deep sump! The combination of high water conditions and the cold led the attempt to be abandoned early on the Sunday morning, by which time the divers had penetrated the sump for a distance of nine to ten feet.

The divers continued to work in the sump for a further eighteen months and reached a point some twenty-one feet into it. All this work however, was, to go to waste during the flood of 10th July 1968 when large amounts of silt were washed back in.

This was the end of attempts to pass the sump by diving and underwater digging, and it was late in 1968 that we decided to attempt holding back the stream by means of dams. These were commenced on 4th January 1969 and many trips were carried out by a small number of regulars to finish a total of six underground dams. They were built with concrete mixed using cave gravel and rapid hardening cement. The cement was transported down the cave in one gallon plastic carboys which were an infernal nuisance to carry. At one stage we were taking down between 1 and 2 hundredweight of cement on each trip.

The typical Tuesday evening session involved hurrying to the Belfry to get changed and descending Cuthbert's at around 7 pm, working for a couple of hours and then retiring to the pub. 'Crange' (Bob Craig) usually led the charge for the pub, regularly breaking the 15 minute barrier from sump to entrance.

On 30th September 1969 after a dry summer, a party went to the Sump Passage to select a site for a dam and noticed that the sump was empty. This at once attracted the attention of the Tuesday night Dining Room diggers, drawn largely from the BEC and the Shepton Mallet CC, who had been excavating a phreatic tube which seemed to be heading over Sump 1. They decided that they would postpone their dig and joined in the sump project. Digging commenced almost immediately and rapid progress was made following the line of the divers' dig. Up to four trips a week were made as it was felt that the onset of rain would mean a rise in water levels and that the soakaway into which the water disappeared would become choked. However, luck was on the side of the diggers as October 1969 was one of the driest on record.

The sump which had long prevented further progress in St. Cuthbert's was finally passed by members of the Shepton Mallet and BEC on Friday 31st October 1969 and the entry in Vol. 6 of the Shepton Hut Log by 'Milche' describes the view over the silt bank as 'momentous'. That breakthrough was halted, perhaps inevitably, by another sump which henceforth became Sump 2 and the site of the original dig, Sump 1.

They had found over 600 feet of new streamway in a high rift passage that headed uncannily straight in the direction of Wookey Hole.

Fig. 20.6
St. Cuthbert's Two Streamway.
Photo by Peter Glanvill

Fig. 20.7
St. Cuthbert's Two Streamway.
Photo by Peter Glanvill

Unfortunately, however, Sump 2 did not prove to be 'relatively short' as had been hoped and was found to be solidly choked. In an article in the SMCC *Journal*, Martin Mills reviewed the options tried, since diving was out of the question:

It appears from the Logs that work was started on Sump 2 in July 1970 by Crange and the 'Tuesday Night Team'. There had been speculation that there may have been a high-level route over Sump 2 and there had been a considerable amount of climbing done in St. Cuthbert's Two; but, in an entry dated 30th June 1970, Crange notes that no such route appeared to exist. Following a number of digging and banging trips, during which an apparent passage above what is now the dig was opened up, Crange notes that there appeared to be a draught from a hole at the end of the passage [in an entry dated 16th November 1971].

It is interesting to note that Tony Jennings made a similar comment on a trip on 12th June 1982. However, on a trip on 14th December 1971 Crange and others were affected by a CO_2 build up and I also noted that the air in the dig on 12th June was not good. I wonder, therefore, whether this draught exists. These digging trips went on until September 1972 and at this stage the passage went straight in to the air bell... Following a large number of dam building trips it was possible to bail the sump during a short evening trip, usually on Tuesday night, and carry out some digging. Although, to a certain extent, this depended on the weather.

It was not until downward progress was made that problems began to occur which led, eventually, to a three and a half-year halt in digging. In an entry dated 2nd September 1972 Crange notes that water appeared to be coming back from Three making bailing difficult. Eventually, Tuesday night trips were not possible and work stopped until March 1976.

Work was restarted on the 31st March 1976 by Gay Meyrick and others and lasted until August that year during which there were 13 recorded trips. Some progress was made but in the end the team was unable to overcome the water problems. Since that time there have been a couple of attempts to make progress with

Fig. 20.8
Argie Glanvill on the
Sour Hall dam.
Photo by Peter Glanvill

explosives although these have had very limited, if indeed any, success. It was not until 1982 that there has been enough renewed interest and support to allow a more serious attempt.

The problems of even being able to bail the sump dig are quite considerable as the preliminary trip on 12th June 1982 proved. There were some 12 people working at the sump on this trip, and we were able to (almost) bail the sump in about 4 hours, during which time it was calculated that some 2,500 gallons of water was bailed behind the 3 dams.

The rest of the story is taken up by Stuart McManus, himself a Water Engineer by profession. His account covers work at Sump 2 from 1982 to 1985:

Further attempts to break through Sump 2 were started in early 1982. Drilling over the top of the sump as previous cavers had tried proved to be very slow going and drill-rods frequently broke. So, it was decided to consider following the stream and bailing the sump out! The approach passage to Sump 2 already had one dam, made out of cement, sand, local rock and sediment in 1970 with a capacity of over 2,000 gallons. It had been used during previous attempts to reduce the sump level sufficiently to bang away at the small chamber just prior to the passage dipping away steeply underwater.

April 1982 saw the construction of another dam called the Sand Bag Dam, which increased the storage capacity by a further 2,000 gallons. More bailing attempts took place in June of that year. It was clear that any attempt on bailing the sump would require good weather, sufficient capacity in all the dams, many buckets and considerable manpower to be successful. The usual procedure was to insert the dam at the Mineries pond some three weeks before-hand, so effectively reducing the amount of water entering the underground catchment area. The Plantation stream on the surface was diverted into the St. Cuthbert's depression so that water could be held behind the dams above the cave entrance to reduce the amount flowing to the underground dams at the Beehive, the Gour Hall and Sump 1 areas.

On the day of the attempt we always started early with the 'dam putting-in' team descending the cave some 2 to 3 hours before the main team to allow the stream in St. Cuthbert's Two to drain away before bailing started. The bailing team consisted of between ten and twelve cavers who would spend the best part of 8 to 9 hours passing buckets between each other until the dams became full, either at Sump 1 or at Sump 2. Then we would release the dams back into the streamway again! The team would leave the cave cold and tired to plan the next bailing operation. Whilst draining the sump in this way to prepare for digging, we realised that the holding capacity of the existing dams would need to be increased and that we would have to think of novel ways of improving the storage space at Sump 2.

During the various bailing epics it was noticed that, once the water had been stopped in the St. Cuthbert's Two streamway, the Sump 2 level would drop by as much as 18 inches over the next 1 to 2 hours. This indicated that the sump was indeed a true syphon with a mud-grit bank on its downstream side through which the water could filter.

A procedure was then adopted whereby, once the streamway had been drained, water held within the dams (that between cave floor and the dam's outlet pipe) was bailed forward into the sump to displace water through the downstream side into St. Cuthbert's Three! This allowed extra capacity of a few thousand gallons to be gained in the dams. The technique worked well and we succeded in lowering the sump still further, though the passage still went steeply downwards. Also, for the first time we could bail for 6 to 8 hours, then dig a little spoil out of the blocked passage for another 1 to 2 hours before the Gour Hall dam upstream overflowed. Very quickly afterwards, the already full dams at Sump 2 would also overflow and bring work to a halt.

It was during one of our many discussions in that much loved seat of learning, the Hunters', that Archimedes' Principle was brought to bear on the Sump 2 problem. As water could be displaced easily, the introduction of empty plastic bottles in the sump at the end of a bailing operation when the dams were released would enable the bottles to displace the same volume of water as their total volume when pushed into the sump pint for pint! This idea had two attractions; firstly, we could increase still further the capacity of the dams at Sump 2 as the bottles would float on the water within the dams, and secondly, we would effectively reduce the actual volume required to be bailed so giving us more time to dig, once the sump was bailed. Over the next twelve months, hundreds of one gallon bottles were taken to Sump 2 to put the principle to the test.

Fig. 20.9
Plastic bottles used for water displacement in Sump 2, 1984.
Photo by Peter Glanvill

Now, as I have already indicated, you would actually need thousands of bottles to displace thousands of gallons of water and, before the first attempt with the 'bottle trick', you had great difficulty seeing the sump let alone getting to it. It necessitated the use of a suspended line along the passage on which the one gallon containers were hung, so that the place became more like a Moroccan bazaar than a cave passage!

It was at the start of 1984 that our minds turned to other methods of bailing the sump; pumps! A trial with two 1,500 gallons per hour double diaphragm hand pumps was carried out in July of 1984 with great success, taking about 5 hours to bail out what had taken 10 cavers 6 to 7 hours before. Another major consideration was the reduction of bad air because we did not need so many people. Only one or two men were needed for the manual pumps.

The combination of hand pump and use of the plastic bottles gave some success but the actual working time for digging was still only an hour or so. It was necessary to pump out the sump as quickly as possible to increase the time for digging and allow the dams to last longer upstream of St. Cuthbert's Two before they overflowed. So, calculations were made on the feasibility of running electricity to Sump 2 to power a 110 volt submersible pump. This was deemed possible and cable was acquired for the 'big push' in the summer of 1985; 1,000 metres in total! All the dams were refurbished and a 6-inch valve installed on the entrance rift dam to ensure better water tightness. The construction of a third dam in St. Cuthbert's Two was also completed, called the Kariba Dam. This dam was destined to hold back between 4,000 to 5,000 gallons of water and was nearly 5 feet high. Again it was made out of cement, sand, rock and sediments with the dam front being backfilled with plastic bags of debris dug out of the sump during the previous year's work.

Fig. 20.10
Tim Large operating a hand pump in St. Cuthbert's Swallet, 1984.
Photo by Phil Romford

Fig. 20.11
Fireman's hose across the Sandbag dam, 1985.
Photo from CDG records

It was at this time during autumn 1984 that preparations were underway for the 1985 British Cave Rescue Council Conference to be held on Mendip. Suggestions from Jim Hanwell that it might be possible to borrow some hose from the Somerset Fire Brigade to carry out an exercise of pumping out Sump 2 using the hose line to drive an air-driven submersible pump, as part of the conference the following May, gave an attractive alternative to the potentially more dangerous use of electricity. It was calculated that 60 fire hoses would be needed to run from the Belfry car-park to Sump 2; a distance of about 1,500 metres with some for spare. The collection of these hoses, their delivery to Sump 2 and subsequent removal is another story. Suffice to say that, without a lot of help and goodwill from the Chief Fire Officer, Nigel Musselwhite, and his colleagues it would not have been possible. A suitable air-driven pump was scrounged from a well-known air tool company, and this with telephone cable, kitchen facilities, and food dumps were all taken down to Sump 2 ready for the 'big push', starting on Saturday 18th May 1985.

With everybody keeping an eye on the weather, the Mineries dam was inserted in an attempt to reduce the water entering the cave. The pushing team was now chomping at the bit! The operation was probably the biggest pumping task ever carried out by cavers at that time; a case of the BEC 'doing it to excess' again!

The capacity of the pump was so high at over 16,000 gallons per hour that the sump was drained to our previous best in under 30 minutes. It was then drained still further to a depth of about 10 feet so that digging could begin, and all within an hour of starting. The pumping was so successful that a plan was put together to keep the operation going every weekend until the sump had been passed. Morale was very high. On the first day over 80 cavers came down to the sump and many more were on the surface. It had taken nearly a year to get all our one gallon bottles to the sump and all were removed in one afternoon!

THE BEC

EVERYTHING TO EXCESS

> 20/5 St Cuthberts
> Morning after the MRO piss-up.
> Mac, Darney & Bob Cork went down at 8.00, Fire Brigade turned up at 10.00, and the pump was started. The sump was bailed by 11.00 and digging commenced.
> Steve, Trebor, Tim Gould & I went down at 11.00 with Chris Castle in ungainly pursuit. At the bottom we manned the Camba dam and organised the phone and the hordes of sight seeing 'Officials'. After a couple of hours, Butch and I were at the digging face. The pump was deafening and ear protectors were needed. Water slowly filled the site from in front of us and so the pump was needed quite frequently. After about 30 mins I took over at the front with Steve behind. The pump worked really well on the water and I dug forward a few feet. The passage was levelling off!
> Knackered after an hour or so I went back to the 'canteen' a few hundred feet back for a breather. We were going to rest for an hour and then go down and take over again...

Fig. 20.12 Extract from Mark 'Gonzo' Lumley's log book, 20 May 1985

Fig. 20.13
Stuart 'Mac' McManus carrying the submersible pump, 1985.
Photo from CDG records

Fig. 20.14
Brian Workman operating the submersible pump, 1985.
Photo: Peter Glanvill

The Sunday also saw another successful pumping session with good underground and surface team support consisting of over 56 cavers helping out below in teams of six. With the pump it was possible to clear the sump and dig at the blockage which consisted mainly of lead tailings from the washing operations of the miners over the centuries. It was, however, necessary to run the pump virtually continuously to remove water that was coming back from the downstream side of the sump. This ominously indicated that it was a deep U-tube.

By the middle of July the sump had been pushed to a depth of 25 feet below normal water level and 65 feet in length, yet still the passage was going down. Time was running out for the team, with the supplier asking about their pump and the Fire Brigade concerned over what they would do if, for example, three jumbo jets crashed on Taunton at the same time with some 4,500 feet of their spare hose underground! The space available for both spoil and water from the sump was also reducing because of the fact that every cubic foot of choke removed meant that an extra cubic foot volume was required in the dams. All this coupled with the mounting cost of hiring and running the compressor 12 to 14 hours per day meant the operation had to be abandoned at the end of July 1985.

So, what of the future prospects at Sump Two? Some people would say that the sump has already reached the magical 480 feet below the surface; the depth at which all Mendip cave passage seem to peter-out and all that is left are small phreatic tubes. But, that well known Mendip saying, 'Caves be where you find 'em', should keep teams trying to overcome the obstacles at Sump Two for sometime to come. The facts remain that the sump level drops 18 inches once the stream is stopped and that there is still over 100 feet of height to drop and 7,000 feet to go to the Twenty-fifth Chamber in Wookey Hole Caves. After all, St. Cuthbert's Two is still the nearest point downstream to Wookey Hole reached by cavers so far. Full details of the saga of exploring and surveying this system have been compiled by Dave Irwin in a monograph for the BEC simply entitled *St. Cuthbert's Swallet*.

It is still possible to attempt a pumping operation using an electric submersible pump, with even bigger and better dams. Alternatively, we could use a syphon device to pulse the sump and remove the large quantities of tailings, but remember what happened last time miners did that! We cannot risk polluting the River Axe at Wookey Hole as they did. Lastly, the old idea of driving a heading over the top of the sump remains a possibility, particularly with new light weight battery driven drills that have come on to the market. Diving the St. Cuthbert's sumps has been out of the question because they are so silted by comparison with the more open ones along the Swildon's Hole streamway. Breakthroughs have only come from well co-ordinated bailing and digging operations in the past thirty years since Don Coase first reached Sump One. And for the future too? We'll see.

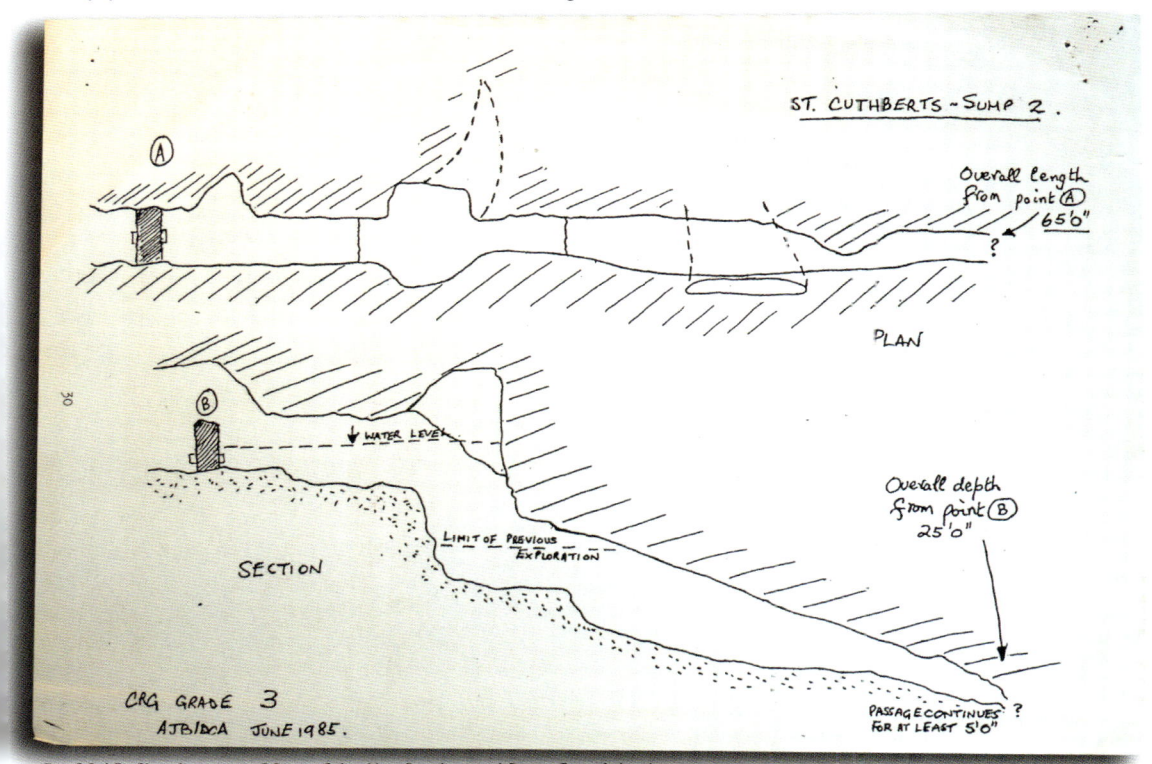

Fig. 20.15 Sketch survey of Sump 2 by Alan Butcher and Dave Cave-Ayland

Fig. 21.1 Divers depart on an underwater mapping mission at Wakulla Springs, Florida. Photo by Wes Skiles

Chapter Twenty One
The Shape of things to Come

by Oliver Wells

This book describes more than fifty years of cave diving from the helmet diving at Wookey Hole in 1935 until the present day. It is to be expected that developments during the NEXT fifty years will take place along two parallel lines. First, and most important of all, human divers will continue to improve their skills. But in addition to this, unmanned vehicles of various kinds will extend the capabilities of the divers.

It is obvious that the sporting aspects of cave diving will never be lost, because no amount of high technology will destroy the opportunity for the enterprising individual to find a successful diving site that has been overlooked by everyone else for many years. Discoveries made by automatic devices will set objectives which the traditional cave diver can try to attain. There will always be the attractive 'man bites dog' aspect of a human diver who succeeds where his automated colleagues have failed.

The idea of automatic devices for cave diving is not new. In the early 1950s, Graham Balcombe told me that to survey underwater, the diver must carry a device containing accelerometers that will plot the journey automatically. I did not believe him when he said this to me, but slowly I begin to see the point. When Sir Robert Davis was asked what he foresaw as being the most promising future developments in diving in general, he answered:

> Television ... Submersible TV did wonderful things on the Affray search and the Comet that fell in the sea at Elba ... (James Dugan, Man Explores the Sea, Hamish Hamilton, London, 1956.)

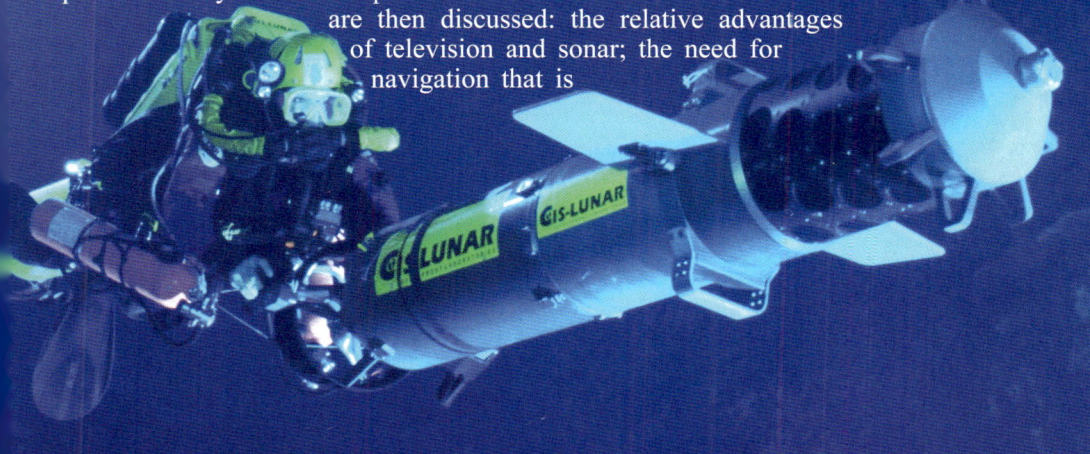

The plan for this chapter is to consider: first, what human divers might do in the next fifty years; next, what might be done with remotely operated vehicles (ROVs) that are controlled by a human operator on dry land. The special features needed for an ROV in a cave are then discussed: the relative advantages of television and sonar; the need for navigation that is accurate enough to retrieve the vehicle in muddy water through squeezes and around corners; operating an ROV without a tether under 'cave radio' control; the various forms of remote sensing that might be useful, and so on. The final question is what can be done with an untethered exploring vehicle that operates beyond the range of 'cave radio' or any other form of direct communication. An example of this approach is the Voyager 2 spacecraft that photographed the planets Uranus and Neptune while operating unattended for many hours under the control of programs stored in a computer. Underwater explorations in caves to distances of tens of kilometres at depths of hundreds of metres might be possible in this way.

To start with human divers, it is remarkable how much progress has been made in the past fifty years in both equipment and skills. Present-day divers swim for distances and at depths that would have been unthinkable only a few years ago. It seems as if the possible limits for human divers have almost been reached. But on the other hand, we felt that we were fairly close to the limits in the 1950s, and you can see what has been achieved since then.

In order to predict what a diver will be able to do fifty years from now, we can consider the life-support, sensing, navigation, communication and propulsion equipment carried by a small manned submarine. The question is whether, over the next fifty years, devices of this kind will be miniaturised, improved, and made cheaper to the point when the individual diver can be instrumented to roughly the same degree, or possibly even more so, as the present-day small manned submarine. If the answer is 'yes,' then we can expect that the cave diver fifty years from now will be enormously more versatile than today in almost every respect. For example, he will no longer use a guide line on the longest dives. Instead of this, he will be pulled at high speed through muddy water using a next-generation scuba-scooter guided by sonar, with the coordinates recorded all the time so that he can know exactly where he is relative to the survey. He will then return to base using an automatic navigational system to show him the way.

Some of the methods that can be used by human divers are as follows. At the present time, cave divers use open-circuit respirators, in which either air or trimix is breathed once and then blown into the water. For the longer dives, respirators are exchanged underwater. On major expeditions, the number of cylinders that must be carried through the cave is considerable. As described in the chapter reviewing the last major push along the River Axe in Wookey Hole Caves during the summer of 1985, 'trimix' is a mixture of oxygen, nitrogen and helium that is carefully chosen for a particular set of diving conditions.

The closed-circuit respirator which at one time seemed so attractive for cave diving (and which was rejected by cave divers in England in the 1960s) has meanwhile been re-worked for military use. This is now called the 'sensor-controlled rebreather', and with present designs will last for six hours down to a depth of 300 metres, when using mixed gases. These rebreathers are described by David Sisman, editor of the *Professional Divers' Handbook*, Submex, London, 1982, pages 166-169, and by authorities in today's technical journals such as J. Rawlins and A.L. Carnegie in *Technological News*, Normalair-Garrett, 1981,

*Fig. 21.2
Bill Stone with
double rebreather.
Photo by Wes Skiles*

concerning 'A short history of diving' and 'Gas control valves for divers' respectively. The breathing circuit, which includes the breathing bag from which the diver breathes, and the soda-lime canister to absorb carbon dioxide, works on the same principle as before. The innovation is in the means used to control the oxygen partial pressure in the breathing mixture. The oxygen, which is consumed in small quantities to sustain life, is contained in one cylinder, while a second gas, which can be either air or trimix, is contained in a second container. The partial pressure of oxygen in the breathing mixture is measured with a sensor. Either one gas or the other is admitted into the breathing circuit in such a way as to maintain a safe situation at all times. This apparatus is hardly suitable for general use, and yet for pushing the limits in cave diving, it does provide an alternative to present-day techniques.

Muddy water presents a serious problem for cave divers. This is bad enough for a walking diver, but when you are swimming, it can be much worse because it is then very easy to lose your sense of direction, to rise or fall in the water without knowing it, and to become disoriented generally. One possible way to solve this problem might be to use sonar, or echo-location as it is sometimes called, to show where you are relative to your surroundings. Here, it is instructive to consider the dolphin, in which sonar has evolved over a period of fifty million years to the point where this animal can swim at high speed through an underwater obstacle course in total darkness. Dolphins can locate small objects at distances of many yards and can catch them for food without seeing them. They can also discriminate between the sizes of the objects which are caught in this way (W.N. Kellogg, *Porpoises and Sonar*, Univ. Chicago Press, 1961). In air and in caves, bats and certain types of bird can do similar things. Sonar is a sense that can be developed to some degree in ordinary people if they are deprived of the power of sight. Oliver Lloyd has described how he went with some friends to Sump 1 in Swildon's Hole, extinguished all their lights, and then came out in total darkness for a distance of over 2,000 feet. In doing this, they found that by talking to one another, they quickly developed the ability to 'sense' where things were. This experiment was repeated by Jim Hanwell who also found that he could sense his surroundings in this way. The ability of blind people to avoid objects in their path by listening to echoes is well-known.

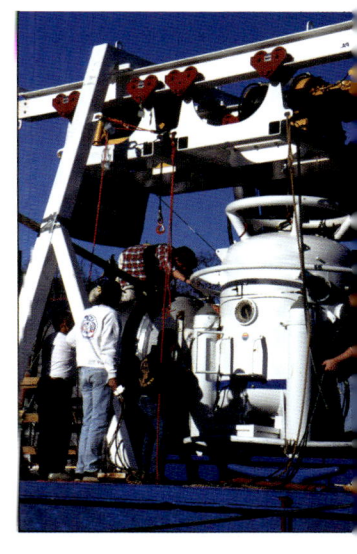

Fig.21.3 Docking transfer bell to decompression chamber, Wakulla Springs, Florida. Photo by Wes Skiles

Ultimately it can be expected that sonar systems in some form will become an essential part of the cave diver's equipment. Tom Cock has described the first steps in this direction:

One of the survey tools we decided to use was a hand held SONAR unit called the Dive Ray. It looks like a yellow gun, 12 inches long. Its range is 99 feet, and the distance readings are displayed at the rear of the gun with a backlighted liquid crystal display (LCD). It is accurate to about 6 inches... surveying a single room underwater cave, we just stood at the central point and beamed readings from all points.

(Thomas Cook: 'The use of the Dive Ray for underwater surveying.' The Northeastern Caver XIII, No. 4, 112-113. Fall, 1982).

Sonar can also be used to measure velocity by the Doppler method. Navigation by Doppler sonar is described below.

Significant advances can also be expected in the use of scooters.

Even at the present time, some amazing distances have been achieved in clear water by cave divers using scooters in Florida and elsewhere. The next step might be to apply the same techniques in muddy water using sonar. A modern scuba scooter can pull a diver at a speed of 2.25 mph (=200 feet/minute) for a distance of 3 miles (=15,840 feet) in about 90 minutes. In comparison with this, Oliver Statham and Geoff Yeadon swam at 40 feet/minute for 150 minutes for the through dive of 6,000 feet length at Keld Head on 16th January 1979. The attraction of scooters is obvious.

The next step is to consider what can be done by remotely operated vehicles (ROVs) controlled from dry land. Many examples are given in recent literature on their use and potential, as in David Sisman's handbook, already mentioned, first published in 1982. Also of interest are papers written by D. W. Partridge on the 'Future development of ROVs' (*J. Soc. Underwater Technology*, Spring 1984), on 'Wide angled, high resolution underwater viewing' (*Underwater Engineering*, February 1985), and how 'UK remotely operated underwater vehicles technology could provide great benefits' (*Petroleum Review*, March 1985). J. A. Adam writes about 'Probing beneath the sea' in *IEEE Spectrum* (USA), April 1985, and J. B. Tucker describes the dives to survey the sunken liner *Titanic* in an article entitled 'Submarines reach new depths' for *High Technology* (USA), February 1986. Back at the beginning of the eighties J. D. Westwood described 'The application of television to remotely operated underwater vehicles' (*J. Phys. E: Sci. Instrum.*, 14, 1981), whilst the vehicle that was eventually used to recover the Air India flight recorders after it had plunged into the Atlantic in 1985 was described by H. R. Lunde, G. A. Reinold, and P. A. Yeisley in Bell Laboratories *Record 59*, No.7, (September 1981). The uses of these devices are well worth reviewing in relation to future cave diving operations.

Fig. 21.4
Remote video grab of diver returning to transfer bell, Wakulla Springs, Florida.
Photo by Bill Stone

It has become an established principle in commercial and scientific diving that the diver must stay out of the water as far as possible whenever danger is involved. In cave diving, a good start has been made in the use of ROVs and it is to be expected that this will be greatly developed in the next few years. ROVs have been used in the Fountain of Vaucluse since 1967. The deepest cave dive so far was by the ROV 'MODEXA 350' to a depth of 315 metres in that same cave in August 1985. This surpassed the depth of 243 metres reached in a truly astonishing 9-hour dive (also in Vaucluse) by Jochen Hasenmayer in September 1983 (*Spelunca*, Oct-Dec. 1983, page 10 and August 1985, page 12). Underwater television was used by Richard Stevenson, Rob Harper and Rob Palmer during an exploratory cave dive in the River Yeo at Cheddar on 26th March 1986. This is surely only the beginning of the use of remotely controlled devices during primary exploration in caves.

ROVs for commercial and scientific diving are not restricted in size as they must be for cave diving. Thus, the ROV 'Scarab' which recovered the Air India flight recorders from the Atlantic in June and July 1985 was developed for underwater cable repair work. This remarkable machine can locate a telephone cable under the mud on the ocean floor from the magnetic field, dig a trench to expose the cable, cut it, attach a rope from above, wait while it is pulled up to be repaired and then replace the cable in the trench and bury it again. All of this is done under remote control. The 'Scarab' is, however, about the size of a large car. The submersible 'Argo' which was used to photograph the *Titanic* in September 1985 is only slightly smaller. In both cases, the vehicle was lowered into the water using winches and cables of the size that one expects to see lifting cars onto an ocean-going ship; except that, in both cases, the cable was thousands of metres long, being more than 2,000 metres for the flight recorders and 4,000 metres for the *Titanic*. Even the 'MODEXA' which reached a depth of 315 metres in the Fountain of Vaucluse was about the size of a motorcycle.

It is clear that miniaturised versions will be needed in caves. Commercial ROVs have tethers about a centimeter in diameter that contain a coaxial cable for the television signal and typically about ten fairly substantial copper wires to carry power for the motors and television lamps. Much thinner cables will be needed, and this implies that the vehicle must operate from batteries in the manner of a scuba scooter, rather than having several hundred watts supplied along power lines from base. This raises problems concerning the rate of power consumption, and the powerful lights presently used for television may not be possible (perhaps the operator will have to settle for one flash picture every second). An optical fibre about as thick as a human hair can transmit television pictures and power over tens or possibly even hundreds of kilometres, and it is conceivable that this might be used as a tether for an ROV in a cave.

Fig. 21.5
The DEPTHX vehicle during transport to a quarry in Austin, Texas.
Photo by Dominic Jonak/Carnegie Mellon University

Fig. 21.6
The DEPTHX vehicle in Poza la Pilita in Rancho Azufrosa, Mexico, during testing of the sample acquisition system.
Photo by David Wettergreen/Carnegie Mellon University

Untethered vehicles that are directly controlled by a signal sent through the water have already been developed for commercial applications as described in David Sisman's handbook, cited earlier, on page 124. Under these conditions, it is almost certainly impossible to transmit a television image, because the data rate that can be achieved through water (even if there is an unobstructed path between the transmitter and the receiver) is so low that it will take several seconds to send a single still picture. In caves, a direct line-of-sight through the water will usually not be available, and a slow method of communication such as 'cave radio' will probably be the best that can be done. This suggests that for use in caves, the vehicle must operate in the manner of the Voyager 2 spacecraft at the outer planets, in which the moment-to-moment decisions are made by a suitably programmed microcomputer, and the messages which are sent by the operator are of a very simple nature, such as:

1) Follow that wall at constant depth.
2) Go as deep as possible.
3) Follow the line of fastest water flow.
4) Move in the direction of the largest sonar offset.
5) Stay still until the water is clear enough to take a television picture.
6) Switch off the motor and drift downstream with the current.
7) Survey accurately within 100 feet of the present position and measure the water flow at every point.
8) Survey the salt-water/fresh-water interface.
9) Measure the tidal flow at that point for the next 12 hours.
10) Come home now, etc.

Each of the above procedures, and others like them, would have its own program stored in a computer on board the ROV. Indeed, even human cave divers very often plan their dives on the basis of simple procedures that are chosen with great care before the dive begins. In addition to the above, the sonar must be continuously active to survey the passages to ensure a safe return.

The final possibility is to design an untethered autonomous vehicle that operates without instructions of any kind. This would be controlled by a computer program that automatically makes choices from a list of options of the kind listed above. This has been described as follows:

> *It's called the high-anxiety-level vehicle. You let it go and hope that it comes back. (Attributed to D.R Blidberg by Adam in 'Probing beneath the sea', 1981, cited earlier.)*

In practice, more than one idea might be combined in a single vehicle. For example, television is excellent in clear water but is useless as soon as the mud rises. An ROV operating with an optical fibre to give a 'live' television image might be programmed to return to base along the same route automatically by sonar if the mud rises or if the fibre breaks. A vehicle might find its way along an underwater passage by sonar using a program stored in a computer, but record a television picture on tape to be viewed later. During the initial stages of an exploration it might be best for the vehicle to keep returning to base to play through its television tape and receive further instructions. And so on. The possibilities would appear to be almost limitless once the basic technology has been worked out.

In case all of this might seem to be too clever to be true, we can point out that most if not all of the techniques needed by autonomous exploration devices in caves have already been demonstrated in commercial and aerospace devices of various kinds. For example, oil tankers which are close to shore can measure their absolute speed from the Doppler shift in the sonar echoes from the sea floor. This suggests a similar system for use underwater in caves. As an alternative, inertial navigation systems used in aeroplanes can keep their accuracy within a drift rate of about 1 nautical mile per hour. Underwater, this can be improved to of the order of a metre by resetting the system every few minutes from the Doppler shift in sonar echoes. In advanced systems, the magnetic compass has been replaced by the more accurate ring laser gyro, in which a sense of direction is maintained by measuring the optical interference between laser signals sent in opposite directions either round paths in a block of transparent material or around coils of optical fibre. To improve vision in muddy water, there is the attractive proposal of range gating, in which the light source is pulsed, and the television camera is switched on briefly when the light wave returns from the chosen distance, thus greatly reducing the scattered light from the water that is just in front of the camera. At the present time, these techniques are too expensive for general use, but it would seem that eventually they will be applied to cave diving.

A 'subject index' on these topics might be compiled from the sources given earlier and the details available on navigation devices. Sonar navigation used by oil tankers is featured in an editorial article entitled 'Getting the drift' (*Exxon Marine*, 27, No.1, Spring 1982); inertial navigation underwater is described by K. Tregonning in the *Offshore Engineer* (January 1983); ring laser gyroscope applications are considered by G. M. Martin in *IEEE Spectrum* 23 (February 1983), and a specification sheet for the 'Trilag' version is available from The Singer Company, Kearfott Division, at Little Falls, New Jersey, in the United States.

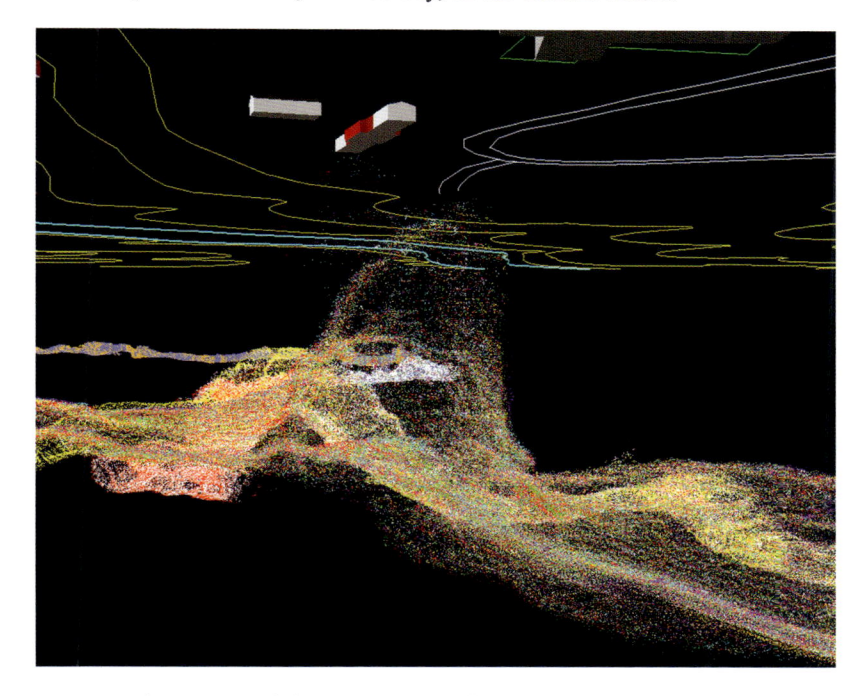

Fig. 21.7
3D computer generated map of the Grand Canyon, Wakulla Springs, Florida.
Photo by Bill Stone

It can be expected that remote sensing will become increasingly important in cave diving. One such technique is penetrating sonar, in which the underwater maps produced by sonar can show details through either mud or rock. Thus, Dr. Harold Edgerton, in his book *Sonar Images*, Prentice-Hall (1986), shows a picture of the the road tunnels under Boston Harbour in the United States that was obtained by low-frequency sonar through the mud and the water from a boat on the surface. The possibility that ROVs might explore by penetrating sonar below the mud floors of cave passages and through the limestone walls is one of the many possibilities that will surely be developed within the next fifty years.

As an example of unmanned exploration, we can imagine that if a free-swimming automated explorer were to be introduced into the resurgence at Wookey Hole, it should have no difficulty in exploring by the submerged route beyond the limit of the tourist area in Nine, but it would need human help to be carried over the dry and flowing sections in Chamber 24. The real payoff will be at the farthest point reached so far. Indeed, one can easily imagine the excitement of using a small and fast vehicle trailing an optical fibre to give a television image as it passes through the small, deep opening that is the present limit of the exploration. Or alternatively, it may be better to use an untethered explorer. If either of these methods is successful and if a major extension to the cave is found, then there will be an interesting contest between the cave divers and the ordinary cavers on Mendip to see who can get in first.

For Swildon's Hole and St. Cuthbert's Swallet, where there is a lot of carrying to be done, and where the final sumps have proved to be too tight for divers to pass through, a miniaturised version will be needed. In these cases, the vehicle will be diving downstream, and television will probably be useless because of the muddy water. Sonar will provide the necessary guidance and surveying information. In the ultimate, such an exploring device can perhaps be made only a few centimetres in diameter, which suggests that explorations can extend not only deeper and farther, but also into smaller passages than can be explored by human divers. And yet I doubt whether with even the most futuristic technology you can satisfy the whimsical request of Derek Ford who wrote in the margin of an early draft of this chapter:

How about an automated explorer less than 1 cm. in diameter. I am interested in speleogenesis?

In summary, it seems that techniques are developing so rapidly that it will be only a matter of time before cave divers explore routinely beyond the farthest points that can be reached today. It is becoming obvious that we are only at the beginning of what can be done in the underwater exploration of caves.

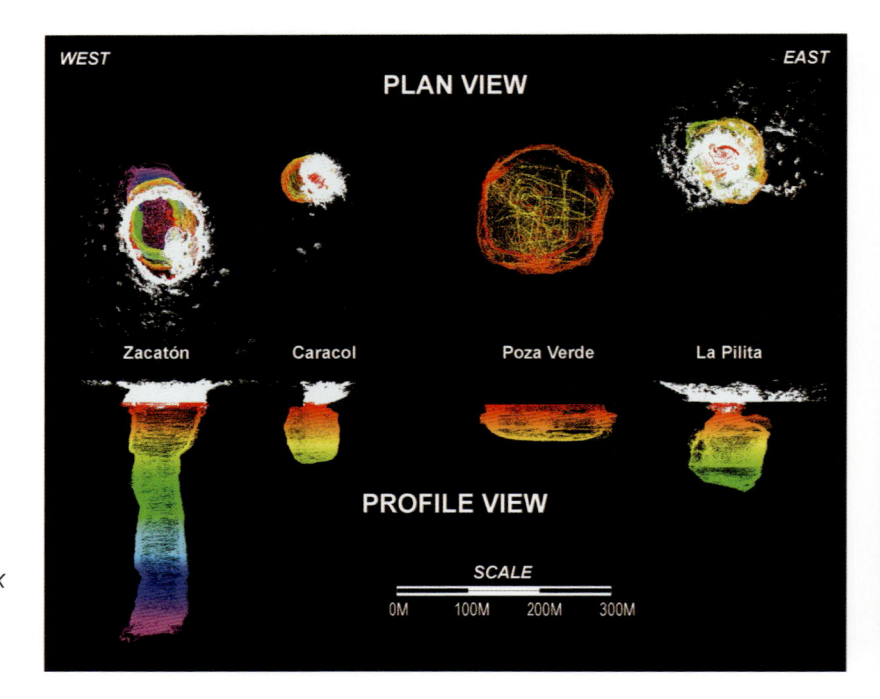

Fig. 21.8
Three-dimensional maps of the cenotes of Rancho Azufrosa, Mexico, created by the DEPTHX autonomous cave exploration robot.
Photo by Nathaniel Fairfield/ Carnegie Mellon University

THE SHAPE OF THINGS TO COME

Fig. 21.9
Divers using long range 'scooters' and rebreathers in Wakulla Springs.
Photo by Wes Skiles

Chamber 3, Wookey Hole, 4 October 1985. Photo by Chris Howes

Operation Jubilee

The 50th Anniversary of cave diving at Wookey Hole on 4 October 1985 was the occasion of a grand reunion of pioneer and current divers.

Maire Trendell (née Urwin)
Phil Collett
Bob Cork
Jack Sheppard
Dany Bradshaw
Rob Parker
Dan Hasell
Julian Walker
Ian Rolland
Graham Balcombe
James Cobbett
Rob Palmer
John Parker
Dave Morris
Brian Woodward
Trevor Hughes
Dave Savage
Tim Reynolds
Phil Davies
Mike Wooding
Ken Dawe
Pete Eckford
Steve Wynne-Roberts

THE CAVE DIVING GROUP
OPERATION JUBILEE

[Page of signatures]

1986-2010

Photo by Gavin Newman

Chapter Twenty Two

Lighter and Faster

by John Cordingley

Photo by Gavin Newman

    I suppose like many events which stick in one's memory, the foundations for this particular trip were laid a long time beforehand. The venue was the Helwith Bridge pub near Settle in the Yorkshire Dales, early in 1976. In those days it was one of THE places to be on a Saturday night if you were a caver. I was still a teenager at the time but I remember the evening in question as if it were yesterday. There we all were, very much the worse for wear after a long session. There had been a lull in the singing in the back room and you could hardly see across to the far side through the smoky haze. Suddenly the door crashed open and someone shouted 'Geoff and Bear have got through Wookey!' The whole pub erupted! We danced around and supped yet more beer – of course I had no idea then of the history or controversy behind this breakthrough, not to mention the wickedly brilliant manipulation by a certain Bristol pathologist! I was just young, daft and carried away on the wave of speleological chauvinism which was rife at the time. Little did I realise that one day I would be swimming along those same magnificent tunnels and have the privilege of coming to know many of the CDG greats who will always be associated with them.

    Wookey Hole is a very special place. It's regarded as the spiritual home of British cave diving and in the 1980s groups of us from the northern wastes were often kindly invited there by local cave divers, so we got into the habit of going to Wookey quite often; petrol was a lot less expensive in those days! In fact I reckon a visit to 24 and back is pretty high on my list of the best of British caving trips. We had also been involved alongside our Somerset Section mates in various expeditions with some of the best deep cave divers in Europe around then, from whom we had gained much technical knowledge which just wasn't available in the UK at the time. The combination of deep diving experience and familiarity with Wookey led to our discussing the feasibility of going to see if the end could be pushed. However, as several of us lived a long way away, the 'siege' approach used by previous explorers was not really feasible. We worked out that it should be possible to do it just on a single trip using various combinations of gases to minimise the possibility of decompression sickness. Andy Goddard and friends had almost managed this not long before and had only been stopped by regulator problems near the start of the final sump. I don't think anyone had been to the end since Rob Parker had shuffled down the steep gravel slope beyond Martyn Farr's limit and there was no true survey of any of the cave beyond 24. The Mendip divers were definitely up for having a go – and so a plan was hatched.

*Fig. 22.1
Mendip Rescue Organization Training Officer Dany Bradshaw on a sump rescue training exercise in Wookey Hole in the 1990s.
Photo from the MRO photo archive*

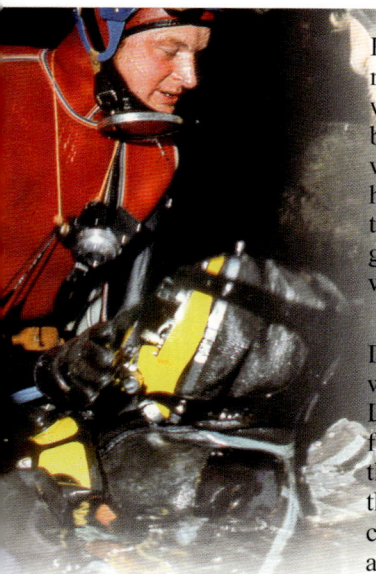

On Friday 14th June 1991, Russell Carter and I were speeding Mendipwards down the M6. The back of his little Peugeot was completely filled with big tanks, stacked vertically. At one point Simon Brooks (also on his way for a weekend's caving on Mendip) overtook us. I remember the look on his face when he first recognised us – then realised what we must be up to. He grinned and gave us the thumbs up before disappearing ahead of our heavily laden vehicle. Several hours later we were mooching around the Wookey Hole car park in the dark, having failed to make a planned rendezvous at the Somerset Section compressor. We got apprehended by the police, who had some difficulty in understanding our explanation of why we were driving slowly round the village with a car full of what looked like bombs! Robin Brown would be diving with us the following day and he and Carol had kindly invited us to stay at their place on the Friday evening. So on the Saturday morning, after a sumptuous breakfast in rather better surroundings than the usual filthy caving hut, this little adventure finally began.

The underwater team consisted of Dany Bradshaw, Robin Brown, Russell Carter, myself, Malc Foyle, Keith Savory and Alan Taylor. We made good progress through the 400 m or so of dives to Wookey 24, where the gear for the final sump was assembled just before the jammed boulder at The Lake. I counted myself lucky that it was going to be me who dived beyond here but several other people in the team could easily have done it instead. There was a lot of 'faffing' needed to get me into the four tanks and all the other paraphernalia but eventually I waved goodbye, sank below that awkward zone at the interface between air and water and then everything clicked into place.

I'm most familiar with the long peaty cobble-floored sumps of the Dales but this is no preparation for Wookey on a good day. The visibility was superb on this occasion and I realised that the underwater part of The Lake is a huge submerged rift. I floated down this until a steep gravel floor forced me leftwards beneath a series of arches to 24 m depth; then the sump reared steeply upwards. There were silt covered boulders with the line occasionally caught on them but soon the tell-tale silvery mirror came into sight and I surfaced in 25. In those days this felt like quite a lonely place but the sections of iron ladder were in situ to enable a heavily kitted diver to get over the large rock rib guarding the final sump pool and I was soon bobbing about in The Well, preparing for a long deep submersion.

I drifted gracefully down towards the bouldery floor at about 10 m depth and immediately managed to foul one of several old loose lines which were floating about in this area. This is one of the things you can just do without before a dive like this. I managed to get free but the visibility had been wrecked and the final stage cylinder of oxygen was clipped to the line by feel. A few fin strokes later I was away from the murk and following a small muddy tunnel which gently rose and ended on the brink of a great inclined rift, soaring down into the depths. The passage is in the form of a bedding plane sloping downwards from right to left. Various people had suggested that there might be an alternative way on in this final sump, so I decided to check the left hand side of this passage all the way down to look for other possibilities but none was found. At -46 m the floor disappeared altogether and I floated out into the roof of a submerged chamber.

Things were going well and I was very confident of reaching the end at this point. After dumping a bit of air from the suit the descent continued. However, at -52 m, a sloping floor of gravel appeared and I immediately spotted loose line strewn around on it. I tied on my own reel to lay line along a separate route to avoid this but straight away some of the loose line blown out in flood from the final few metres of the sump managed to wind its way around one of my head torches. Fortunately it did not take long to get this off but it caused silt clouds to roll off ahead of me. I drifted rapidly further down the slope and at -54 m briefly emerged from the silty water and realised I was at the far end of the chamber at the start of the final gravel slope down. I had an excellent view down this and in crystal clear water I could see where the gravel deposits made the passage too small ahead at about 60 m depth. The clouds rolled in again and I had another encounter with loose line. Again this was sorted without too much trouble but the visibility was now zero. I was unable to see any instruments and could no longer verify the contents of the tanks so there was nothing for it but to negotiate the coils of old line back up the sloping chamber floor and begin the long climb out.

The visibility soon improved again and it was possible to take bearings, so I became absorbed in trying to make a survey back to 24. It's a big place in there and it was not possible to survey everything just on this one visit but I did my best and managed to collect enough data to produce a half reasonable drawing for the CDG *Newsletter*. Decompression stops in The Well were a dark experience and I remember having occasional problems following the line back down the shaft on the downstream side of the 25 airspace. Eventually I was doing more decompression in the great submerged rift below The Lake and finally surfaced to tell the tale to the lads. We had not quite made the end but we would easily have done if the accumulation of loose line below 50 m had not resulted in a long delay. We had also managed to do a survey of what was then still quite a remote place and showed it was possible to get to the end on a single trip, so everyone was fairly pleased on the journey out. The only fly in the ointment was that by the time we surfaced in the show cave it was pretty certain we would be too late for the pub. But this was where Dany, bless him, really starred. He shot off like a terrier and somehow managed to generate a couple of crates of beer in double quick time. Talk about pleased; we almost kissed him! (I said almost . . .)

After this trip it was obvious that the way on in Wookey really was in the deep zone – and I was certain that I had seen a definite blockage on that occasion. We reckoned the answer would be to make repeated visits and hit it on the right day, when the level of the gravel floor was lower. To do this was beyond our time resources; Malham Cove Rising had gone big not long before, the Peak Cavern system was occupying a lot of our energies and the long underwater passages of Kingsdale were calling. So we remained in the northern wastes and never went back to Wookey but it's been great to watch the progress made by our friends in the Group over the years since then. My main memories of that trip are not so much what we achieved but of an enjoyable experience in the company of some very competent people who were great to spend time with; there is nothing quite like a Mendip welcome!

Fig. 22.2
John Cordingley, Kingsdale, Yorkshire Dales, just before the first UK 1 mile in and out cave dive. 17 April 1998. Photo by Susan Hoyle

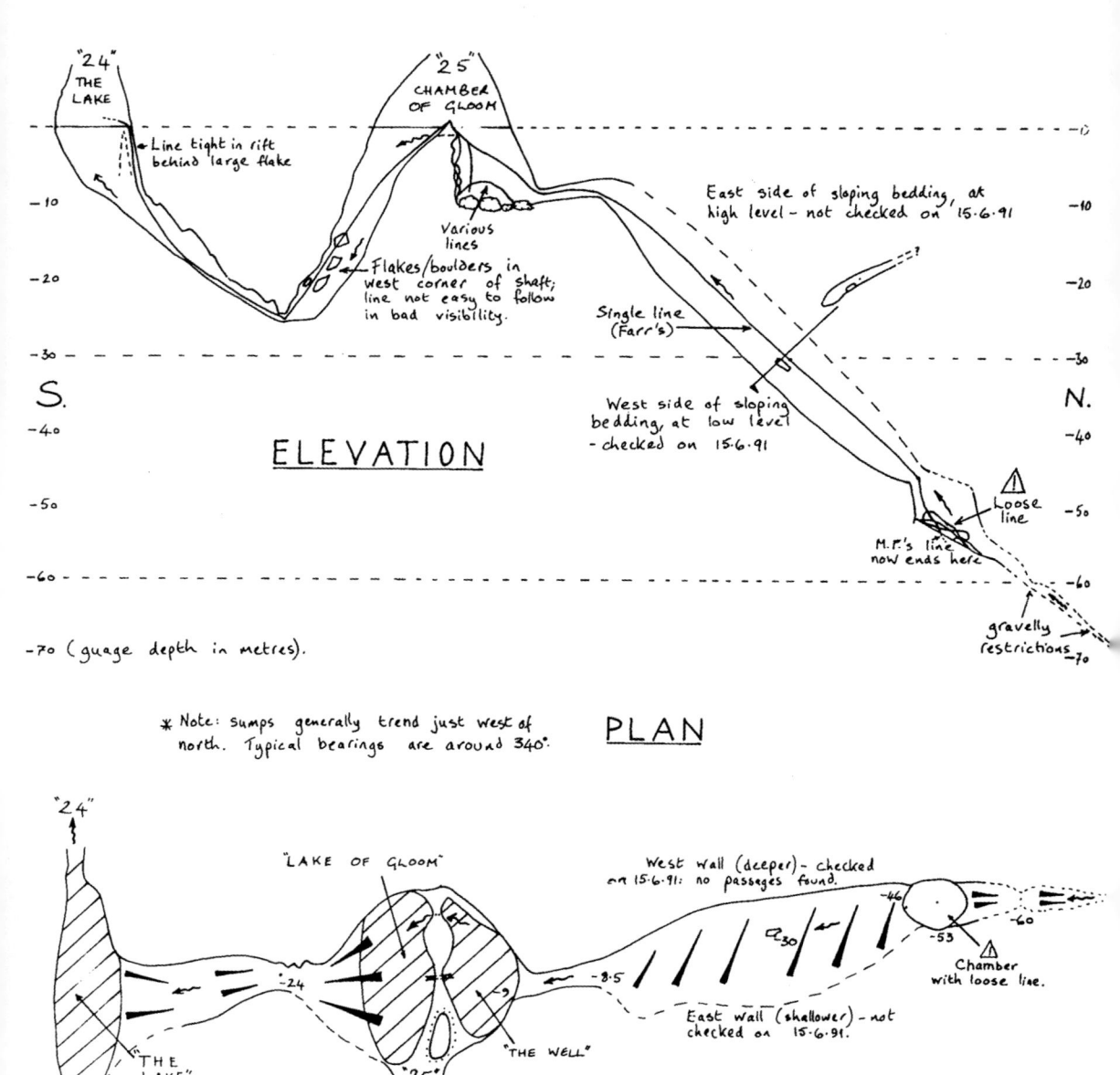

Fig. 22.3 John Cordingley's survey of Wookey Hole beyond Chamber 24, 1991

The water level at the Wookey entrance is at 230 ft O.D., the top of the aven in 22 at is around 360 ft O.D. and the end of Hallowe'en Rift is at 410 ft O.D., making a vertical separation between the two points of only 50 ft. According to the cave surveys, the end of Hallowe'en Rift is only 165-195 ft horizontally from the top of the aven, making it the most likely location for achieving a dry connection with the further reaches of Wookey Hole.

Fig. 23.4
Alex Gee, Hallowe'en Rift, 2009.
Photo by Mark 'Gonzo' Lumley

Fig. 23.5
Trevor Hughes, Caine Hill Shaft, Priddy, 2009.
Photo by Mark 'Gonzo' Lumley

Hallowe'en Rift is possibly an old resurgence or a large high-level relict passage that has been truncated by surface erosion, in the same way that the Wookey's Ninth Chamber was entered from above. The main passage is filled almost to the roof with calcite and mud infill, so appears as a low bedding about 15-20 ft wide. A shallow trench was excavated on the left hand side heading at an angle downwards of 2 or 3 degrees. This ends after about 120 ft at a flat-out wormhole dig. A side passage dug by Trevor Hughes and Tony Jarratt goes off to the right before turning to the left and running roughly parallel to the main passage. The walls, floor and roof of this passage are liberally coated in calcite. The route ends in a squeeze into a small decorated chamber at the base of a 20 ft aven with no obvious continuation[35].

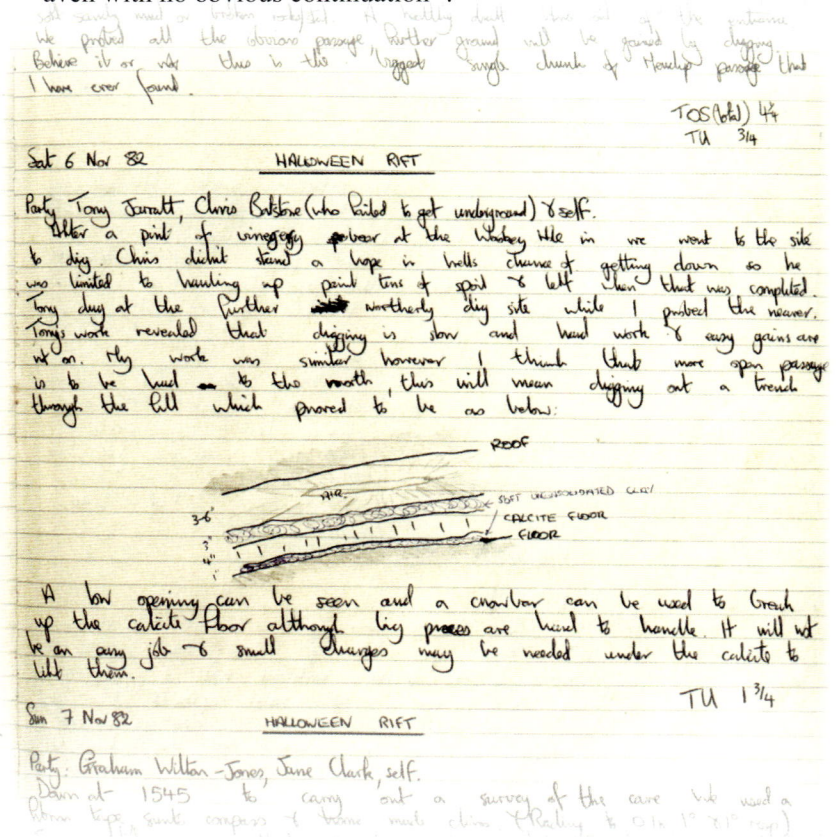

Fig. 23.6 Extract from Trevor Hughes' logbook.

At the time of writing, the terminal sump in Wookey Hole appears, despite the heroic efforts of several individuals over the years, to be all but impassable with present technology. The main hope of further progress, therefore, may lie in the discovery of a high-level bypass. Whilst much effort has been expended by previous generations of divers to achieve this (and no doubt efforts will continue to be made by current and future divers), I believe the best way to facilitate further progress would be to make a surface connection with the farther reaches of the cave. This would allow access for many more cavers and significantly increase the exploratory work. CDG *Newsletters* are full of accounts of sites such as 'small passage pushed until blocked by boulder' and tenacious Mendip diggers could be guaranteed years of happy and almost certainly productive work above the water table.

At around the same time that I was climbing the avens in Wookey 22, I was approached by Dr Roger Stenner who asked if I would assist with his ongoing studies of cave water chemistry. So, along with Tim Chapman and Clive Stell, we collected water samples, that Roger had analysed at Bristol University. The results of these studies were published in the *Belfry Bulletin* and *Cave and Karst Science*[36]. One of the conclusions of this work, based on the difference in the levels of magnesium in the water in the Static Sump (separating Wookey 22 and 23) and the main River Axe in the cave, was that the water in the Static

[35] The dig was restarted in 2009 and has received the blessing of Professor Derek Ford!

[36] T. Chapman, A. Gee, A.V. Knights, C. Stell & R.D. Stenner; Water studies in Wookey Hole Cave, Somerset, UK., *Cave & Karst Science*, Vol 26 (3) pp 107–113 (1999)

Photo by Martyn Farr

Sump comes from the same original source, but possibly divides from the main flow upstream of the terminal sump at an unknown junction. This would imply that there could be a way on beneath the surface of the Static Sump. Although other simpler explanations for the difference in the water chemistry based on the greater relative contribution of percolation water in times of low flow have been put forward, Roger's suggestion would be worth further investigation.

I conclude this essay with a cautionary tale of the effects that outside events can have on the mind of the cave diver. During 1997-98, I was under a great deal of mental stress due to business problems. Foolishly I thought this would have no impact on my caving activities and continued to dive. As a result, during a routine dive to 22, I started feel extremely claustrophobic and uncomfortable whilst halfway between 20 and 22. The consequent rush of adrenaline prompted my breathing and heart rate to increase out of control. In a blind panic, I somehow reached airspace in Wookey 22. I still don't know how I got there – I suppose that it must have been instinct and good luck. After getting myself together, I re-entered the water and endured a most unpleasant dive out. I have not cave dived since. I also experienced similar symptoms shortly after entering a dry cave a few days later. I then gave up caving for other pursuits.

I feel that adding the pressure of cave diving to the high level of stress I was experiencing at the time, was the cause of this episode – one I was lucky to survive! I also believe that this could be the cause of other accidents, where competent divers have been found drowned, with fully-functioning equipment and no discernable problems.

I was uncertain whether to disclose the circumstances surrounding my retirement from cave diving, but with the encouragement of my peers, I have done so in the hope that it may spare someone else a similar experience (or worse). In writing this chapter, I have also found that revisiting these (mostly happy) events has been most therapeutic and it has ignited a desire to exorcise old demons. This has rekindled the old caving itch. To that end I have recently returned to caving after a break of 12 years and, along with others, will be reviving the quest for a dry route into the further reaches of Wookey Hole. Despite my enthusiasm for this, I don't think I'll be the only ex-sumper to feel a little sad if our magnificent 'private playground' of the last 75 years is opened up to the 'Mongol Hordes'[37].

---

[37] These climbs were re-visited in October 2009 by Claire Cohen, Duncan Price and Phil Rowsell who used the existing ropes and belays left by Alex Gee. The karrabiner on the top belay was found to be almost completely corroded through after 11 years in the cave (but only after Phil had climbed up on it!). The way on, although promising requires significant enlargement to make progress.

Jon Volanthen, Wookey 14, 2009
Photo by Gavin Newman

Chapter Twenty Four

Digging Deeper

*based on
Mike Barnes,
Pete Bolt,
Tim Chapman
and Clive Stell*

Fig 24.1
Clive Stell in Wookey 21.
Photo by Tim Chapman

John Cordingley's slick and successful investigation of the deep upstream terminal sump of Wookey Hole in June 1991 described in Chapter 22 marked the beginning of a new era in the exploration of the cave. His visit (and Andy Goddard's attempt twelve months earlier) demonstrated that a well-organised and highly motivated team could make an Alpine-style assault on the end of the cave. By the 1990s the use of mixed gases and drysuits was becoming more common and digging underwater had become a routine means of forcing previously impassable sites.

On 25th September 1993, Robin Brown, Malcolm Foyle, Tom Chapman, Mike Thomas, Rich Websell and Keith Savory went into Wookey Hole where:

Uneventful progress was made to 24. MSF and RAB continued to 25, removing loose line en-route. Reaching the Chamber of Gloom, RAB sorted out the lines in The Well to a depth of 18 m. MSF then checked the line to a depth of 30 m. In good visibility the divers regretted not having the gear for a push.

However, the principal protagonists in the next phase of exploration were active in the main feeder, Swildon's Hole. Mike Barnes takes up the story:

My own interest in the Wookey Hole/Swildon's Hole connection was first aroused after moving to Somerset as a teenager. Early trips through Sump 1 in Swildon's without any sort of protective clothing proved that cave diving wasn't for me; far too cold. However, a book in the school library showed (told) that the Swildon's water eventually surfaced at Wookey Hole, and that the way on beyond Chamber 15 was unexplored. Can it be that simple I thought, get some diving gear and go exploring new cave? Maybe cave diving was for me after all.

Sadly, the book proved to be well and truly out of date. After subsequently hearing about 'Sump 25', 'cave diving depth records', 'mixed gases', 'decompression schedules' etc, etc, I realised it wasn't going to be that simple after all. Still, never one to be defeated, the obvious solution now seemed to be to get into the area of undiscovered cave that must lie between the two systems via Swildon's Hole. Once work had started producing money, the dive kit was purchased and the end of Swildon's was soon reached at the gloomy looking Sump 12. A lot of effort was put into trying to dig a way through this sump over several years, but with little success. During this period, I teamed up with Pete Bolt, a most determined and experienced caver based in Cardiff. Eventually, we both got fed up with the long trips underground with little or no progress. There had to be an easier way.

Pete Bolt continues:

Swildon's 12 proved to be well silted. We spent a considerable number of dives digging at the bottom of Sump 12 to finally emerge into a tight rift that appeared to be an extension of Sump 12a. This was confirmed by diver's air rising into 12a. The difficulty of passing this sump was also indicated by the flood water marks around Sump 12, a sign of serious constrictions below. The most promising find was an unexpected one; a strong draught through the narrow connection between Sumps 12 and 12a, felt on a day of wet weather and rapidly falling air pressure. It was also the first and only time I felt a good draught at the cave entrance. The source of this draught was, unfortunately, not sought at the time although on later trips a possible rift above Sump 12a was noted.

It was not until 24th July 2004 that, after a series of arduous underwater digging sessions, Phill Short was able to pass the tight rift described above and establish an underwater connection between sumps 12 and 12a. The way on (into Sump 12a) was pushed to a gravel choke 25 m in at a depth of 13 m. This is a similar point to that first reached by Martyn Farr 30 years earlier.

Fig. 24.2
Tim Chapman in the 'Wookey Window'.
Dinas Rock Silica Mines.
Photo by Mike Barnes

Mike Barnes takes up the story:

At this time, I was living near Wells and would pass Wookey Hole most days on my way to work. As a diabetic, deep diving was out of the question. If my blood sugar were to drop below a certain level, unconsciousness, and death, would certainly follow. On a dive of any length, such as a deep dive with associated decompression, I would never be able to maintain a safe blood sugar level. However, whilst waiting at a T junction for the car in front to move, the thought occurred to me that no matter how deep I was, I could surely eat a Mars bar underwater and keep my blood sugars at a safe level.

The first attempt at this nearly killed me. It had seemed that the best approach would be, after removing the valve, to take a large bite, put the valve back in, and then slowly chew the chocolate whilst breathing. After taking several relaxed breaths and opening the wrapper, a large piece was bitten off in between Chamber 19 and 22. It immediately became apparent that trying to hold a half Mars bar and a demand valve in one's mouth at the same time could prove to be one of the more unusual fatalities in cave diving. After much choking and spluttering, the chocolate was spat out and normal breathing eventually resumed. Far better was to take a small bite, eat very quickly, then breathe. A later concern was the use of oxygen on the decompression stage. Would the fat and sugar in the chocolate, mixed with the oxygen, cause an explosion inside me. Luckily, this didn't happen!

Once Pete and I had made the decision to switch our attention to Wookey, the first problem was getting rid of Martyn's and Rob's line. In 1991, John Cordingley found it had been swept back into a dangerous tangle at the base of the wall at -50m. The first attempt [26th March 1995] didn't go very well as

Fig. 24.3
Mike Barnes at Stoney Cove, Leicestershire in 1997.
Photo by Nicky White

much of the line was found to be covered in a very fine silt. After removing just a few loops, I found myself in very poor vis, still surrounded by much loose line. This dive was done on a very weak Nitrox mix, but because of the depth, narcosis soon started to manifest itself. Rather than risk entanglement, I turned tail and ran. A subsequent dive, using Trimix, showed there to be a lot more loose line than previously thought. With a clear head, and the major advantage of a pair of good scissors, the remainder of the line was quickly chopped up into small pieces in situ, whereupon it harmlessly floated up and out of the way. These early dives were reminiscent of Martyn's first dives here. The return and decompression were made in much reduced, almost zero, visibility.

Having cleared the way on, it was now Pete's turn to step into the breach.

Pete Bolt continues:

Exploring the bottom of Sump 25 was my first technical dive. Mike had very thoughtfully set the dive up for me; bottles, gas mixes and decompression schedules. With a rich helium mix I had the luxury of exploring the bottom of Sump 25 with clear headed comfort. The chamber at -50 m was easily reached. Mike had cleared the old line, although the odd length still protruded from the silt at this point. Descending across the chamber the ongoing route started as a rift and descended into a low wide passage where it split into two. To the left a descending low arched passage appeared that fitted Rob Parker's 1985 description. Ahead, I could look through a low silted passage into what appeared to be something larger ahead. I pushed forward through the silt to emerge into a small square passage, about 1 m by 1 m. Delighted that I was into something new I continued only to promptly run out of line. With nothing to belay the line on I buried the reel into the silt and, after a difficult turn around, returned in zero vis through the constriction and back up through the chamber to commence my decompression.

Mike Barnes was next to dive:

At long last, the open passage of my schoolboy vision was in front of me. If only it didn't have to involve so much equipment! On a set-up trip from Chamber 9, Pete and I departed with, between us, 11 tanks, 2 tackle bags, and a line reel. I was starting to think there had to be an easier way! The dawn of the next push came on 7th October 1995. With excellent support from Pete, and other members of the Somerset section, Chamber 24 was soon reached. Here a possibly serious problem was overcome using a somewhat crude method. My blood testing equipment, which had been carried in and stored in a waterproof bag, failed to work in the dampness. Pete, with a PhD in micro-electronics, solved the problem by putting it into an empty saucepan on the stove and gently cooking it to remove the moisture. Concerns about, at best, sugar level problems underwater, or at worst, tea tasting of melted plastic were soon forgotten. The device worked again and the blood sugars were perfect. During the kitting up process, with several helpers eagerly helping me strap on what seemed like a mass of equipment, I was thinking, 'What have I got myself into'. However, once in the water, it started to feel like just any other dive.

With lots of helium in the bottom mix, a very clear head showed there to still be some loose line below the wall. This was soon cleared, and the end of Pete's line reached at -62 m. The low section at -60 m was found to be easy to negotiate. After attaching the new reel, the passage ahead was found to be in beautiful, clean, solid limestone with superb vis. It was a total contrast to the murky, horrible stuff further upstream at Swildon's. Sadly, it ended all too soon at -65 m. Here, a low arch about 15 cm high blocked the way on. Removing my helmet, I was able to push my head in far enough to see the gravel starting to rise in a more open section on the other side. Was this the elbow?

Several things kept us from having another look for almost a whole year.

Fig 24.4
Pete Bolt in Porth yr Ogof, 1983.
Photo by Chris Howes

Pete's dig in South Wales had gone and is now better known as Ogof Draenen. This was obviously proving to be a worthy distraction from the cold, deep water of Wookey. Also, I'd been doing some depth training in Dorothea Quarry when a faulty valve over-inflated my drysuit at depth. After shooting to the surface from -69 m, my lungs felt peculiarly stretched. But having been spared a burst lung, or even decompression sickness, I was advised to not dive for three months. It was 5th October 1996 when the next attempt was made. With Pete now almost bored with wandering off down vast unexplored passage, with no decompression to worry about, it was down to me to continue with the dive. Using the same gases and gear as the last dive, it was once again such a relief to finish kitting up and start the dive. Sadly, at -60.8 m, the line plunged into the sand floor in a passage now only 10 cm high. The level of the floor had changed, presumably because of winter floods. A possible way on was seen to the right, but with a fin of rock in the roof, and the position of the sand, it didn't look like entry would be easy. On surfacing in 25, Tim Chapman was on hand to help with the kit. He seemed disappointed at the news, but there was a strange look of optimism in his eyes.

After these dives, Pete and I decided to abandon the project. It seemed as though the only way a diver might get beyond this point would be to strike lucky and turn up when the currents and winter floods had opened a route through.

On 20th July 1996, Mike Thomas (supported by Robin Brown and Malcolm Foyle) undertook a dive in the sump between 24 and 25 to search for any other possible ways on. In excellent visibility and with powerful lighting no ways on were found save for an underwater route around the wall in Chamber 25. Unfortunately the minimum depth was only 2 m so it would not help decompression profiles very much but would save future visitors from having to climb over the wall after deep dives in the end sump.

Tim Chapman now takes up the story:

Back in 1996 I had only been a member of the Somerset Section of the CDG for a short time, when I sat cold and alone on the wall in Chamber 25 separating the Lake of Gloom from the terminal sump of Wookey Hole. I was looking down at the bubbles surfacing and wondering what account of the way on Mike Barnes was about to give me. I had seen the survey and read the stories in 'The Darkness Beckons' and now we were living our own dreams.

Two close friends, Clive Stell and Jonathan Edwards, and I had been caving and cave diving for a number of years and at that time had been doing increasingly deep and technical dives. Doing training at Dorothea Quarry, North Wales, we had increased our knowledge and experience of mixed gas diving. Holidays to the Dordogne had proved that we had the ability to do long deep mix dives; we just needed a project in the UK to concentrate on.

Unfortunately for Mike his gallant efforts in the Terminal Sump had not resulted in a way on and he decided to move on to other projects. It therefore fell to us to take up his gauntlet. Our diving philosophy was always to keep things simple and light weight, streamlining kit configurations and the number of cylinders. So in the summer of 1997 we decided to take our own look at the end of Wookey Hole…

In early 1997, after a prolonged dry spell, Tim and Clive took up the challenge in the course of two trips on the 1st and 9th February using compressed air at depth to probe the end. On the first trip, Clive reached 61.5 m depth whereupon progress was then halted by an extremely tight constriction formed by the gravel banks. The following weekend Tim kicked through the constriction and gained slightly roomier passage beyond. Unfortunately the whole sump appeared to be full of relatively coarse gravel and the existing line was found to be buried. Apart from the help of Jon Edwards for the first trip, the two divers took turns in helping one another.

Mike Thomas undertook an inspection of the roof of the terminal sump on 18th May 1997 helped by Robin Brown and Phill Short. Nitrox-28 was used for the main part of the dive, limiting the safe maximum depth well short of the end of the sump. The rift beyond The Well was pushed to -3 m in the roof whereupon the visibility was lost due to the silt coating the walls, floor and ceiling. A search along the main line to a depth of 40 m (the maximum safe depth of the gas mixture) led to the conclusion that there were lots of areas worth investigating and it would need several dives to complete the search.

Meanwhile Tim and Clive set their sights on finding a way past Rob Parker's limit in the bedding plane at 60 m depth. Their activities concentrated on a week in August 1997 during which four Trimix dives were made into the final sump. A kit dump was established in Wookey Chamber 24 in July and early August which involved each of the divers undertaking a number of dives to Chamber 24.

The emphasis throughout was on reducing the kit to a minimum. Each dive had a relatively short bottom time (between eight and ten minutes), but it was hoped that streamlining of the rig would enable each diver to make progress. The lean Trimix-19/35 contained in a pair of sidemounted high pressure 7 litre cylinders allowed the divers to use that gas to inflate their dry suits, rather than using a separate cylinder of argon. This reduced the kit requirements further. Travel and initial decompression mix (Nitrox-50) was breathed from a chest-mounted 232 bar 12 litre cylinder which was staged at 18 m depth. Decompression from 9 m depth to the surface was conducted using Nitrox-80 out of a staged 232 bar 12 litre cylinder. Decompression was chilly, but this did not represent a serious problem.

It became clear that the gravel slope at 60 m depth was subject to very considerable movement depending upon water flow. When the gravel has banked up after high flow, attempts to pass the various squeezes were considered to be futile and hazardous.

On 19th August both divers made visits to the 'end' and examined two possible 'window' slots. The line leading to the main route forward was embedded in deep gravel and that option was impassable without very substantial digging. A low arch on the right was tightish and looked to continue in similar fashion. On the left of the main passage was a slightly more open squeeze beyond which the roof seemed to remain relatively high. This was kicked through by Tim but the reel was left for the second diver. Clive spent a few minutes tidying up the main line and tying on the line reel. He then forced a way through the gravel in the left hand squeeze and exited leaving the reel for another time. Although passable, there was no possibility of securing the line on the other side of the restriction. The fine gravel floor and the smooth ceiling did not offer any belaying opportunities. Lead weights would simply be carried down the slope with the moving gravel.

Two more dives on 21st and 22nd August conducted alternately by Clive and Tim showed that the way on was unexpectedly tight. The lack of belays for the line was a cause of considerable concern. On going through the window slots, it was noticed that the gravel slope immediately filled up the excavated section once the diver's body had passed the low section. It was therefore necessary for the diver to dig his way out on the return. In low visibility, the possibility that the line had been dragged into a different (and impassable) slot left the diver questioning his sanity as he dug out his exit route. A maximum depth of 66.2 m was eventually attained on the final dive by Tim but it was felt that the line was now leading the wrong side of a roof projection. The line was reeled in and an alternative window examined but in the poor visibility due to the digging it was not possible to see if the route continued. Again, there was concern that the line had dragged down into a lower section of the bedding plane.

Further Trimix dives were planned during the summer, but were postponed following the loss of a bag between Chambers 9 and 22 containing decompression schedules and a dive computer. Work resumed in the autumn following its recovery. The initial dive on 25th October was limited to an examination of the chamber at 50 m depth owing to a valve failure. Clive tied off a line reel and ascended to a dark slot in the roof. This led into a small chamber above the main chamber but no way on could be found.

The next dive was delayed until 2nd November due to heavy rain at the end of August. Tim was supported by Jon and Clive but the lack an air cylinder in Chamber 24 combined with the rich travel mix used on the dives (50%) meant that the diver had to switch to bottom mix on the way through to Chamber 25. This disrupted the decompression calculations and the run times for the dive into the terminal sump. On reaching the first of the squeezes at 60 m depth Tim realised that the gravel had receded leaving considerably greater headroom, although the main way straight on was still blocked. In the short amount of time available, the diver looked into both left and right hand slots and both looked as if progress would be possible.

December was one of the wettest on record which resulted in no pushing dives being made in Wookey until the middle of February 1998. Some equipment had been left in the cave from the previous year, including a bottle of Trimix and a bottle of travel mix. Tim and Clive therefore planned a light trip to the end, using the remaining equipment and an additional bottle of Trimix, to see the changes brought by the winter floods. Clive's dive was planned for 14th Feburary but whilst kitting up in Chamber 24 it was found that the pillar valve in the travel mix had corroded and most of the gas had leaked out. The cylinder was new in May 1997 but had spent seven months of its short life in Wookey. The dive was therefore postponed and the opportunity taken to partially reline the Wall Bypass in Chamber 25.

A week later, on the 21st, Tim and Clive returned and a dive was undertaken using the equipment left the previous week with an additional cylinder of travel/decompression mix. More Trimix was also taken in for future dives. Tim utilised three high pressure 7 litre cylinders, two containing Trimix 19/35 Trimix and one with Nitrox-42. The dive plan was based on the previous year's tables, modified so that a separate decompression mix was not required.

An uneventful descent was made in surprisingly poor visibility, staging the travel mix at 28 m depth (not the planned 30 m). The diver quickly gained the canyon at the far side of the 50 m chamber and was disappointed to find that the main line was buried at 59 m depth, graphically illustrating the amount of shifting of gravel that occurs. The diver continued to explore after digging out the line and tying on a fresh real. The line reel left last summer could not be found. No possible continuation could be found to the right of the pendant (Barnes/Bolt route); exploring to the left of the pendant (the same route pushed during 1997), space was found, but no continuation. A return was made to the surface after a dive of 46 minutes.

Two years were to pass before Tim made another attempt on 9th June 2000. Again, an ultra light-weight approach was adopted using two 7 litre cylinders of Trimix 21/15 for the terminal sump and one 6 litre cylinder of Nitrox-44 for decompression. To reduce the carrying, the three cylinders that were to be used in the terminal sump were also used as the 'bail out' for the preceding sumps.

Using the Nitrox-44 for the dive from 24 to 25, this was staged at 12 m depth beyond Chamber 25 where the line was completely buried under many inches of sediment. Some time was taken to pull it out and continue, and the line down to 40 m depth was as usual rather loose. At 50 m depth the gravel floor was reached and by 60 m depth the large pendant to which the line is belayed was found to be well-buried. The squeeze to the left of the pendant was completely blocked but to the right a very low passage could be seen to continue. With little run-time remaining the diver pushed a short way into the very low continuing passage but it soon became impassable. Run-time was then reached and the dive was terminated at 61.5 m depth. The overall terminal choke appeared to be more blocked up than ever and there was still no sign of the dive reel and line left in 1997. The current was very strong and could be felt from about 50 m depth.

Decompression was carried out from 12 m on the Nitrox-44. In the shaft a powerful up-welling current was also observed which was physically lifting the diver; this had not been noted on previous dives. Interestingly it was also observed that the wall separating the two pool areas of Chamber 25 was covered by 4 inches of water although no perceptible flow was observed.

With such a disappointing outcome, Clive and Tim made no further dives, and it was not until Gavin Newman's filming dives in 2003 described later that anyone went to the 'end'. It was at this point that Mike Barnes returned to the fray, escorting Jo Wisely on a dive upstream on 21st August 2003:

JW wanted to dive to the end of the terminal sump which she duly did, 60 m depth achieved on air. However, on the way in, after an absence of 6 years, I noticed several of the fixed aids in poor condition. The bolt holding the hand line in Chamber 22 appeared very corroded, as did the belay holding the iron ladder leading into the upstream sump in 22. This observation was proved when it snapped as JW climbed down, fortunately only falling into the water. Also the line upstream from the lake in 24 appeared to be in a very dangerous condition.

Over the course of several trips the following month remedial work was carried out on the entry to the 22 Static Sump and a dive platform to assist in kitting up was constructed. This wasn't purely out of generosity to visiting divers, however, as Mike had a grand scheme in mind:

Wookey was filed away in my mind under 'if only', as I moved on to other things. But the vision of that rising passage kept gnawing at my mind. I knew that with sufficient investment of time, money and enthusiasm, along with a few clever ideas, it could go. But how? An estimation of the amount of gravel that needed moving was 120 cubic metres. At the far end of Wookey, it wasn't going to be easy.

One of the oldest and simplest techniques used in salvage work is an airlift, a most effective and powerful tool. Compressed air injected into the bottom of a submerged open-ended tube rises up the tube creating a strong suction at the base. This can remove large quantities of sediment very quickly. It does, however, have one big drawback when used underground: a continuous supply of air is needed to run it, far exceeding anything that could be supplied by a normal diving cylinder. The consumption of an average-sized airlift seemed to be about 80-100 cubic feet per minute [cfm] – a typical 7 litre cylinder only holding 50 cubic feet. Nigel Brock of the National Diving Centre suggested simply making a smaller lift. After some experimental dives at Vobster Quarry using a variety of tubes modified into airlifts, I discovered that with a 5 cm diameter pipe, we would only need 11-12 cfm. Whilst this is still a reasonable quantity of air to pump to the far end of the cave, it suddenly appeared to be a feasible prospect.

After many hours of thinking about the possible problems and ways to overcome them, the plan is as follows[38].

The airlift itself will hang vertically down a short 3 m wall at -46 m, (it is this short drop which traps the gravel in the first place). From here a length of flexible hose goes to the working end at the first squeeze where a simple frame keeps the end of the pipe in the gravel. At the top of the wall, another length of flexible hose directs the gravel away to where it will be dumped. This section of passage is up to 1 m high but about 10 m across, sloping at an angle of 40 degrees with plenty of space to deposit the spoil. Because of the quantity that needs moving, the intention is to operate the air lift without the need for a diver to be present. A test

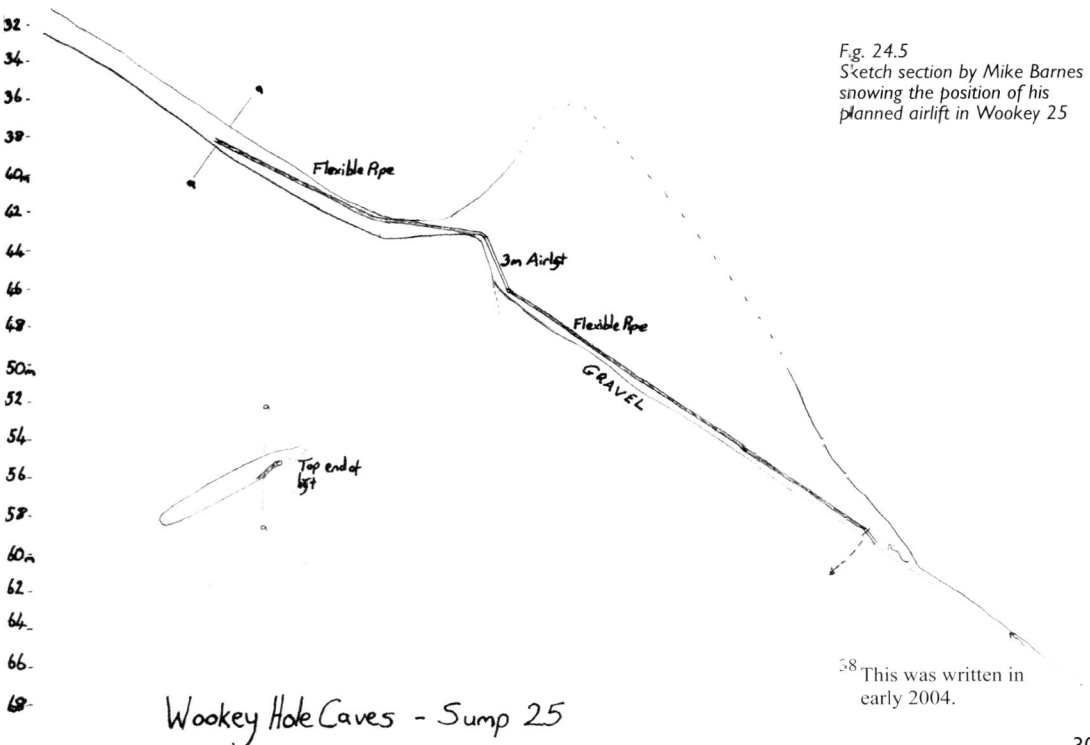

Fig. 24.5
Sketch section by Mike Barnes showing the position of his planned airlift in Wookey 25

[38] This was written in early 2004.

Wookey Hole Caves - Sump 25

in one of the approach sumps showed that as gravel is removed from the bottom of the slope, a continuous avalanche from above feeds the ever-hungry air lift. Wookey Hole Caves are offering their full support and have agreed to the location of an electric air compressor in Chamber 9 at the end of the show cave. From here approximately 1 km of air hose will be laid to the air lift itself. As operation will only be possible at night when the cave is closed to the public, the digging may take up to 6 weeks once lifting begins.

The team hope to have all the pipework in place and be up and running by early summer [2004]. Several tasks have already been undertaken to facilitate the operation. These include installing some rigid ladders on climbs and the construction of a dive platform in 22. With the kind cooperation of a dry-ish summer offering good visibility, the search for the elusive Chamber 26 can begin.

Access to Wookey Hole Caves was suspended during the autumn and winter of 2003-4 due to a change-over in public liability insurance for cavers, so it wasn't until the spring of 2004 that cave divers were able to ply their trade once more in the River Axe.

A trip to 24 on 29th February by Mike Barnes and Rich Dolby was reduced to another session working on the 22 Static Sump platform after Rich's ears failed to clear properly due to bad nasal congestion. The pair were back in the cave on 7th March 2004 for another go:

A working trip, the intention to dive upstream from 25 and reconnaissance with regard to the imminent arrival of MB's home-built airlift, and hopefully the sponsorship to provide the anticipated 1000 m+ of pneumatic hose needed to drive the new toy.

...

MB stayed in 24, placing various bolts & making other improvements (including a fixed ladder for easier access to the rift/traverse before Sting Corner) whilst RD dived on into the lake upstream in 24. Two breaks in the line, the first at 15 m depth which was repaired, and a second at 24.7 m depth, which could not be fixed and therefore prevented access to 25. Much disappointment as he had been looking forward to diving to 25. Still, a productive 6 hour trip though, with further surveying carried out in much of 24 apropos the length of pneumatic hose required through this part of the cave.

Will return in the near future to re-line the sump from 24 to 25 with some nice blue polypropylene line.

Meanwhile another diver had his sights on the 'end' and everything was about to change…

Photo by Martyn Farr

Chapter Twenty Five

Wet to Twenty-Four

by Duncan Price

When Wookey 22 was discovered in April 1971 the flow of the water was lost and instead an abandoned fossil passage was followed to regain the river in Chamber 24. Chapter 13 tells the full story of these discoveries and it was not until the early part of the 21st century that this missing section of the River Axe was finally located.

The first efforts in finding the course of the river from Wookey 24 to Wookey 22 took place on 12th August 1984 when Rob Harper undertook an investigation of the point at which the River Axe sinks just downstream of Sting Corner. In drought conditions he found a tight rift which widened out at depth. Rob returned a week later in the company of Simon Brooks when he dived the passage wearing two small 4 litre cylinders and reached a depth of 2.7 m before the passage became too narrow. The way on was seen to continue tight with the widest point being at different levels.

Rob came back to the area on 27th March 1993, where (after looking at Pleasant Valley Sunday) he and Tom Chapman examined the boulder pile just downstream of Sting Corner and found a small static sump. On 7th April 1994, Mike Barnes and Rob made the first dives in this new sump using single kit. On closer examination it was found to take a fair current and proved to be the downstream continuation of Sting Corner sink. On that occasion it was pushed for 4m in a tight rift. The route continued and the line reel was left in place but it was not until January 1995 that it was revisited. The water levels were now sufficiently high for the flow to be reversed with a stream draining all the 23-24 sumps. Mike dived the tight rift in a very strong current to check on the flow at the bottom of the rift. This as expected was the same. He followed the rift at floor level for about 10 m to where it started to widen out. However on a base-fed line Mike felt unhappy and returned.

Mike Barnes came back to the site on 10th June 1995 when the water level had dropped about 1 m and a waterfall was falling into the pool. On descending with a full line reel he found the situation somewhat unnerving as everything was a mass of bubbles. After 12 m the passage had widened to about 1.5 m and started to descend. He dropped down the narrow rift to a depth of 3.5 m. Due to the lack of belays Mike returned removing the line.

A year later conditions were more favourable and with water levels very low, Mike Barnes soon reached his previous limit laying a new line in this tight rift. Progress at floor level was easiest and after 5 m a route up the rift was found surfacing in a wider continuation with fast flowing water. This was found to sump again after 30 m. Pete Bolt and Tim Chapman managed to squeeze over the top of the sump without diving but this was only due to the very dry conditions. Tim and Mike returned on 21st September that year and took advantage of the drought conditions to bypass Sump 1. Mike tied on a reel to a submerged boulder and set off. The rift continued as before for only a few metres before it dropped down a small pot. The passage was found to be a much higher and narrower rift and continued descending, the roomiest section again being at floor level. At a depth of 12 m, progress was only possible by traversing across the rift following the flow. Mike had three lengths of plastic pipe to jam across the rift for belays but these were soon used up, so at a depth of

14 m, Mike cut the line and exited. The passage was seen to continue but the route would require very careful lining to be safe. Tim Chapman, Jon Edwards and Clive Stell looked at this area in the spring of 1998, but did not progress beyond Mike's limit.

Fig. 25.1
Mike Thomas and Mike Barnes, Wookey Resurgence in 1996. Photo by Martyn Farr

The next phase of exploration of the course followed by the River Axe was to change location to the vicinity of Wookey 22. Although the area had been thoroughly searched during the 1970s, the discovery of the flood overflow route via Chamber 23 had lessened the impetus to try for a direct route.

On 13th July 2001, Mike Thomas and Malc Foyle planned a trip to 22 to test some of Malc's diving kit that had not been used in a while. Both divers planned to surface in the left hand sump pool in 22. The visibility was very good and Mike was swimming well off the line using Malc as a 'lighthouse'. Progressing from the Edmunds Chamber junction to surface in 22, Mike decided to swim up the far left hand wall. On reaching a depth of 3 m he swam over a flat sandy area with obvious flow marks, and on further investigation a sloping bedding plane passage heading off into the distance was found. Upon surfacing Mike rather excitedly called Malc over and positioned him at the start of the passage again to act as a beacon. As Mike had no line reel and was approaching thirds, only a brief examination was possible. He went in about 5 m and confirmed the passage continued but it was low; about 0.5 m high and about 5 m across. Mike struggled to get into this passage due to the flow coming out, but thought that this had to be the water that disappears at Sting Corner in Wookey 24. The divers headed back to Chamber Nine and a return was planned – and then it rained!

It was not until 24th February 2002 that Mike Thomas was able to return to the fray. Even though the visibility was only about 3 m he decided to have another look at the 22 inlet passage. A part-used 6 litre cylinder was used to get to 22 leaving the diver with two full 7 litre cylinders for exploration. The flow coming out of the passage was very high and Mike just managed to pull his way into the low section and jam himself between the roof and the floor. Twenty metres of progress was made at a depth of 3 m before Mike retreated due to a complete lack of line belays and continuing low passage. The line was removed on the way out. While slowly exiting to the 22 sump pool one regulator started breathing wet and on stripping the second stage back in the pool, debris was found under the exhaust diaphragm. This was removed and the diver returned to Chamber 9 with all valves working.

Another dive was planned for 17th March but conditions in the cave were still very poor. Mike took some lead weight belays through to the start of the passage in 22 and after stashing these he had a good look around. He found that the floor at 6 m depth gave a little more room to manoeuvre.

Better conditions were present on 13th April when Mike Thomas got in the water early in order to beat a 'cast of thousands' who were also planning to dive. Mike swam to 22, again using a part-used 6 litre cylinder which was dropped in the sump pool. This time he followed the floor of the inlet passage at 6 m depth making 20 m of progress until it got too tight and he became forced back

Fig. 25.2
Helen Rider leaving Chamber 3
Photo by Martyn Farr

up to 3 m depth. Another 15 m of line was laid until the bedding again got too constricted. Mike could not find anywhere to tie the line off so it was removed again. From his limit of 35 m from Chamber 22, he could see another 8-10 m, but it showed no sign of getting bigger. On returning to the 22 sump pool he had problems with debris under both exhaust diaphragms again and both regulators had to be stripped and cleaned before returning to Chamber Nine.

The inlet passage starts low (about 0.5 m high) at the beginning, decreasing to about 0.3 m at the furthest point with a lot of loose material on the floor and roof of the passage – this was what was causing Mike problems with wet breathing valves.

On 2nd May Mike installed a permanent line between the main dive line in Wookey 22 and the start of the 22 Inlet Passage – a distance of 27 m at a depth of 3 m. This trip was followed up by a series of dives by Pete Mulholland and Andy Stewart who pushed the ongoing tight inclined bedding passage to a distance of 80 m from the line junction. In order to make the route safe, they installed some plastic stemples along the bedding in order to keep the line in the widest part of the passage.

At the invitation of Pete Mullholland and Mike Thomas, I visited the 22 Inlet Passage on 10th May 2003. It was the day of the CDG AGM and dinner held in Somerset that year and there were a lot of divers in the cave. Clad in a wetsuit with a pair of 12 litre cylinders, I squeezed my way to the end of the line with a fresh reel but found that there was ample line already on the reel left in situ. Another 20 m of line was laid out in a low, inclined bedding until air margins forced a retreat. The line reel was jammed across the passage which appeared to continue in the same fashion as far as I could see.

No further visit was made until 19th June 2004 when I adopted a drysuit and smaller 7 litre cylinders with a 12 litre stage tank to get me to 22. Some plastic stemples were installed along the previous section of line that I had laid in order to safeguard my return. Only a further 5 m of line was laid to reach a 1 m high cross-rift where the reel was belayed. The passage continues small ahead and is in the flow but I felt that more air reserves were required. Even though the end is only just over 100 m from 'safety' in Chamber 22 the passage is very awkward to traverse and (as Mike Thomas found) quite aggressive towards one's diving gear.

Whether the flooded passage between Chamber 22 and Chamber 24 can be fully explored by cave divers remains to be seen, but it would be interesting to be able to submerge beneath the River Axe at the resurgence and surface in the grandeur of Wookey 24. Well over 100 m of 'missing' streamway remains to be entered in the phreas between 22 and 24 with the passage probably continuing north from 22 in a straight line course for 24 with no let up in its awkward nature.

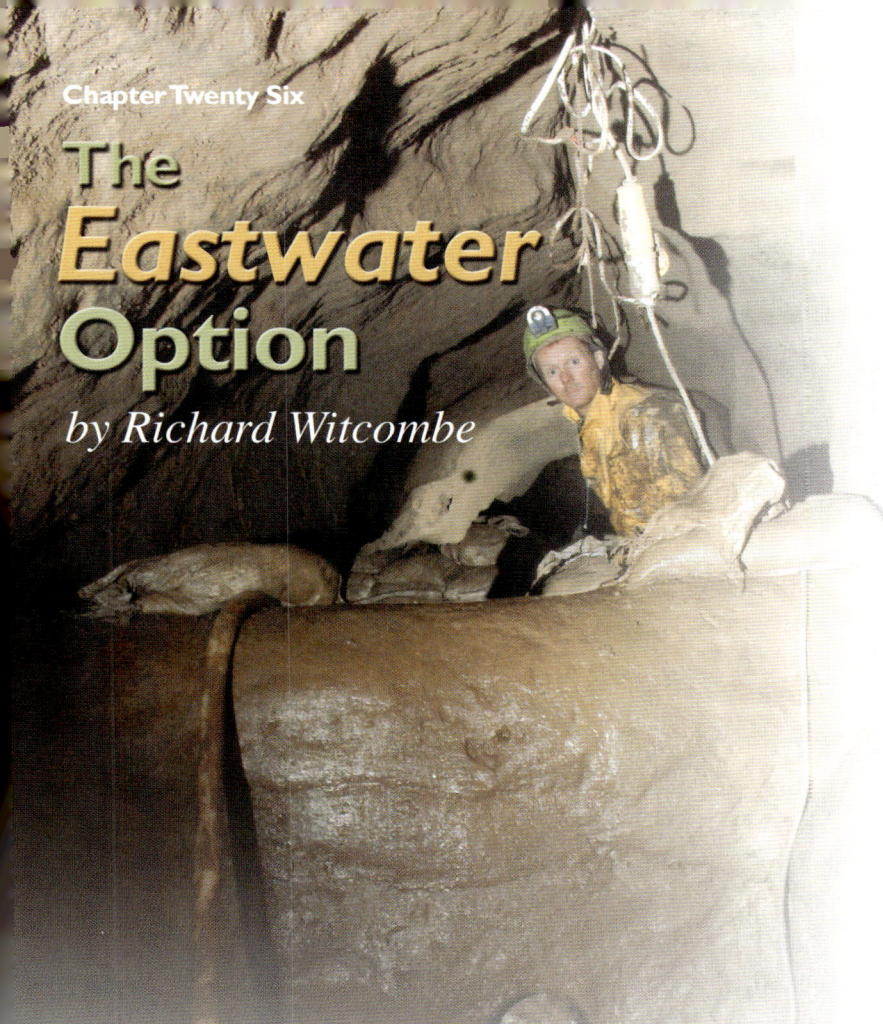

Chapter Twenty Six
The Eastwater Option
by Richard Witcombe

*Fig. 26.1
Phil 'MadPhil' Rowsell working above the terminal sump dam, 2009.
Photo by Tom Clayton*

*Fig. 26.2
In the Canyon, Eastwater Cavern.
Illustration by Reg Balch*

In the century-old search for a through route from the Priddy swallets to the Wookey Hole resurgence, Eastwater Cavern has tended to be the poor relation in comparison to its larger neighbours, Swildon's Hole and St. Cuthbert's Swallet, and yet water tracing experiments have shown a flow through time of only 16 hours from the Eastwater sink to the Wookey entrance. Eastwater Cavern is situated a mile east of Priddy Green, and was dug open by Herbert Balch and his friends from Wells and Wookey Hole in 1902 in the hope of finding a second entrance to Swildon's Hole where the landowner was preventing access. The explorers were young members of the Wells Natural History and Archaeological Society which later gave birth to the Mendip Nature Research Committee, Mendip's oldest caving club. Beyond the entrance boulder ruckle, they found a cave of steeply descending bedding planes and high rifts, unadorned by any significant stalagmite formations. Although the cave quickly reached the then considerable depth of 420 ft – thought by Balch to be nearer 600 ft – the spiralling passages had made very little progress either southwards towards Wookey or westwards towards Swildon's. In the first decade of the twentieth century, attempts were made to extend the system by pushing tiny passages at the bottom of the cave and by digging in the high-level 380 Foot Way which seemed to carry the stream further to the west, but no significant progress was made.

Fig. 26.3
At the Mouth of the Swallet. Forming a chain gang to pass down kit, 1913, H.E. Balch. Wells & Mendip Museum Collection

Fig. 26.4
Denis Warburton surveying with his trusted ship's compass, 1960. Photo by Jim Hanwell

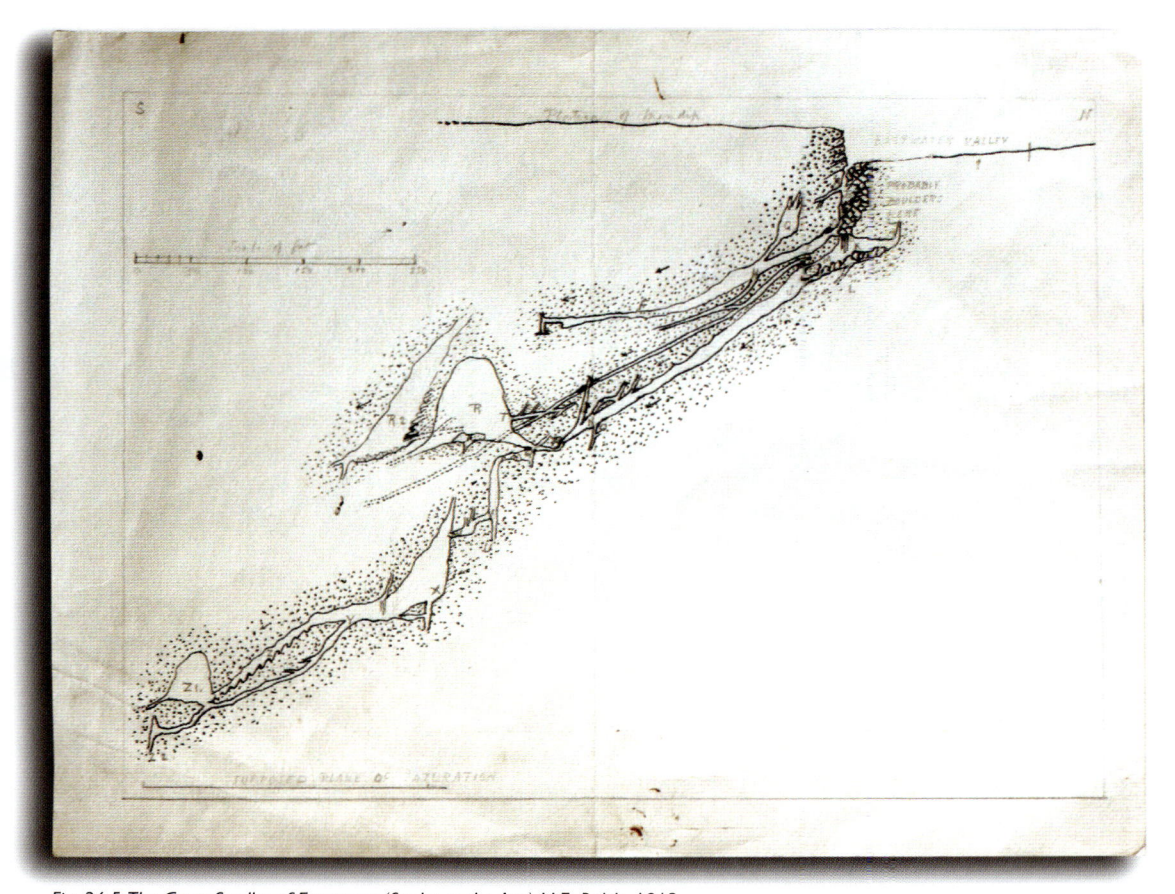

Fig. 26.5 The Great Swallet of Eastwater (feeder to the Axe) H.E. Balch, 1913

The cave remained a popular sporting trip and in 1933, MNRC member Charles Wyndham 'Digger' Harris from Wells used his long legs to climb up from the bottom of the Second Vertical to reach an inlet passage which now bears his name. Wessex cavers continued to make small finds in the cave, but neither Dolphin Pot opened by Paul Dolphin, Colin Low and Norman Paddock in 1940 nor the spectacular 185 ft deep Primrose Pot, first descended by Howard Kenney and Vincent Stimpson in 1950, took the cave beyond its southern limit. Harris's Passage had closed down at a boulder choke, but in 1951 Johnny Ifold of the BEC managed to dig through this to enter a short series of unstable passages, soon christened Ifold's Series. The early 1950s saw two Wessex members, Denis Warburton and Alan Surrall, complete a very detailed survey of the cave which undoubtedly prompted future generations to examine its various extremities in the hope of finding a way onwards towards the resurgence.

Nearly thirty years went by before anyone took a serious look at the nooks and crannies of the Ifold's Series, but in the Spring of 1983 two BEC members decided to probe a low tube at the far end. On an uncertain date but probably 28th May, Keith Gladman and Andy Lolly broke into several hundred feet of roomy inclined bedding and rift passages after a total of only six hours of excavation. This very significant extension was dubbed the West End Series from its position relative to the rest of the cave, and spawned a host of London names for the new discoveries.

Klondike fever soon gripped the BEC and over the next six months dozens of trips involving dozens of cavers saw the Series extended to many thousands of feet. As well as beautifully decorated passages such as Regent Street with its 'Crystal Palace' and 'Serpentine' lake, a sinuous, sporting streamway was found which took the cave into deeper territory via the 75 ft deep Gladman's Shaft

Fig. 26.6
Formations in Regent Street,
West End Series, 1984.
Photo by Phil Romford

and the 40 ft deep Lolly Pot. Beyond the latter, 100 ft of fairly miserable passage ended in a body-sized tube with a howling draught whistling in. It was half-full of water and with a high tide mark suggesting recent sumping, it deterred all but the most determined cavers. In October 1983, Tony Jarratt, Tim Large and Barry Wharton dug through this squalor to enter 50 ft of inclined tube given the ironic title of Blackwall Tunnel. In June 1984 a joint BEC/WCC team inspected the site. The trip resulted in Pete and Alison Moody undertaking a banging project along the most promising lead – a typically tight Eastwater bedding plane. With the draught blowing strongly inwards it was possible, after one charge, to return half an hour later, clear the debris and then fire a second. On 15th September, after the first charge of the day, the combined team saw Alison squeeze through into a large cross passage with several leads. A further charge was laid before exiting. The first follow up was on 27th October and pushing trips over the next few weeks resulted in the Moodys, Tony Jarratt and others making several important finds. The complex Charing Cross area was explored in October, the aptly named and flood prone Chamber of Horrors and the roomier Jubilee Line were explored on 10th November, and a week later, the Blackwall breakthroughs culminated in the discovery of a major fossil inlet terminating in the magnificent 120 ft high Cenotaph Aven.

The Gladman and Lolly pitches carry a small, intermittent stream and this was nearly the cause of a major disaster on 2nd March 1985. An exploration party coming in found only 2 inches of airspace in the Blackwall Tunnel 'grovel' but carried on to Cenotaph Aven to alert a bolting party there to the danger of the rising water. Discarding all the climbing kit, a dash for freedom was made. The airspace was now unusable but Pete Watts of the Wessex managed to force the squeeze underwater and proceeded to cobble together a dam and start frantic bailing. The trapped team were lucky to find a small hole 30 ft away into which they could bail, and after two hours' combined work, all were able to escape and make their way out of the cave in a very chastened state. The site remained sumped for over two months.

Photo by Phil Romford

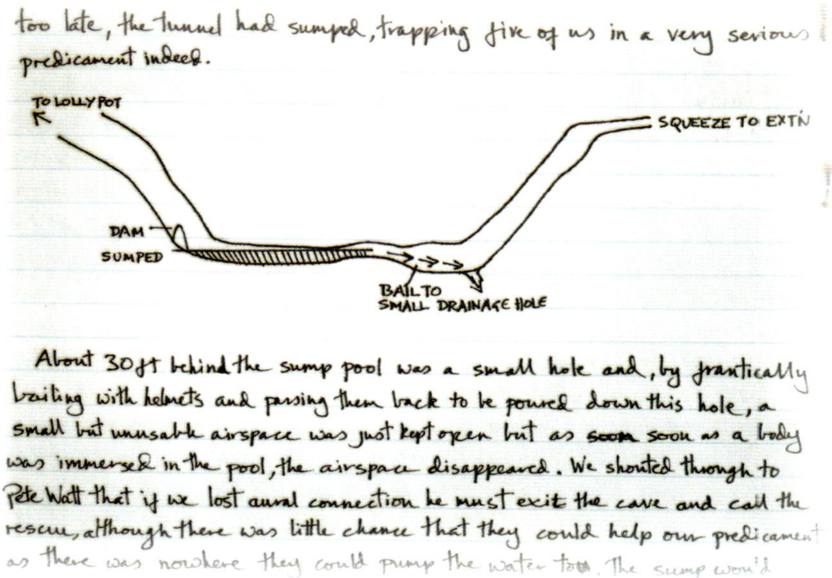

> too late, the tunnel had sumped, trapping five of us in a very serious predicament indeed.
>
> [sketch: TO LOLLY POT / SQUEEZE TO EXTN / DAM / SUMPED / BAIL TO SMALL DRAINAGE HOLE]
>
> About 30 ft behind the sump pool was a small hole and, by frantically bailing with helmets and passing them back to be poured down this hole, a small but unusable airspace was just kept open but as soon as a body was immersed in the pool, the airspace disappeared. We shouted through to Pete Watt that if we lost aural connection he must exit the cave and call the rescue, although there was little chance that they could help our predicament as there was nowhere they could pump the water to. The sump would

*Fig.26.7
Trapped in the West End Series.
Extract from Mark 'Gonzo' Lumley's logbook for 2 March 1985*

The new series was heading steadily southwards towards Wookey Hole, well beyond the boundaries of the 'old' cave, but the pushing trips were getting increasingly arduous and the momentum inevitably slowed. Then in 1986 a disagreement between local farmers and the Nature Conservancy over the scheduling of Sites of Special Scientific Interest in the Priddy area led to Eastwater Cavern being closed.

Peace eventually returned to the hills and a rejuvenated Wessex digging team returned to the far end of the Jubilee Line in late 1988. On 10th December, following a number of digging trips, the Moodys, Murray Knapp and Rob 'Tav' Taviner pushed a squeeze to discover 500 ft of new passages, including a section carrying the main Eastwater stream. Although 8 ft in diameter, the stream passage – Blackfriars – proved disappointingly short. The water emerged upstream from an impenetrable crack and the mud-coated tunnel terminated after only 50 ft in a gloomy sump. Further back in the new extension, a window in the passage wall had been passed from which emanated the enticing sound of flowing water, but this was left for another day.

Three weeks later the same team was back in what had been christened the Southbank Series, and the window in the wall yielded 250 ft of very narrow ascending passage. The dimensions of Lambeth Walk meant that the stream level was mostly impassable with caver progress only possible by high-level traversing. It seemed likely that the inlet led up to an eventual connection with the heavily choked Morton's Pot dig at the end of the 380 Foot Way, the scene of intermittent work since the days of Herbert Balch. The diggers also looked again at Tooting Broadway, the wide abandoned streamway at the most southerly point of Southbank. Various difficult leads were pushed but only 50 ft of new passage was found. The return trip was typically grim with Murray Knapp getting grit in his eyes in one of the ducks and having to go straight to hospital on leaving the cave.

This remote series at a depth of nearly 500 ft saw very few trips in the 1990s, but Alison Moody and Phil 'MadPhil' Rowsell went back to the end on 16th August 2002 and pushed on for 80 ft to a static sump, Canary Wharf. On 3rd June 2003 Alison returned to Southbank accompanied by Simon 'Nik Nak' Richards, Laura Trowbridge and cave diver Phill Short. Phill entered the Blackfriars sump feet first and forced a 45 degree slope to a levelling off at a depth of about 6 ft. Here he felt around with his feet, concluding that the way on was totally blocked with stream debris. He then tackled the Canary Wharf pool, again feet first, but found it completely choked when he was in as far as his neck. The diving possibilities in Eastwater were thus soon exhausted, but the same could not be said for the digging options.

It seemed clear from the polythene bag and timber debris found in Southbank that the ascending inlet of Lambeth Walk must bring in water from the Morton's Pot dig, and renewed efforts were made in the 1990s to make the connection. BEC diggers, including Adrian Hole, Tony Jarratt, Graham 'Jake' Johnson, Vince Simmonds, Rich Blake and Trevor Hughes, installed an overhead cable system for spoil removal and excavated a 20 ft pot punningly christened A Drain Hole. Renewed BEC digging in the new century was twice thwarted by major flooding problems but with fresh impetus provided by 'MadPhil' Rowsell the campaign was rewarded in 2003 by the discovery of the two short Pointless Pots. Then, by following a very narrow rift cut into a steeply descending bedding plane, the diggers slowly engineered their way on downwards, finally linking up with Lambeth Walk in October 2004. This very hard won passage, a tribute to the persistence of diggers over two decades, was named The Technical Masterpiece, and it was to play a useful role in the next phase of the Eastwater story.

Fig. 26.8 Graham 'Jake' Johnson working in the Morton's Pot dig, 2003. Photo by Sean Howe

A re-survey of the cave's far reaches by 'MadPhil', Emma Heron and Kevin Hilton in 2003/4, had shown that the Chamber of Horrors was the deepest point in the system and that the Blackfriars sump was in fact 'perched' about 30 ft above this level. This prompted an assault on the sump by a combination of bailing and digging. Starting in 2004, 'mud-crete' dams were constructed and after many hours hard work the diggers were able to lower the water level sufficiently to allow a short digging session. The opening up of the Technical Masterpiece later that year transformed the situation. A power cable was laid all the way from the entrance to the sump, and a generator on the surface not only powered a semi-submersible pump but also provided lighting. The digging team, who styled themselves the 'Fat Belly Boys', could now empty the sump in just over an hour.

A mud-choked tube was uncovered and as excavation proceeded, this was protected by a metal cage to reduce the potential for sudden slumping and limit the infilling between trips. The tube was finally passed on 8th July 2007. Beyond the 15 ft of excavated passage, was a 20 ft long chamber ending in another sump. This too was heavily choked. Dam capacity thwarted further digging in Sump 2, and so an attempt was made at engineering a by-pass, but this proved painfully slow and enthusiasm waned.

It was already known that the sump dig was at much the same level as another Southbank site only 20 ft away, the Pea Gravel Dig, which had completely separate drainage. On a preparation trip for a memorial dig at the sump during the weekend of the late Tony Jarratt's wake, 15th and 16th November 2008, it was discovered that much of Sump 1 had fortuitously washed through into the Pea Gravel Dig, lowering the water level by over 3 ft, and more importantly creating a natural soakaway independent of Sump 2. This permitted Sump 2 to be pumped back to the main dam above Sump 1 in stages, with the contents then being allowed to drain away via Pea Gravel. A pumping and digging frenzy ensued and after a long and grim campaign a further breakthrough was made on the 10th May 2009. After reaching a depth of 10 ft the diggers had excavated ahead for 30 ft and then followed a tight tube upwards into open passage. The breakthrough duo, 'MadPhil' and Tom Clayton, returning on 12th May, explored about 200 ft of passage, but disappointingly the phreatic tunnel hit a major joint after only 100 ft and dropped steeply for 45 ft to another sump at the lowest point in the cave – 498 ft deep.

Before Sump 3, a narrow side passage was followed upwards to a constriction in a tight phreatic bedding plane passage heading toward the Dark Cars and Primrose Pot zone. On 30th May this point was passed and another 60 ft or so of ascending passage was explored until it again became too tight. The new sump was dived by 'MadPhil' on 31st May and after a number of attempts a committing constriction was reached at a depth of 15 ft. Further dives are planned in attempt to pass this squeeze and the team will also be re-examining Tooting Broadway (only accessible in extremely dry weather) for other higher level downstream options.

Fig. 26.9
Phil 'MadPhil' Rowsell entering Sump 3, 2009.
Photo by Tom Clayton

Eastwater Cavern has now been extended to a length of over 10,000 ft and as the terminal passages are doglegging steadily southeast and southwest by turns, it could yet prove to be the dark horse in the Wookey Hole catchment area. Meanwhile, the plateau between the North Hill swallets and Wookey Hole has not been neglected. In 1973 a mechanical excavator was used to break into Wookey 9 from the surface and in the early 1990s Wessex member, Dave 'Tuska' Morrison, developed the technique to clear vast amounts of spoil from deep depressions over a single weekend. Several sites on the plateau south of the Hunters' to Priddy road have been dug in this way, including Eighteen Acre Swallet above the far reaches of St. Cuthbert's Swallet, the well decorated 1200 ft long White Pit south of Swildon's Twelve, and the immense 180 ft deep shaft of Templeton Pot just north of Ebbor Gorge. So far none of these caves has gone deep enough to intercept the main upper Axe streamway, but as Jim Hanwell speculated many years ago, digger may yet one day meet diver somewhere deep beneath this quiet, unspectacular farmland.

Fig. 26.10
Templeton Pot, 2006.
Photo by Clive North

Chapter Twenty Seven

Making 'Wookey'

by Gavin Newman

Photo by Gavin Newman

It is a well known fact that water and electronics don't mix. It's also a well known fact that caves trash nice shiny diving kit – we are talking real caves here not the Florida/Mexico holiday diving variety. So two years ago when I decided to make a film about Wookey Hole Caves in Somerset it was fairly predictable that firstly, it wasn't going to be easy and secondly things probably weren't going to go to plan.

The idea first reared its head as so often is the case during alcohol-fuelled conversations in the pub. Back in 1985 my good friend Leo Dickinson had made a film about Rob Parker's attempt to push on beyond Martyn Farr's limit at the end of Wookey. I knew Rob at the time as we were both members of the South Wales Caving Club, but not well, and so I wasn't involved in the film. I was only just venturing into the world of cave diving and was concentrating much more on still images than film work. I later got to know both Rob and Leo very well and they became my mentors in their respective areas of expertise. Rob taught me to cave dive, initially to join him and others on an expedition to Northern Spain in 1986. The expedition needed a vehicle to carry all the diving equipment and I owned a Range Rover at the time so I was co-opted onto the team!

At the end of Leo's 1985 film Rob swims off alone into the darkness to try to push the end of the cave and we never see where he actually went. Small underwater video cameras did not exist then, and lights were large and heavy. It simply was not practical for a diver, making what was one of the first sport dives ever on mixed gases, to film himself at the same time.

After Rob's death on a cave dive in the Bahamas my quest began with the simple idea to go and have a look at the end of Wookey for myself. I had no thought that I was going to push it further but I wanted to see where Rob had been and if possible film it so others could see. Talking through the idea with Leo, he suggested that we re-edit his old film, adding in footage of the end of the cave and making it into a tribute to Rob, as in spite of several attempts no-one had since managed to push the cave any further. In fact due to the changing nature of the gravel choke at the end of the cave, no-one had even reached Rob's line reel which, like Martyn's before him, remained buried somewhere in the gravel. So the idea was born that I would try to get footage from as deep a point at the end of the cave as I could and re-edit it into Leo's old film. That at least was the plan but plans have a habit of developing...

Before going to the end of the cave I needed to refine all the filming equipment that I would use so I started by filming in the easily dived sumps leading from the end of the show cave. The summer of 2001 was unusually dry and it was also during the foot and mouth crisis that closed many of the caves on the hills above Wookey. Whether the lack of cavers in the feeder caves made the difference is hard to tell, but the visibility in the cave during those few months was extraordinary. It very quickly became apparent that with the combination of the fantastic conditions and modern video equipment we were getting footage far better than anything in the earlier film. I soon decided to reverse the plan and make my own film that would include footage from Leo's film.

It was a daunting prospect. I was a professional cameraman and stills photographer but all my filming work had been done for other people's productions, I just shot what they asked me to shoot. My cave diving work was almost entirely based around stills photography and apart from small projects at college, I had never made my own film. Not being made as a commission, the film had no budget, but fortunately many of the Somerset cave diving community rallied round and gave an extraordinary amount of help without which the project would have never even got started. This was not going to be an exploration cave diving film as there was apparently nothing left to explore in Wookey, but I wanted to make a film that would appeal to a wide audience and maybe try to explain why people go cave diving, as well as showing as best as I could the spectacular cave that lies beyond the show cave. I approached the owners of Wookey Hole Caves who were very supportive and gave us unlimited access to the caves and, with the full support of the Cave Diving Group, the project was underway.

I had decided to base the story around the exploration of the caves right from the earliest cave men through to Rob's exploration and my attempt to film where he had been. The trouble with cave diving is that it is not a great spectator sport – in order to engage with the audience, I decided to use a presenter. Not a super-hard cave diver, but someone who could experience the cave for the first time and convey their own feelings and emotions on a level that the audience would relate to. This was going to be a difficult balancing act; to find someone experienced enough to be safe in the environment and happy to make the dives, but also sufficiently unfamiliar with conditions at Wookey Hole to be able to convey those experiences to the audience.

*Fig. 27.1
Mike Thomas,
Chamber 24.
Film Still by Gavin Newman*

Roger Whitehead was a diver I had met a couple of years earlier whilst filming sharks in Africa, a safe competent diver who had done a certain amount of caving and continental cave diving but never anything like Wookey. British, but now living permanently in the US where he's a lecturer in psychology at Denver University, I hoped that his academic background would bring an interesting angle to the question – why explore caves? I approached Roger and he agreed to take on the role of presenter – although he was obviously nervous about the dives involved.

Meanwhile filming began in earnest, working most weekends throughout the summer to fit in with work commitments and the availability of assisting divers both in front of and behind the camera. Because of his own commitments Roger was only available for a three-week period, so we had to shoot much of the background material without him – that meant filming most of the big scenic sections of the cave and then inter-cutting Roger's sequences later.

It has often been a feature of cave films that they are not filmed where they say they are. To make filming easier, locations are substituted for harder ones and a certain amount of poetic licence is used. I decided early on that we would NOT do this. With the equipment available today, there was no excuse not to film everything where we said it was, so that's what we set out to do. We succeeded apart from three shots. Two are 5 second close-up shots, which were filmed in different parts of the cave from the sequence in which they appear and were not visually location-specific in any way. The third was an explosion sequence. This appears in the film as Sump 1 in Swildon's Hole, which is a very easy place to get to, but not very diplomatic to blow up. At the time, however, part of our team was exploring and blasting at Sump 12 in the same cave, an altogether much harder place to get to. So we dragged all the filming gear down through Sump 1, our actual location, and on down to Sump 12 where we filmed the explosion. The picture was flipped in the computer and run in slow motion and 'hey presto!' we had our Sump 1 explosion. I doubt if there are many film productions that deliberately use much harder locations as a double for easier to reach ones!

Fig. 27.2
Jo Wiseley and Roger Whitehead in Chamber 3, dressed in 1930s gear when playing the parts of 'Mossy' Powell and Graham Balcombe.
Photo by Gavin Newman

Part of the Wookey story involved the first dives done in the caves using Standard Navy Diving Dress. As Sid Perou had done earlier in his series of films illustrating the history of cave diving, we wanted to recreate these dives and film the divers underwater in the very passages they explored. With the help of the Historical Diving Society, Roger and Jo Wisely were able to make a series of dives from the original dive base in Chamber 3, recreating the 1935 dives of Graham Balcombe and 'Mossy' Powell. The Historical Diving Society put a lot of effort into ensuring the equipment was as it should be and with Leo filming on the surface and me filming underwater we managed to get a real flavour of what those early dives must have been like.

Another reconstruction deals with the first dive in caves using open circuit scuba. One of the Cave Diving Group members, Pete Mulholland, was to play the diver and we had borrowed an ancient and rather questionable twin hose regulator and mask for him to use. Throughout the sequence a suitably hidden stage bottle with trusty Poseidon regulator was there for safety, as was a safety diver, but when the mask started to disintegrate and the regulator flooded some way out from Chamber 9 the look of terror on the 'lost' diver's face didn't require much acting! Pete kept swimming and I kept filming and the sequence is one of my favourites in the film.

Fig. 27.3
Mike Thomas leaves the water in Chamber 22.
Film Still by Gavin Newman

The filming proved to be the easy bit of the project, whereas getting everything and everybody to the locations proved far harder. I wanted to really show the size of the large chambers in the cave but to do this needed some serious lights. I'd managed to get hold of some 270 W arc lights that produce the equivalent of 1200 W of halogen light. They were perfect for the job apart from the 36 V battery power required to run them. Three gel cell type car batteries would give us approximately 30 minutes of light and with two light heads we required two sets of batteries.

Fig. 27.4
Mike Thomas and Steve Marsh, still wearing diving gear, stagger across Chamber 22.
Film Still by Gavin Newman

The sumps from Chamber 9 to Chamber 24 are constantly changing in depth and buoyancy control of a pack of three car batteries all soldered together became something of an art. In the end most of us adopted the same technique and gave them all to our tame gorilla, Andy Stewart, who abandoned all attempts at buoyancy control and just proceeded to walk along the floor of the sumps with a battery pack in each hand! The lamp units themselves presented the opposite story. At £3,000 each I wasn't going to risk getting them even slightly wet so I commandeered a large Perspex dry box that had been made by Pete Scoones for the original Wookey film. Due to its size and its tendency to try and kill the diver manhandling it, it became known as the Perspex coffin. The amount of lead needed to sink it made it unmanageable on the surface but the cave dictates at least three carries between sumps to get to Chamber 24… this was nobody's favourite load!

Roger made several dives beyond Chamber 9 where the show cave ends, including one trip to the end of Chamber 24, at the end of which he crawled back to the car on hands and knees and announced he had officially retired! He is the first to admit he's not a British-style cave diver but he put a supreme effort into the project and got further than I ever expected of him and his reactions were everything I could have hoped for.

Fig. 27.5
Mike Thomas and Steve Marsh approach Sting Corner in Chamber 24.
Film Still by Gavin Newman

Personally my goal had always been to go as far as I could in Rob's footsteps and bring back images from as deep as possible. I had never been beyond Chamber 24 before so to hope to get to the end and bring back the images on the first dive was with hindsight rather ambitious. From all the reports I knew the end of the cave was very small so I decided to take a smaller camera and housing than I usually use and built a neutrally buoyant rig including the lights that I could use on the end of a pole to film myself. I reached the end surprisingly easily and started to dig my way down the gravel-filled slot that Rob had followed. At -64 metres I couldn't get any further; the gravel and the visibility was closing in behind me and the decompression was building up fast so it was time to leave. I'd opted to make the dive on air and in a wetsuit to reduce the number of bottles and extra equipment required, so narcosis was a major factor at the bottom. I pointed the camera in what I thought were all the right places and then headed back. Thirty minutes of chilly decompression beneath Chamber 25 was followed by the short dive back to Chamber 24 and the rest of the waiting team who were fast asleep in 'bivvie' bags.

Fig. 27.6 Mike Thomas and Steve Marsh, Chamber 24. Film Still by Gavin Newman

Once back on the surface we were able to review the footage and discovered that the water pressure had activated the camera housing's auto focus button, causing the focus to hunt in and out constantly. All the footage I'd shot above -50 metres was fine but all the footage of the end was unusable. It quickly became obvious that I'd have to go back again.

Several weeks later I was back in Chamber 24 with the camera's auto focus button firmly disconnected. Again the trip to the end went smoothly and all appeared to go to plan until halfway back to Chamber 25 I realised that the wide angle lens from the front of the camera housing was missing.... I spent most of the decompression time more worried about how I was going to tell the team that we might have to do yet another dive than how I was going to tell Leo that his £300 video lens was somewhere at the bottom of Wookey...

The video was perfect, so clear and sharply focussed this time that you can see exactly where the lens falls off the camera and ruins everything...

Attempt number three was to be the make or break, although I had already broken more than enough. I decided to take the risk of taking my larger and infinitely more expensive camera housing, and hope that digging through gravel with it would not scratch the expensive glass lens. It really was third time lucky. The dive went like clockwork. I reached the end easily and spent five minutes digging around in the terminal slot getting a variety of shots, and headed for home. Familiarity with the location meant that in spite of the narcosis and being blinded by the filming lights shining in my face I was able to get the shots I needed. It was to everybody's relief when we checked the tape that evening and everything had finally come together. It was not however until much later that we were to find out just how significant that dive was to become.

I completed the initial edit of the film and showed it at various caving and diving events where it was received with great enthusiasm. It was at a Dive Show in London that Rick Stanton saw some excerpts from the film and asked if he could have a copy.

Rick is generally considered to be the best cave diver of his generation. A pioneer of new rebreather technology, he is also a natural cave diver who really understands caves and how they form. This is something that is so often lacking in the new generation of open water technical divers moving into cave diving these days. Proving this, Rick has on numerous occasions over the last few years found the way forward in cave systems that other divers have declared finished.

However, based on the reports from Rob Parker and several other divers who had been there on mixed gases since Rob, even Rick had considered Wookey a finished cave until he watched the film footage from the very end. The film clearly shows a gravel mound in the centre of the passage suggesting a split in the water flow. Rob's old line leads into the gravel slot on the left but there was a possible space on the right.

Rick decided to take a look using a new very low profile side-mounted rebreather rig that he had built especially for such caves and swam straight through the gravel constriction that had always been considered as the end of the cave. Nobody was more surprised than Rick, who had not even taken a line reel. Now beyond the squeeze and back in wide-open passage the story was set to continue.

Fig. 27.7
Traversing over the River Axe in Chamber 24.
Photo by Martyn Farr

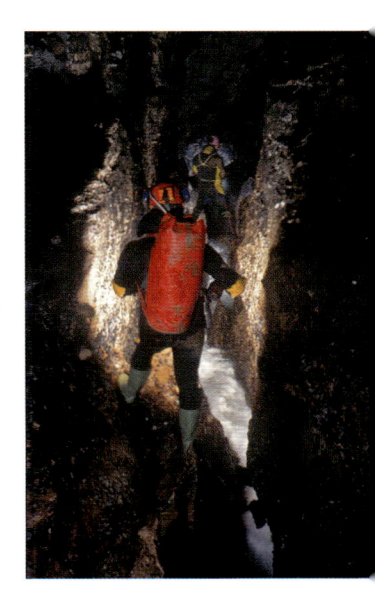

My film was proving to be out of date even before it was finished so the only thing to do was continue filming and update the original edit. But the cave proved to go deeper on every dive; this was now firmly side-mounted rebreather territory only. I could film as far as Chamber 25 although beyond was to prove more difficult. But the film had been born of the idea to film Wookey to the bitter end and we weren't going to be stopped now. I designed and built a miniature helmet-mounted camera unit connected to a waist-mounted digital recording unit that Rick wore on his final dive into the cave. It doesn't get much more extreme than digging through boulders at -70 m and reaching an eventual depth of 90 m in a second boulder choke.

I got a better ending to my film than I could ever have hoped for and in my own little way helped to find the way on in Wookey after 18 years. The film is now finished and I start the hardest job of all in trying to get it broadcast. An early edit of the film won the best adventure film prize at the Kendal International Mountain Film Festival and was only pipped to the grand prize by 'Touching the Void', which I couldn't really complain about! I took on the roles of director, producer, scriptwriter, cameraman, lighting and editor out of necessity and without excuse made the film I wanted to make. It has been a long and sometimes painful process and I personally made over 50 trips into the cave, but I could never have made the film without the totally unselfish support of a lot of people who helped out and the best gratification of all is that now it's finished, they still think it was all worth it...

Fig. 27.8
Chamber 25:
The Lake of Gloom -
the furthest upstream
airspace yet reached in
Wookey Hole.
Film Still by Gavin Newman

Fig. 27.9
The 1985 limit of exploration showing Rob Parker's line (left) disappearing into the gravel floor at -60 m with a whaleback of gravel in the centre and the suggestion of a passage to the right which turned out to be the way on.
Film Still by Gavin Newman

Film Still by
Gavin Newman

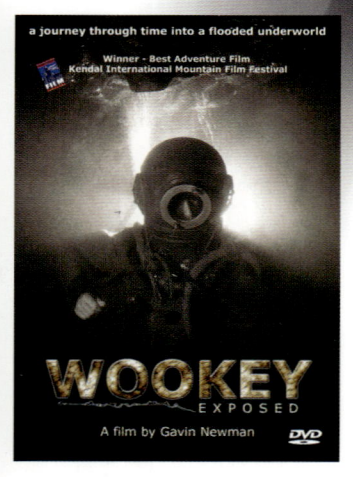

Fig. 27.10
The completed DVD
(available from
www.wookeyfilm.com)

Chapter Twenty Eight

Full Circle

by Duncan Price

Photo by Mark 'Gonzo' Lumley

In 1991 Michael 'Trebor' McDonald produced the third edition of the *Somerset Sump Index*. Building on the text written by Phil Davies (1957) and Ray Mansfield (1964) 'Trebor' wrote in his introduction to the entry for Wookey Hole:

If the Cave Diving Group was considered to have been conceived in Swildon's Hole, then it was most certainly nurtured and cradled in Wookey Hole, where many great diving scenes have been enacted, from light-hearted water romping to toil and triumphant discovery – but also to grim tragedy. The present extent of exploration represents the greatest achievement of the CDG in Somerset, although Gough's Cave in Cheddar is running second. The Wookey explorations which started in 1935, are still progressing, albeit at a slow and intermittent pace. The sharp end of the cave deep beyond Wookey 25 represents the very forefront of diving techniques and human endeavour. It is felt that the equipment side of the equation has to catch up with the human side. Divers are perfectly capable of diving and operating at such depths and remoteness, but the use of cylinders is, or has become, awkward and impractical. New equipment such as rebreathers will have to be perfected. It is poignant to note that Wookey explorations commenced with rebreathers and we have come full circle to start considering their use again for deep, remote cave diving.

…in retrospect, these words were somewhat prophetic!

By the turn of the 21st century most people had written off the upstream end of Wookey Hole as impassable. Several divers had put in concerted attempts to dig their way through the blockage that had first defeated Rob Parker in 1985 and no one had ever got as far as him. Gavin Newman had made a film of the cave up to the limit of exploration and most cave divers were content to see the 'end' from the comfort of their sofas. It was, however, the ability to inspect the limit of the cave at leisure which led Rick Stanton to comment to John Volanthen that 'something wasn't right' about the shape of the gravel deposits on the passage floor and that there might be a way around the gravel blockage on the opposite side from the line. A plan was hatched to go and have a look…

Rick Stanton hails from Epping Forest in Essex and wanted to be a cave diver ever since he saw *Underground Eiger* (the film by Sid Perou of Oliver Statham and Geoff Yeadon's traverse from Kingsdale Master Cave to Keld Head) on TV in 1979. Rick recalls that it was his mother who suggested that he might be interested in the programme. Shortly afterwards, Rick went to Aston University where he simultaneously joined the caving club and sub-aqua club so as to fulfil his ambition.

It was almost on the other side of the world, in Peru, that Rick met up with Ian Rolland on an Army Caving Association expedition in 1987. Ian was also a cave diver and had supported Rob Parker on his 1985 dive. Ian was very active in South Wales where he was busy pushing cave passages beyond sumps in Ogof Daren Cilau. Rick was invited to join Ian on his explorations and they were very successful in discovering the Inca Trail (named after their Peru trip) extensions in the cave.

I knew Ian as well, since we were both members of the Chelsea Speleological Society. Although I was born and brought up in Somerset, in the village of Chilcompton – only 12 miles from Wookey Hole – I had taken up scuba diving in my teens before being introduced to caving by a school friend during the summer vacation after my first year at Exeter University in 1983. During my postgraduate studies at Birmingham University I had spent far too much time digging in Agen Allwedd where Ian had helped out on occasion. One thing led to another, and I started cave diving as a means of getting to promising dig sites beyond sumps under the watchful tutelage of Ian Rolland and Martyn Farr.

By 1990, I had moved to Coventry where I had a job with a major chemical company. Rick also worked in Coventry as a fire-fighter (he frequently turned up at my employers whenever the alarm went off). Ian suggested that Rick and I should do some caving together, and we subsequently undertook the excavation of the Pwll-y-Cwm resurgence to the Llangattock caves in the Clydach Gorge in order to establish a convenient 'backdoor'. This facilitated further cave diving discoveries in Daren Cilau.

John Volanthen and I first met in October 1998 when we went for an evening trip in Ogof Capel. My diving watch was still set to British Summer Time and we ended up rushing out of the cave only to find that we had an extra hour of drinking time. John comes from Brighton and first started caving with the Scouts in 1982. The surname 'Volanthen' is unusual, being an anglicised form of the family name of 'Von Lanthen' which was held by John's Swiss grandfather. As a consequence John is frequently known by the affectionate nickname of 'Volleyball.' My dive logs show that whilst I did a fair amount of cave diving with Rick or John, it was not until the Wookey Hole dives described here, that the three of us got to work together.

In 2002, Rick and John combined forces in order to push on from Rob Palmer's upstream limit in Gough's Cave. This required the use of a rebreather to ease the logistics and Rick was well-placed to be an expert in such technology, having been part of the 1998 explorations to Wakulla Springs in Florida led by Bill Stone. The diving there had used sophisticated computer-controlled mixed-gas rebreathers, sonar underwater mapping devices mounted to long range diver propulsion vehicles and submersible decompression chambers borrowed from the marine oil exploration industry. It will not surprise readers to learn that Bill Stone is currently developing autonomous robotic underwater cave mappers for use in the sub-glacial lakes of Antarctica and beneath the ice-covered oceans of Europa, one of Jupiter's moons. Wookey Hole Cave Divers get everywhere!

Rick's rebreather, dubbed the RS2000, was less sophisticated, but nevertheless, most effective. He had put this together from various second-hand components during 2000. The RS2000 did the job in Gough's Cave and Rick was able to surface in a blind airbell in the middle of the boulder choke at the end of Sump 3 first reached by Palmer in 1990.

The proposed dive at the end of Wookey would require a more streamlined rebreather than the one used at Cheddar. Rick devised a novel side-mounted rebreather which could be worn like conventional cave diving cylinders rather than more traditional rebreathers which are commonly worn on the back (or chest). Interestingly, the concept of mounting a rebreather under one's arm pit was not new to cave diving as Steve Wynne-Roberts had used his innovative ATEA/SEBA side-mounted oxygen rebreather in Swildon's Hole, Stoke Lane Slocker and other caves during the 1960s (Chapter 10). Rick's unit was capable of operating at depths approaching 200 m. Two galvanic cells were used to monitor the breathing loop for oxygen content and a constant mass-flow orifice matched the addition of this gas closely to the diver's metabolic rate so that only small corrections were required. With the appropriate choice of make-up gas (air or Trimix) the wearer could stay underwater at almost any depth for several hours.

The genesis of this device, however, was far more interesting. In order to test the layout of the carbon dioxide scrubber and counterlung, Rick constructed a prototype which used a plastic sweet jar (of the sort that you see on the shelves of traditional corner shops) as a proxy for the scrubber. Despite not having any medium for the removal of CO_2, this device was successfully tested in Coventry swimming baths. Both John and I were building similar rebreathers and there was a bit of a competition between us as to who would dive theirs underground first. In the end we declared it a three-way draw.

Meanwhile, back at Wookey Hole Caves, Vic Cooper and David Haselden donated some cylinders containing suitable gas mixtures for a dive and cached these in 22 during the autumn of 2003 awaiting the push dive. Unfortunately diving access to Wookey Hole Caves was then suspended while public liability insurance cover for caving activities in the UK was resolved.

Fig. 28.1
Rick Stanton with prototype rebreather, Coventry swimming pool, February 2004. Photo by Duncan Price

With the cave reopened for exploration six months later, we were ready to go. John and I planned to do a set-up dive at Wookey on 31st May 2004, but were informed en route that permission to dive had been refused on the grounds that it was a Bank Holiday and the show cave did not want divers in the way of the crowds of tourists. John and I went to Stoke Lane Slocker instead and were defeated by a broken line in Sump 6. The next day, Tuesday 1st June, Nick Lewis, Ian Pinkstone, Laura Trowbridge, John Volanthen and I assisted Rick and a modest pile of gear into the cave. Collecting the gear cached in 22, good progress was made to 24 where Ian and I also had plans to go as far as 25 in support. Unfortunately the guide line was broken at the elbow of the sump from 24 to 25, and only Rick and I had sufficient gas supplies to reach Chamber 25. Here I did a quick bounce dive to a depth of 50 m to see the 'end' for myself before Rick overtook to explore further. I retired to the dismal gloom of the airspace above to wait alone for Rick's return and assist him as required. An hour later, Rick surfaced to say that he'd passed a low arch off to the right of the line, down a gravel slope and through a squeeze to reach the elbow of the sump into an ongoing phreatic tube with an ascending rift in the roof. The arch passage had bypassed the squeeze encountered by Rob Parker in 1985. Rick had reached a new depth of 70 m but had not taken a line reel as he had not expected to pass the old terminus so quickly and easily. Luckily water conditions had been very clear.

With the way on apparently open, plans were made to put Rick and John in the sump simultaneously. This was because they hoped to ascend to further dry passage and explore yet another of Wookey's dry chambers: No. 26! However, during a set-up trip made by John and Ian

Pinkstone on 6th June they found that the guideline between Chambers 22 and 23 had been removed and several of the cylinders left in 24 had been emptied! Mike Barnes also had his sights on the end of the cave and was planning to employ an airlift (driven by over a kilometre of pneumatic hose) to excavate the gravel squeeze at a depth of over 60 m (Chapter 24). Feeling that his project was being 'poached' by Rick after he had put in a lot of preparatory work, Mike had carried out a deliberate act of sabotage. Over the next two weeks the damage was put right and sufficient equipment was carried to Wookey 24 to allow both two divers to pass the squeeze and explore the passage beyond. On 18th June, Rick and John were efficiently escorted to Chamber 24 by Nick, Ian and I (Laura having dropped out in Chamber 9 due to ear clearing problems). Rick led off followed by John – the latter using a small chest-mounted rebreather, which although less streamlined than Rick's was easily manoeuvred in small spaces. The roof-rift was examined, but appeared to close down at -48 m. Line was laid down a 2 m diameter tube in a north-easterly direction but after a couple of turns the route ended in what appeared to be a pot of about 3 m in diameter. The floor of this pot was at a depth of just over 70 m, but was full of large angular boulders through which the current seemed to rise. This was 70 m beyond Parker's 1985 limit. Ian had to leave the cave early, meaning that Nick and I had to do most of the work of getting the gear out.

The next trip took place on 3rd July. Again, Rick and John were put in the water, supported by Andy Chell, Jon Beal, Gary Jones, Nick and myself. Although the aim of the dive was to take another look at the terminal choke with a view to calling it a day the divers surfaced optimistic that with the right gear (lump hammer, crowbar and lifting bag) the blockage could be shifted. Rick's shift pattern as a fire-fighter now dictated that the next trip took place on a Monday so that only Andy, Jon and I were available. Even so, on July 19th a cut-down team sent Rick and John off and sat down at the campsite for a series of hot drinks to wait. An hour later, Rick returned to say that the electronic oxygen display on his rebreather was malfunctioning and he'd been unable to get to the end. The problem was traced to a new battery which was packing up below exactly 50 m depth. John came back after two hours away, having managed to pull the key boulder out only for it to slip back into place. With John and Rick off to the France for a cave diving expedition, the sherpas took a much needed summer break.

It was business as usual on September 5th, with the regulars – Andy, Jon and I – joined by Pete Mulholland and Jo Wisely in support of Rick and John. Rick was delayed in Chamber 9 because he had to fix the oxygen supply regulator for his rebreather, but the divers set off from 24 in good time. The boulder at the end of the sump was moved aside enough to enable John to back through with Rick waiting on the downstream side. Mindful of the awkward return, John confirmed that the way on was open beyond and, with some difficulty, squeezed back to join Rick.

Just over two weeks later the team from 3rd July escorted the divers to Chamber 24 once more. Gavin Newman was also in attendance with a video camera to record the proceedings. Rick and John were interviewed before they set off and the sherpas sat down for a hot meal at camp. Rick

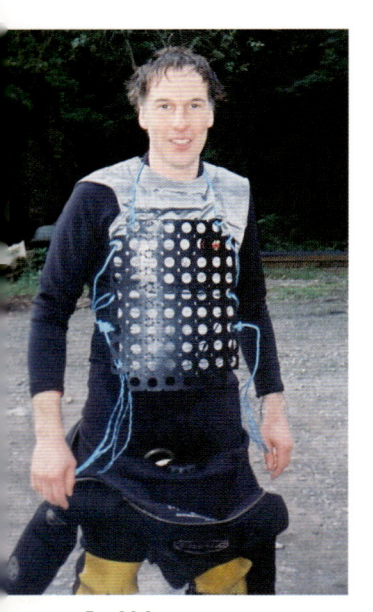

Fig. 28.2
Rick Stanton modelling the latest in thermal protection, 19 July 2004.
Photo by Duncan Price

and John came back after a couple of hours, very cold but happy to have explored another 70 m of passage which ascended to the lip of another pot at -59 m where it was deemed prudent to return. Significantly, Rick and John had reached -75 m and -76 m, surpassing the previous record of -72 m achieved in Speedwell Cavern, Derbyshire. Worryingly though, Rick was low on the Trimix required to run his rebreather. When he looked at his contents gauge, he was heard to remark 'I've got some gas left.' When asked to elaborate on this, he explained that it was 'enough but not plenty!' Gavin had him repeat this for the benefit of the video camera and more interviews followed before the team made a speedy exit to the usual drinks (at Rick and John's expense) in the Wookey Hole Inn where they chatted to the owner's son, Gerry Cottle Jr., about their escapades.

Another visit to Wookey Hole took place on 29th September 2004. Ever mindful of the need for good public relations, the team was met by the press at the cave. Radio and TV interviews were given and celebratory champagne was drunk (with the divers in their gear) in Chamber Three. After the toast of 'Chamber 26 – and beyond!', and with Rick and John still being fêted by the media, the support divers slipped into the water for a slightly inebriated swim…(I do admit to being more inebriated than the others as I had downed Andy, John and Jon's bubbly since they don't drink).

The last push of 2004 had to be aborted owing to high water conditions. A set-up trip had taken place on Saturday 23rd October in preparation for the push on the following day. I had cached some cylinders of Trimix for a tourist dive through the gravel squeeze at -65 m earlier in the year and carried a large cylinder of Nitrox-40 and a smaller tank of pure oxygen to get me from Chamber 24 to Chamber 25 and back (including decompression). High water levels made the planned dive impossible and further activities would have to wait until the following spring.

In comparison to the drama of 2004, the continuing underwater exploration of Wookey Hole in 2005 was of a different character. Trips in 2004 had been on a somewhat impromtu basis and of an alpine style whereby the team turned up (often midweek) and took all the gear (save some of the cylinders which seemed to live in the cave) in and out on every push. At the start of the 2005 'Wookey Season' a well-planned list of dates for trips was agreed between the lead divers (Rick and John) and the 'National Union of Rebreather Porters' (as I dubbed the support team). By request, the 2005 season's dives were to take place at weekends and thus had to fit in around Rick's shift pattern as a firefighter. This meant that unlike 2004 there was often an audience of people who could be persuaded to help carry equipment up to the dive base in Chamber 9.

It was not until 14th May 2005 that my deep dive beyond Chamber 25 eventually took place. Supported only by John Volanthen from Chamber 9, we made good time to Chamber 24. Wearing a thick wetsuit, I wore two sidemounted 7 litre cylinders of Trimix plus another 7 litre cylinder of Nitrox-40 and a further 7 litre cylinder of oxygen. The intention was to dive from Chamber 24 though to Chamber 25 and drop off the oxygen at -6 m beyond this, then carry on down to -25 m on the Nitrox before staging this tank and continuing on the helium mix to pass the squeeze.

Fig. 28.3
Duncan Price, post diving to -70 m, 14 May 2005.
Photo by Antoinette Bennett

Fig. 28.4
John Volanthen after helping Duncan (above), 14 May 2005.
Photo by Antoinette Bennett

Rather than a pure tourist trip, the dive was being undertaken to ensure that the route though the gravel-filled arch was still passable for future dives by John and Rick. Ear-clearing problems delayed my exit from 24, but by the time I had dropped off the cylinder of Nitrox, I was on a mission to get as far as I could. Viewers of the film by Gavin Newman of the old end of Wookey will have seen Rob Parker's line disappearing into the sand. Actually being there (with a clear head) is like being part of cave diving history.

Just before Parker's limit a newer, thin line led off to the right through a low arch and then descended down a mobile gravel slope. Even in a very streamlined rig, I was aware of the gravel running down behind me when suddenly I popped out in a little chamber at the base of the slope with the bank of gravel behind me. Above my head was a blind ascending rift that had been probed by Rick in 2004, whereas ahead and to the right, the cave passage continued as a narrow rift to the next restriction. I had no intention of seeing this as I was already at nearly 70 m depth – in a wetsuit! Turning around, I had to clear the gravel that had slumped in behind me in order to get out but the rest of the dive was uneventful, if a little chilly. Chamber 24 was reached after only an hour's absence and we managed to exit with all the spent cylinders from the day's dive after a 5 hour trip.

Another major push took place just over a month later on 16th June. Rick and John were supported by a strong team of Jon Beal, Andy Chell, Tim Morgan, Gavin Newman, Charlie Reid-Henry, Laura Trowbridge and myself. I was roped into transporting a cylinder of oxygen to Chamber 25 for decompression purposes. Gavin had also come along with an underwater video camera to shoot John and Rick as they set off. Tim took the opportunity to visit Chamber 25 using my gear. Although the gravel squeeze was open, there was some concern that the boulder choke beyond might have moved. In the event, the way on was still passable allowing Rick and John to progress beyond the lip of the pot at -59 m. They dropped down a steeply inclined shaft to -63 m, then again to -75 m where the passage turned into a meandering rift as the floor fell away. John stopped at -78 m but Rick continued alone and followed the bottom of the rift at a depth of 90 m for a short distance before prudently returning to base. Back in Chamber 24 the sherpas were having a miserable time of it. The air quality was poor and none of the lighters would work. Eventually, Tim soaked the cloth from my camera gear in methylated spirits and managed to spark it alight after drying the lighter on Gavin's film light. It is questionable whether Tim really needed to have used all of the rag, but at least everyone could have a brew. By the time that John and Rick returned the sherpas were keen to get out. That it was also a Sunday night, and everyone had to work the next day, didn't help matters.

The next trip took place on 2nd July. Because of the depth and distance from base John and Rick elected to dive separately from now on as it was clear there was not going to be an easy route up to dry passage. The two had continued diving together both in the forlorn hope of this possibility and as a contingency against damage to equipment during transport through the cave. Due to the gear-intensive nature of each venture, should there be a problem with one set of equipment on the approach, then the 'mission' could still go ahead with the other diver. With only

Fig. 28.5
Charlie Reid-Henry,
12 June 2005.
Photo by Antoinette Bennett

Fig. 28.6
Jon Beal, 12 June 2005.
Photo by Antoinette Bennett

Fig. 28.7.
Andy Chell, 12 June 2005.
Photo by Antoinette Bennett

Fig. 28.8.
Gavin Newman, 12 June 2005.
Photo by Antoinette Bennett

Fig. 28.9
Rick Stanton at dive base in 24 with Charlie Reid-Henry and Gavin Newman, 2 July 2005.
Photo by Martyn Farr

one exploration diver the logistical mountain was now somewhat eased. John was going to use two rebreathers – his chest-mounted unit that had been employed on previous dives and his own side-mounted rebreather, like Rick's, but smaller. Gavin Newman was keen to get some footage of the cave beyond Parker's limit so Rick was put into the sump with the aim of filming up to the dug out boulder choke. Martyn Farr was also present to take some still photographs so the dive base was all rather crowded. Unfortunately John was unable to make progress through the gravel squeeze with both rebreathers and has to turn back. Rick got to the second squeeze but on surfacing found that the camera had turned itself off. Not a successful trip and a re-think was needed.

A few weeks later on July 23rd everyone was back for another go. Technically it was Rick's turn to dive, but John had streamlined his gear and was ready for a second attempt so it seemed fair to let him have another chance. Gary Jones and Gavin Newman had helped set up equipment in the cave mid-week prior to the trip and a smaller team of Jon, Charlie, Rick and I put John into the water for the push. Clive Westlake joined the sherpas in 24 after a couple of hours to help carry out the gear. With lower water levels than in June the sumps from 24 to 23 were largely open and there was improved air quality at camp. Rick was rather bemused to find himself one of the support team for the first time. John was able to push on from Rick's limit, laying another 30 m of line, initially going up to -80 m, and then descending again to -90 m in a chamber where the water appeared to rise through boulders in the floor. The dive took over four hours and John was adamant that he'd gone as far as he felt he could, and that it was Rick's turn.

The divers took their summer break and reconvened at Wookey on 17th September in preparation for another solo push – this time by Rick, who had only just returned from a cave diving trip to France. David Haselden and Brian Judd turned up to help Rick, John, Charlie and I with the set-up trip to 24, but in the event both Brian and Charlie retired early with various problems giving the smaller team a bit of a challenge to transport all the extra gear to the end.

The next day the gang was back at Wookey with Martyn Farr and Gavin Newman on hand to take surface photos and shoot more video. Other familiar faces were on hand – Antoinette and Harriet Bennett, Helen Rider, Jackie Ankerman and Lee Hawkeswell – to assist, this being the final scheduled exploration of the year. Rick was launched into the water at Chamber 24 and

Fig. 28.10
Rick Stanton, about to dive on 2 July 2005.
Photo by Gavin Newman

Fig. 28.11 John Volanthen, Chamber 24, 2 July 2005.
Photo by Gavin Newman

Fig. 28.12 John Volanthen and Rick Stanton, with dressers Jon Beal and Andy Chell, Chamber 24, 2 July 2005. Photo by Gavin Newman.

was accompanied by John Volanthen to about -40 m in the final sump. Here John left a back-up rebreather for Rick to decompress on in an emergency. Rick also carried a small 'helmet cam' built by Gavin to film some of the passage en route to the end. This time everything worked. Rick reached John's previous limit and confirmed that the route ahead was blocked by boulders. So far in, and at such a depth, the prospects did not look hopeful. However, Rick reckoned that if a diver could pull a few of the rocks aside, it might be possible to squeeze through by removing most of his equipment. The choke however, is unstable with friable walls. A very low route ahead could be seen to continue for a few meters beyond this point – so it was not too inviting. Whether this will ever be attempted by Rick, John, or anyone else remains an unanswered question.

Fig. 28.13 Jon Beal and Duncan Price in, Chamber 9 after the final dive of the 2005 season, 18 September 2005. Photo by Martyn Farr

Fig. 28.14 John Volanthen and Rick Stanton, Chamber 9, 18 September 2005. Photo by Martyn Farr

Rick's last dive on 18th September 2005 was nearly five hours in duration, of which one hour was spent beyond the 1985 limit, and required two rebreathers, a heated undersuit and endless patience by the sherpas: Jon, Brian (who came in behind the others on his own) Charlie, John and myself. Some equipment was left behind and removed later on a clean-up trip under high water conditions in December 2005. The final sump in Wookey Hole had been extended upstream by a modest 230 m since Parker's dive and remains (now at -90 m) the deepest cave dive in Britain. Gavin Newman was able to use the footage from the 2004 and 2005 campaigns to provide an up to date conclusion to his *Wookey Exposed* film.

A return match is not being actively considered, but is also not discounted. The technology exists to enable a diver to operate at such depths, but the physiological problems of extended decompression in such conditions, to say nothing of the psychological barriers of the dangerous and unstable squeezes at such great depths, now present new challenges to future explorers. Wookey Hole remains the unconquered Everest of British cave diving.

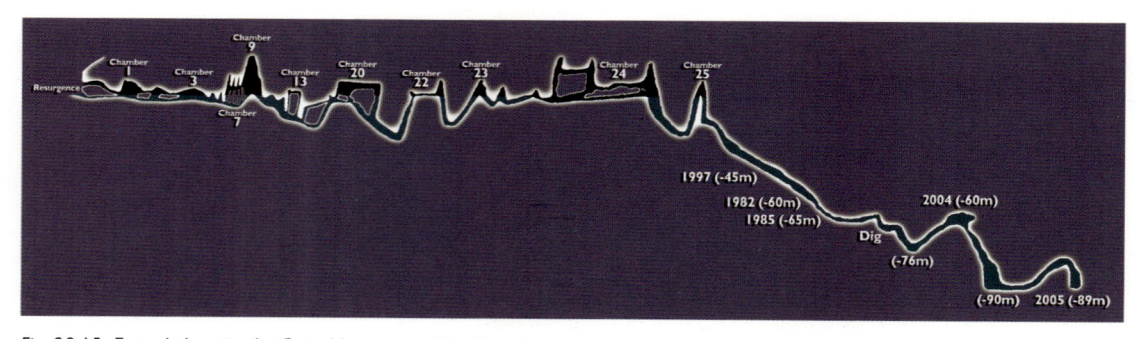

Fig. 28.15 Extended section by Gavin Newman and Jon Volanthen

Fig. 28.16
Rick Stanton sets off from Chamber 24 on 16 June 2005 followed by John Volanthen.
Film Still by Gavin Newman

Fig. 28.17
The way on into new territory as revealed by Gavin Newman's video footage (Chapter 27), filmed by Rick Stanton using a helmet-mounted camera on 18 September 2005.
Film Still by Gavin Newman

Fig. 28.18
Rick Stanton films himself squeezing though the 2004 breakthrough. A 12 litre cylinder pushed ahead of the diver gives scale to the passage.
Film Still by Gavin Newman

Fig. 28.19
Illuminated only by the diver's helmet-mounted lights, the cave continues...
Film Still by Gavin Newman

Fig. 28.20
Even if this was in a dry cave, it would not be appealing!
The excavated squeeze at -70 m.
Film Still by Gavin Newman from footage shot by Rick Stanton, 18 September 2005

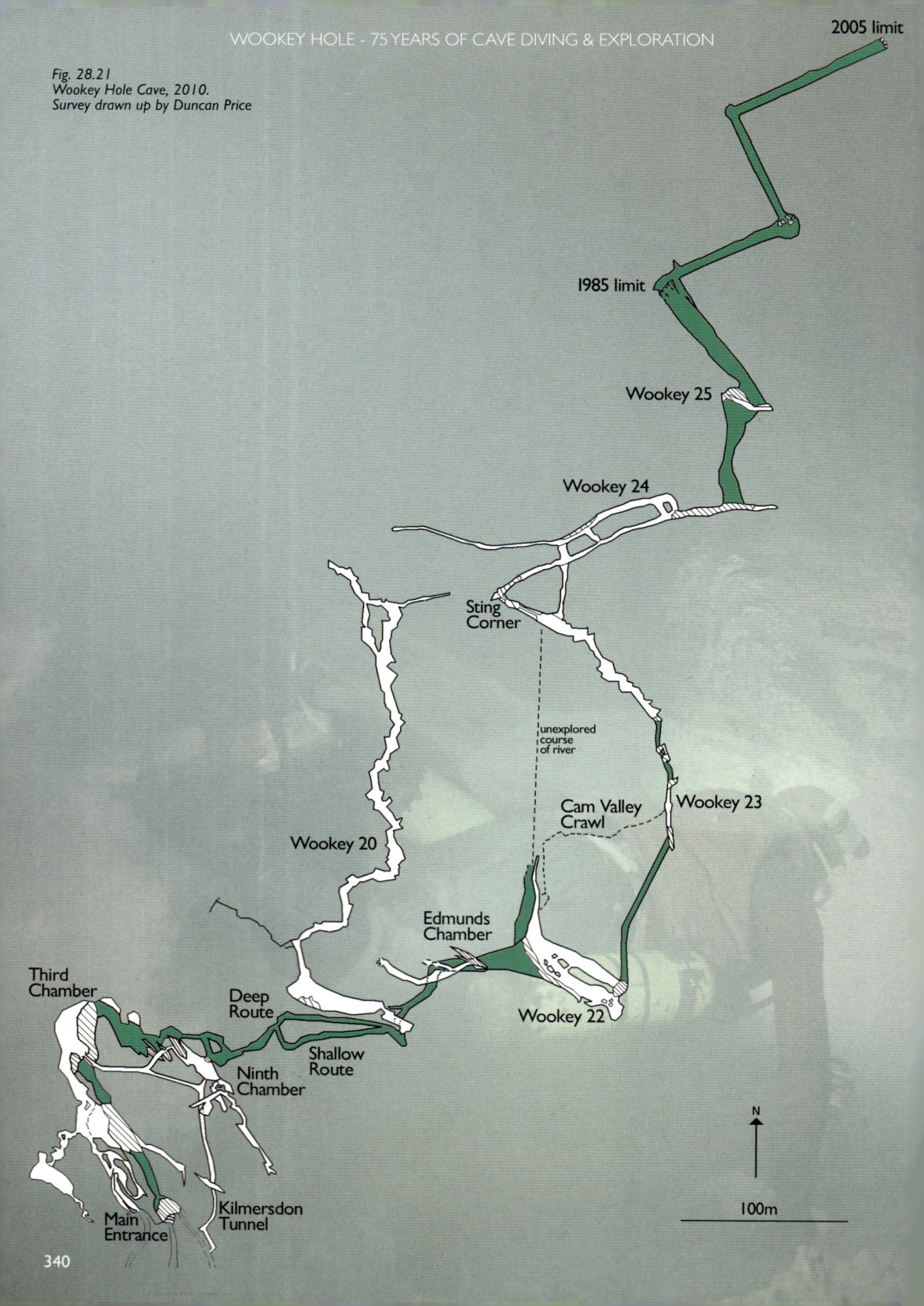

Fig. 28.21
Wookey Hole Cave, 2010.
Survey drawn up by Duncan Price

Acknowledgements

At the end of his Introduction written in August 1987, Jim Hanwell lists the large number of people who gave invaluable help in the preparation of his original 50th anniversary book. They were, in alphabetical order: Graham Balcombe, Bruce Bedford, Barney Butter, Fred Davies, Judy Hanwell, Chris Hawkes, Peter Haylings, Olive Hodgkinson, Graham Jackson, Eric Lewis, the Main family, Ray Mansfield, Frank McBratney, Ken Pearson, Brian Prewer, Jack Sheppard, Willie Stanton, Alf Stapleton, Dave Turner, Rich West and Linda Wilson.

To this list must now be added the authors of the chapters which bring the Wookey story up to date, the photographers who have contributed material, especially Martyn Farr, Peter Glanvill and Gavin Newman who provided many images from their extensive collections, and those individuals who have helped the editors to bring the combined volume to successful publication, including Alan Butcher, John Buxton, Bob Gannicott, Rob Harper, Mike Hearn, Chris Howes, Trevor Hughes, Stuart McManus, Alison and Pete Moody, Phil Romford, Phil Rowsell, Tony Setterington, Bill Stone, John Volanthen, David Wettergreen, Martin Whiteley and Brian Woodward.

The following have played a particularly important part in the project:

Martin Grass who took responsibility for publicity and financial aspects of the project

Ric and Pat Halliwell who assiduously proof read the text and provided necessary revisions

Mark 'Gonzo' Lumley who applied his artistic and professional skills in design, layout and photographic reproduction with a commitment which went well beyond his original remit

Tim Reynolds who strove to bring the original history to publication and helped the present editors resurrect the project

Nick Williams who smoothed the path of discussions between the editors and the British Cave Research Association who funded the book

Clive Westlake who read the text with a diver's eye and ensured that it made technical sense

The Wessex Cave Club who provided a base from which to conduct editorial meetings and library and internet facilities for research purposes

The Wookey Hole Caves owners and management who continue to allow cave divers to push the frontiers of exploration and science in the far reaches of their beautiful and historic cave system

Finally the three editors appreciated the support and patience of **Andrea**, **Antoinette** and **Geraldine** during the lengthy gestation of this book – now it is done, you can have us back!

Photo by Gavin Newman

Bibliography

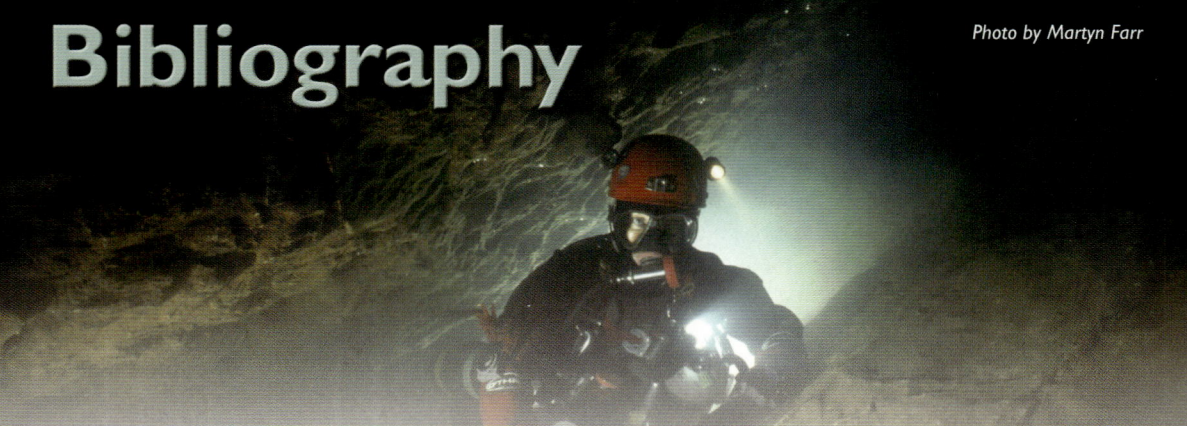

Photo by Martyn Farr

The following publications afford particular historical and technical details to the stories told herein. They represent only a fraction of the literature on Mendip caves and cave diving in general, of course, but are chosen for broadly covering the subject of this book and for providing further reading in themselves. Certain widely known periodicals and newspapers mentioned in the text have been omitted from the list below.

'Alfie' (Collins, S. J.), *Reflections* (Barton, 1971)

Baker, E. A., Caving: *Episodes of Underground Exploration* (Chapman & Hall, 1932. Republished S. R. Publishers, 1970)

Baker, E. A. and Balch, H. E., *The Netherworld of Mendip* (J. Baker & Son, 1907)

Balch, H. E., *Wookey Hole: its Caves and Cave Dwellers* (Oxford University Press, 1914)

Balch, H. E., *Mendip: the Great Cave of Wookey Hole* (3rd Edition John Wright & Sons Ltd, 1947)

Balch, H. E., *Mendip: Cheddar, its Gorge and Caves* (2nd Edition John Wright & Sons Ltd, 1947)

Balch, H. E., *Mendip: Its Swallet Caves and Rock Shelters* (2nd Edition John Wright & Sons Ltd, 1947)

Balcombe, F. G. and Powell, P. M., *The Log of the Wookey Hole Exploration Expedition 1935* (Balcombe, 1935. Reprinted by the Cave Diving Group, 2009)

Balcombe, F. G., *'Cave Diving'* in *British Caving* edited by C.H.D. Cullingford (Routledge & Kegan Paul, 1953)

Balcombe, F. G., *A Glimmering in Darkness* (Cave Diving Group, 2007)

Barrington, N. and Stanton, W., *Mendip: The Complete Caves and a view of the hills* (Barton, 1977)

Beck, H. M., *Gaping Gill: 150 Years of Exploration* (Robert Hale, 1984)

Bonnington, C., *Quest for Adventure* (National Geographic, 2000)

Boon, J. M., *Cave Diving on Air* (Cave Diving Group, Technical Review No 1, 1966)

Boon, J. M., *Down to a Sunless Sea* (Stalactite Press, 1977)

Burgess, R. F., *The Cave Divers* (Aqua Quest Inc., 1999)

Casteret, N., *Dix Ans Sous Terre* (Perrin, 1933). First published in U.K. as *Ten Years Under the Earth* (J. M. Dent & Sons, 1939)

Cousteau, J. Y. and Dumas, F., *The Silent World* (Hamish Hamilton, 1953)

Coysh, A. W., Mason, E. J. and Waite, V., *The Mendips* (4th Edition Robert Hale, 1977)

Davies, P. (Ed), *A Pictorial History of Swildon's Hole* (Wessex Cave Club, 1975)

Davis, Sir Robert H., *Deep Diving and Submarine Operations* (St. Catherine Press, 1935)

Dawkins, W. B., *Cave Hunting* (Macmillan, 1874. Republished E. P. Publishing Ltd, 1973)

Drew, D. P., *Aspects of the limestone hydrology of the Mendip Hills* (PhD thesis, University of Bristol, 1967)

Drew, D. P., *'Cave Diving'* in *Manual of Caving Techniques* edited by C.H.D. Cullingford (Routledge & Kegan Paul, 1969)

Dugan, J., *Man Explores the Sea* (Hamish Hamilton, 1956)

Edgerton, H., *Sonar Images* (Prentice-Hall, 1986)

Ellis, B. M., *Surveying Caves* (British Cave Research Association, 1976)

Empleton, B. E. et al, *The New Science of Skin and Scuba Diving* (Association Press, New York, 1962)

Exley, S., *Caverns Measureless to Man* (Cave Books, 1994)

Farr, M., 'Cave Diving' in *Caving Practice and Equipment* edited by D. Judson (British Cave Research Association, 1991)

Farr, M., *Diving in Darkness* (Wild Places Publishing, 2003)

Farr, M., *The Darkness Beckons* (Diadem, 1991)

Farr, M., *The Great Caving Adventure* (Oxford Illustrated Press, 1984)

Farr, M., *Wookey/the Caves Beyond* (Redcliffe, 1985)

Farrant, A. R., *A walker's guide to the geology and landscape of Western Mendip* (British Geological Survey, 2008)

Ford, D. C., *Aspects of the geomorphology of the Mendip Hills* (DPhil thesis, University of Oxford, 1963)

Gemmell, A. and Myers, J. O., *Underground Adventure* (Dalesman, 1952)

Haldane, J. S. and Priestley, J. G., *Respiration* (Clarendon Press, 1920)

Hanwell, J. D., 'Eighty years of British Caving' in *The Alpine Journal* Volume 80 No 324 (Alpine Club, 1975)

Howes, C., *To Photograph Darkness* (Alan Sutton, 1989)

Irwin, D. J., *St. Cuthbert's Swallet* (Bristol Exploration Club, 1992)

Irwin D. J., *Mendip Cave Bibliography and Newspaper Catalogue* (2nd Edition Mendip Cave Registry, 2005)

Irwin, D. J. and Jarratt, A. R., *Mendip Underground* (4th Edition Bat Products, 1999)

Irwin, D. J., Moody, A. A. D. and Farrant, A. R., *Swildon's Hole – 100 years of exploration* (Wessex Cave Club, 2007)

Johnson, P., *The History of Mendip Caving* (David & Charles, 1967)

Larson, H.E., *A History of Self-contained Diving and Underwater Swimming* (National Academy of Sciences: National Research Council, Washington, D.C., Pub. 469, 1959)

Lavaur, G. de, *Caves and Cave Diving* (Robert Hale, 1956)

Lloyd, O. C., *A Cave Diver's Training Manual* (Cave Diving Group, Technical Review No 2, 1975)

Mansfield, R. W., Reynolds, T. E. and Standing, I. J., *Mendip Cave Bibliography and Survey Catalogue* (Cave Research Group, Publication No. 13, 1965)

McDonald, M. C. and Price, D. M., *Somerset Sump Index* (Cave Diving Group, 2008)

Newson, M. D., *Studies of chemical and mechanical erosion by streams in limestone areas* (PhD thesis, University of Bristol, 1970)

Kellogg, W. N., *Porpoises and Sonar* (University of Chicago Press, 1961)

Miles, S. and MacKay, D.E., *Underwater Medicine* (Aldard Coles. 1962)

Pyatt, E. C., *A Climber in the West Country* (David & Charles, 1968)

Rose, D. and Gregson, R., *Beneath the Mountains* (Hodder and Stoughton, 1987)

Savory, J. (Ed), *A Man Deep in Mendip* (Alan Sutton, 1989)

Shaw, T. R., *Mendip Cave Bibliography, Part II Volume 14 No 3* (Cave Research Group, Transactions, 1972)

Sims, M., *Shepton, The Diving Club* (Shepton Mallet Caving Club, in preparation)

Sisman, D. (Ed), *Professional Divers' Handbook* (Submex, 1982)

Smith, D. I. and Drew, D. P. (Eds), *Limestones and Caves of the Mendip Hills* (David & Charles, 1975)

Stone, W. C. (Ed.), *The Wakulla Springs Project* (United States Deep Caving Team, 1989)

Stone, W. C, am Ende, B. and Paulsen, M., *Beyond the Deep* (Warner Books, 2002)

Terrell, M., *Principles of Diving* (Stanley Paul, 1967)

Thomas, A. (Ed), *The Last Adventure* (Ina, 1989)

Verne, J., *Vingt Mille Lieues sous les mers* (Musée des familles, 1870).
First published in U.K. as *Twenty Thousand Leagues Under the Sea* (Sampson Low, 1873)

Yeadon, G., *Line Laying and Following* (Cave Diving Group, Technical Review No 3, 1981)

Waltham, A. C. et al, *Karst and Caves in Great Britain* (Chapman and Hall, 1997)

Wells, O. C. et al, *Scanning Electron Microscopy* (McGraw-Hill, 1974)

Witcombe, R. G., *Who was Aveline Anyway?* (2nd Edition Wessex Cave Club, 2008)

Yeandle, D. W., *The Adventures of Another Pooh* (Writers Club Press, 2002)

Young, D., *The Man in the Helmet* (Cassell, 1963)

Most Cave Diving Group publications and records have not been cited because all have some bearing upon the subject of this book and are too numerous to list in full. CDG *Letters to Members, Reports, Reviews, Sump Indexes* and *Newsletters* in particular are key sources of information. There is also a considerable body of literature on diving itself, of course, that is relevant to the technical adaptations made by cavers. *The Cave Diving Group Manual* edited by Andrew Ward, Colin Hayward et al. and published by the CDG (2008) represents the definitive text on British techniques. In the course of compiling Chapter 21 especially, professional periodicals such as *High Technology* (USA), *Offshore Engineer, Petroleum Review, Underwater Engineering*, and the *Journal of the Society of Underwater Technology* have been used. References have also been made to the Royal Navy's *Diving* Magazine. Company literature and equipment specifications have included Bell Laboratories *Record*, Normalair-Garrett's *Technological News* and many other sources.

The local cave guidebook, *Mendip Underground* by the late Dave Irwin and the late Tony Jarratt (Bat Products, 1999), is currently the best source of information about the caving clubs and organisations on Mendip holding detailed records of exploration over the years. Throughout the period concerned, these clubs have published much material, such as Bristol Exploration Club *Belfry Bulletins* and *Caving Reports*, Mendip Nature Research Committee *Reports*, Shepton Mallet Caving Club *Journals*, University of Bristol Spelæological Society *Proceedings*, Wells Natural History and Archaeological Society *Reports*, and Wessex Cave Club *Journals* and *Occasional Publications*. The popular cavers' magazine, *Descent* (Wild Places Publishing), contains numerous references of relevance to the exploration of Wookey Hole Caves. Much of this literature is available in Bristol Central Reference Library, whilst the University of Bristol Spelæological Society and Wells and Mendip Museum libraries contain additional materials and manuscripts. The latter's archives include rare documents, original photographs and the first cave diving breathing apparatus called the 'Bicycle Respirator'. Wookey Hole Caves Museum has a display of old diving equipment and the first sump rescue apparatus on loan from the Mendip Rescue Organization (now Mendip Cave Rescue). Over the years, numerous editions and revisions of brochures about Wookey Hole Caves have been available to tourists. All have included relevant information about the history and exploration of the Great Cave so that accounts of cave diving must be widespread in thousands of homes. Much of the hitherto unpublished material used in this book, on the other hand, remains the personal property of the contributors.

The Mendip Cave Registry and Archive hosts a plethora of references, photographs and cave surveys, not only relating to Wookey Hole Caves, but the whole of the Mendip caving area and its environs. The bibliography is continuously updated and available for search online at www.mcra.org.uk.

It is also worth noting that the Great Cave and divers have featured in many broadcasts and films over the years. These special archives contain such notable British firsts as Graham Balcombe's BBC Radio broadcast from underwater in 1935, and the national screening of live cave diving scenes filmed by submersible cameras directly to millions of BBC Television viewers around the country in 1986. Episodes drawn from the history of cave diving have been re-enacted and filmed underground in both Swildon's Hole and Wookey Hole Caves by Sid Perou during 1987. His television series, called *Hidden Depths*, was broadcast nationally by the BBC in 1989. Recently, as described by Gavin Newman in Chapter 27, a film was made covering the exploration of Wookey Hole from prehistoric times to the present day. These productions make a fitting epilogue to *WOOKEY HOLE – 75 years of cave diving & exploration* which all began with Graham Balcombe's first ever radio broadcast publicising the pioneering cave dives at Wookey Hole in 1935.

Index

NB All Wookey Hole chamber and passage names appear under Wookey Hole Cave(s)

Adam, J.A., 280
Adcock, Neil, 206
Admiralty Experimental Diving Unit, 61
Allen, Paul, 177
Alum Pot, 61, 65, 67, 71
Anderson, Rob, 293
Ankerman, Jackie, 336
Arkenside, Mark, 1
Army Caving Association, 330
Ashwick Grove Springs, 175
Atkinson, Paul, 244
Avon Gorge, 10
Axbridge Caving Group (ACG), 169
Axe, River, iv, vi, ix, x, xii, 2, 16-19, 27, 33, 34, 37, 39, 59, 63, 69, 71, 72, 75, 84, 91, 94, 98, 125, 133, 142, 143, 178, 187, 205, 214, 220, 234, 251, 255-257, 261, 262, 264, 265, 278, 299, 308-310, 312

Badger Hole, 19, 73
Baker, Ernest, 2, 8, 10, 144, 219
Baker, Tom 'Jumbo', 55, 176
Balch, Herbert, 2, 4, 6, 7, 10, 17, 27-29, 48, 73, 83, 92, 94, 144, 152, 163, 172, 176, 200, 219, 313, 317
Balcombe, Graham, v, vi, xi, xii, 2, 9-13, 15-22, 25-27, 29, 32, 33, 47, 48, 50, 54, 55, 57, 59, 61-66, 68, 69, 72, 75, 76, 78-81, 83, 84, 88-92, 96, 119, 120, 122-125, 128, 129, 131, 143, 149, 155, 160, 170, 176, 277, 286, 323
Balcombe, Mavis, 57, 66, 90
Ballard, Geoff, 252
Barnes, Mike, 301-304, 306, 308-310, 332
Barrington, Nick, 193
Beal, Jon, 332, 334, 337
Beck, Howard, 3
Bee, George, 241, 244
Bennett, Antoinette, 336
Bennett, Harriet, 336
Bennett, Roy, 267
Bert, Paul, 59
Bevan, John, 134, 139
Bickell, Brian, 256
Bishop, Martin, 196, 208, 250, 251
Blackdown, 8
Blake, Rich, 296, 318
Blue Holes, Bahamas, 245, 253
Bolt, Pete, 301, 303, 309
Boon, Mike, xi, 125, 144, 147, 151-159, 161, 162, 168, 169, 172, 195
Boreham Cave, 202, 204, 208
   Sump 8, 202
Bowden-Lyle, Sybil, 85, 86, 106, 135
Boycott, Tony, 229-231, 230, 231, 233, 235
Bradshaw, Dany, 178, 231, 251, 286, 290, 291
Bristol Exploration Club (BEC), 8, 69, 76, 82, 103, 120, 134, 143, 147, 252, 265, 267, 273, 294, 315, 318
Bristow, Colin, 167
British Broadcasting Corporation (BBC), 28, 39, 41, 54, 74, 100, 121, 191, 192, 213
British Spelaeological Association (BSA), 16, 65, 66

Broadbent, Dave, 71
Brock, Nigel, 307
Brooks, Andrew, 187, 188, 191-193, 195, 196, 198
Brooks, Norman, 130, 191, 222
Brooks, Simon, 236, 251, 290, 294, 309
Brown, Frank 'Mac', xi, 20, 47-50, 54, 55, 71
Brown, Robin, 290, 301, 304
Bufton, Bill, 20, 31, 32, 50
Burrington Combe, 8, 69, 175, 227, 293
Burwood, Charles, 16, 32, 33, 44, 47
Butcher, Alan 'Butch', xii, 265, 266
Buxton, Audrey, 122, 125, 128, 135, 143
Buxton, John, 109, 119, 120, 122, 124, 126, 127-129, 131, 133, 134, 136, 139, 141-143, 147, 152, 166, 177, 178, 181, 266, 294

Caine Hill Shaft, 297
Carnegie, A. L., 278
Carter, Russell, 290
Casteret, Norbert, 3, 119
Causer, Dave, 144
Cave Diving Group (CDG), xi, 9, 17, 26, 55, 61, 69, 72, 73, 75, 76, 82, 84, 85, 89, 90-92, 94, 103, 109, 111, 113, 120, 121-124, 126, 128, 130, 131, 133-135, 138, 139, 143, 147, 152, 159, 160, 161, 167-172, 175, 177, 179, 181, 187, 191, 192, 196, 197, 199, 200, 202, 203, 237, 253, 265, 289, 291, 294, 298, 304, 312, 322, 323, 329
Cave Rescue Organisation (CRO), 61
Cave Research Group (CRG), 72, 175
Cerberus Spelaeological Society, 277
Chambers, Keith, 155
Channel Four, 264
Chapman, Lillian, 7
Chapman, Tim, 296, 298, 301, 304, 309
Chapman, Tom, 301, 309
Charterhouse-on-Mendip, 51, 71, 163
Cheddar, ix, 17, 47, 50, 55, 73, 84, 143, 152, 175, 179, 191, 227, 230, 280, 329, 330
Cheddar Gorge, ix, 47, 191, 227
Chell, Andy, 332, 334, 335, 337
Chelsea Speleological Society, 330
Cheramodytes, pseudo. O.C. Lloyd, 168, 170, 175
Churcher, Bob, 197, 201, 203
Churchill Rocks, 10
Clayton, Tom, 319
Cleave, Noel, 161, 177
Clegg, Alan, 144, 147, 168
Club des Sous l'Eau, 61
Clydach Gorge, 330
Coase, Don, 69-75, 79-87, 90, 119, 120, 123, 131, 135, 265
Cobbett, James, 16, 187, 190-193, 196, 197, 200, 286
Cohen, Claire, 299
Collett, Phil, 184, 187, 190-192, 196, 199, 200, 286
Collins, Stanley 'Alfie', 143, 144
Comheines, Georges, 61
Condert, Charles, 58
Cook, Tom, 279
Cooper, Aldwyn, 199-201
Cooper, Vic, 331
Cordingley, John, 289, 291, 301, 302

345

Cork, Bob, 178, 251, 286
Cornwell, John, 144, 148
Council of Southern Caving Clubs (CSCC), 171, 172
Cousteau, Jacques-Yves, ix, 61, 100, 123, 200
Cow Hole, 8, 47, 51
Coysh, A. W., 75
Craig, Bob 'Crange', 267-269
Crooks' Rest, 73, 84, 109, 136, 137, 139, 148
Cross Swallet, 71
Cross, Dr. Maurice, 261, 262
Crossley, Geoff, 234
Crowther, Robert, 224, 226
Cueva de Vegalonga, Northern Spain, 204
Cullingford, Cecil, 72

Dan-yr-Ogof, 228
Darbon, Frank, 130
Davies, Fred, xi, xii, 136, 144, 147, 149, 151-153, 155-161, 172, 177, 266
Davies, Mel, 191, 194, 215, 216
Davies, Nell, 86, 87
Davies, Phil, 17, 72, 102, 104, 124, 128, 129, 133, 134, 137, 138, 141, 147, 150, 152, 153, 286, 329
Davies, Robert 'Bob', xi, 57, 84-88, 100, 103, 109, 110, 116, 119, 120, 122-127, 131, 142, 143, 146, 152, 181, 184, 236, 249
Davies, Tom, 201
Davis, Sir Robert, 16, 22, 32, 42, 61, 72, 73, 125, 277
Dawe, Ken, 135, 136, 137, 141, 152, 286
Dawes, Len, 130, 131, 134, 155, 169
Dawkins, William Boyd, 18, 30, 219
de Graaf, Brian, 132, 134, 136-139, 141
de Graaf, Valerie, 135
Denayrouze, Auguste, 58
Devenish, Luke, 98, 112, 119, 125, 128, 134, 135, 137-139, 163, 177, 179, 201, 227
Dickinson, Leo, 255, 260, 264, 321
Diepolder Sinks, USA, 257, 258
Dinas Rock Silica Mines, 302
Disappointment Pot, 61, 67
Dolby, Rich, 308
Dolphin, Paul, 68, 69, 315
Dordogne, 304
Dorothea Quarry, North Wales, 304
Drake, Bob, 178, 179, 255
Drew, Dave, xi, 164, 169, 172, 174-176, 183
Duck, Jack, 15, 21, 94, 221
Dumas, Frédéric, 123
Durston, Jim, 201, 229, 231, 232, 235-237, 249
Dwyer, John 'Half-Pint', 84

Eastwater Cavern, 2, 4, 8, 19, 27, 69, 313, 314, 320
    380 Foot Way, 313, 317
    A Drain Hole, 318
    Blackfriars, 317, 318
    Blackwall Tunnel, 316
    Canary Wharf, 317
    Cenotaph Aven, 316
    Chamber of Horrors, 316, 318
    Charing Cross, 316
    Dark Cars, 319
    Dolphin Pot, 69, 315
    Gladman's Shaft, 315
    Harris's Passage, 315
    Ifold's Series, 315
    Jubilee Line, 316, 317
    Lambeth Walk, 317, 318
    Lolly Pot, 316
    Morton's Pot, 317, 318
    Pea Gravel Dig, 319
    Pointless Pots, 318
    Primrose Pot, 315, 319
    Regent Street, 315, 316
    Second Vertical, 315
    Southbank, 317, 318, 319
    Sump 1, 319
    Sump 2, 319
    Sump 3, 319
    The Canyon, 313
    The Technical Masterpiece, 318
    Tooting Broadway, 317, 319
    West End Series, 315
Ebbor Gorge, 48, 83, 135, 177, 223, 320
Eckford, Pete, 286
Edgerton, Dr Harold, 283
Edmunds, Colin, 170, 199-203, 206, 208, 225, 237, 241, 244, 252
Edwards, Jonathan, 304
Eighteen Acre Swallet, 320
Elliot, John, 195, 196
Ellis, Bryan, 191
Ennor, Nicholas, 19
Equipment
    Aflo (Aflolaun), 57, 58, 63, 104, 106, 107, 111-116, 119, 126, 128, 129, 137, 143, 161
    Amphibious Tank Escape Apparatus (ATEA), 61, 153, 157-162, 331
    Aqualung, ix, 61, 110-115, 119, 123-125, 128, 139, 142, 143, 146, 154, 168, 181, 184, 245, 256
    Bicycle respirator, 13, 26, 66, 120, 149, 155
    Copwac, 82
    Cowsack, 82
    Davis Submerged Escape Apparatus (DSEA), 61, 156, 157, 161
    Goon Suits, 134, 153
    Normalair Sump Rescue Apparatus, 150, 168
    Nyphargus, 154, 156, 158, 162
    Port Party (P-Party), 61, 84, 110, 114, 124-129, 131, 141, 142
    Remotely Operated Vehicle (ROV), 277, 280, 281, 282
    Scooter (Diver Propulsion Vehicle), 278, 280, 281
    Sefus Suit, 104, 136
    Self Contained Underwater Breathing Apparatus (SCUBA), 58, 125, 264
    Siebe, Gorman Amphibian Mark One, 61, 121
    Siebe, Gorman Amphibian Mark Two (SGAMTU), 61, 121, 139, 153, 156, 158, 161
    Sladen Suit, 121
    Submarine Escape Breathing Apparatus (SEBA), 153, 157-162, 331
    Tacwack, 82
    Universal Breathing Apparatus (UBA), 123, 125, 131, 139, 141, 144, 161
    Whodd-Whodba, 63
Esser, Paul, 196, 197
Exley, Sheck, 257

Farr, Martyn, xi, 3, 187, 200, 202, 206-208, 227, 228, 230, 231, 234, 237, 239, 241, 244-246, 251, 253, 256, 289, 302, 321, 330, 335, 336

Ferraro, Mike, 170
Fleuss, Henry, ix, 59
Ford, Derek, 112, 115, 122, 143, 172, 174, 175, 284, 298
Ford, Trevor, 120, 128
Fort Bovisand, Plymouth, 256, 261, 264
Foyle, Malc, 290, 310
Frost, Frank, 20, 31, 55, 56, 155, 172

G. B. Cave, 71, 163
Gagnan, Emile, 61
Garrett, Mike, 175
Gee, Alex, 293, 299
George, Charles, 132
Gemmell, Arthur, 61, 293
Giles, Tony, 196
Gilmore, Chris, 183
Gladman, Keith, 315
Glanvill, Angie, 269
Glanvill, Peter, xi, 227, 247, 249, 251, 261
Glencot Spring, 35, 220
Goddard, Andy, 289, 301
Gouffre Berger, France, 147, 161
Gough's Cave
    Lloyd Hall, 179
    Sump 3, 330
Gower, 10
Goyden Pot, 65, 66, 67
Graham, Colin, 170, 195
Grass, Martin, v, 179
Green Lane, 216
Gregson, Richard, 255
Grosart, Christine, 189
Grosvenor, Tom, 84
Grotte de Saint Hélène, France, 119

Hades Caving Club, 294
Hague, Brian, 240
Hainsworth, Reg, 61
Haldane, John Scott, 61, 73
Hall, Molly, 25
Hallowe'en Rift, 236, 252, 293-298
Hanwell, Jim, iv, vi, xii, 1, 13, 57, 101, 134, 138, 157, 163, 165, 171, 172, 191, 194, 200, 222, 223, 231, 272, 279, 320
Hanwell, Judy, xii
Harlech Television (HTV), 200, 253, 255, 261, 264
Harper, Rob, xi, 199, 205, 225, 235-237, 247, 280, 294, 296, 309
Harptree Hill, 8
Harris, Charles Wyndham 'Digger', 8-11, 15, 19-21, 24, 25, 31, 33, 35, 39, 51, 54, 55, 160, 315
Harvey, Chris 'Zot', 187, 190, 193
Harvey, Peter, 68, 69, 91
Haselden, David, 331, 336
Hasell, Dennis 'Dan', 76, 93, 94, 100, 113, 131, 135, 137, 141, 165, 173, 178, 203, 286
Hasenmayer, Jochen, 257, 280
Hassall, John, 52
Hawkes, Chris, xii
Hawkeswell, Lee, 336
Haylings, Peter, xii
Health and Safety at Work Act, 253
Helwith Bridge Inn, 289
Hensler, Eric, 176
Hepste Resurgences, 187

Heron, Emma, 318
Hewer, Tom, 163
Hill, Arthur, 91
Hilton, Kevin, 318
Historical Diving Society, 323
HMS Poseidon, 61
HMS Vernon, 126, 139
Hodgkinson vs. Ennor lawsuit, 219
Hodgkinson family, 18, 224
Hodgkinson, Gerard 'Wing Co', 48, 72-74, 76, 81, 85, 101, 122, 167, 177, 178
Hodgkinson, Olive, xii, 18, 98, 100, 101, 144, 147, 167, 178, 193, 223
Hodgkinson, William, 6
Hole, Adrian, 318
Holtzendorff, Louis, 257
Howes, Simon, 196
Hughes, Trevor, 199, 205, 235, 236, 247, 248, 252, 286, 294, 297, 318
Hunt, Sir John, 170
Hunters' Lodge Inn, 20, 50, 204, 234, 236, 270, 320
Hurtle Pot, 128
Hyaena Den, 18, 27, 219, 220
Hy-Mac excavations, 214

Ifold, Johnny, 315
Independent Cave Diving Group (ICDG), 171, 172, 176, 177
Independent TV (ITV), 152
Ingleborough Cave, 3, 67
Ingram-Marriott, Gordon, i, 76, 77, 84-87, 90, 103, 105, 123, 147, 230
Irwin, Dave 'Wig', 221, 225, 267, 275, 293

Jackson, Graham, xi, xii, 200, 201
Jarman, Bob, 144
Jarratt, Tony 'JRat', 293, 296, 297, 316, 318, 319
Jeanmaire, Mike 'Fish', 169, 170, 187, 190, 267
Jennings, Tony, 269
Johnson, A. C., 89
Johnson, Graham 'Jake', 318
Johnson, Peter, 170
Jones, Gary, 332, 336
Judd, Brian, 336

Keld Head, 61, 63, 65, 67, 68, 121, 125, 197, 202, 204, 280, 330
Kellogg, W.N., 279
Kemp, Dennis, 130, 134, 155
Kenney, Howard, 315
Kenney, Richard, 163
Kingsdale Master Cave, 330
Kingston, Phil, 267
Knapp, Murray, 317
Knibbs, Tony, 225
Knockshinnock Colliery disaster, 59

Lake District, 9, 11
Lamb Leer Cavern, 7, 8
Lambert, Alexander, ix, 59
Lancaster Hole, 147, 168
Lander, Jack, 69
Lane, Barry, 267
Langstroth Cave, 161
Large, Tim, 316

Larson, Howard, 58
Lavender, Annie, 207, 210
Lawrence, Eric, 54
le Prieur, Yves, 61
Lewis, Eric, xii
Lewis, Nick, 331
Little, Bill, 169
Lloyd, Oliver, xi, 115, 119, 120, 126, 127, 131, 135, 147, 149, 150, 154, 163-169, 171, 175-179, 182, 183, 188, 190, 191, 193, 194, 196-204, 206, 227, 255, 279
Lolly, Andy, 315
Long Churn Cave, 67, 71
Longwood Swallet, 191, 222
Lord, Pete, 209
Low, Colin, 315
Lucy, George, 82, 84, 106
Ludwell Cave, 119, 175
Lumley, Mark 'Gonzo', v
Lunde, H. R., 280

Madame Tussaud's, 18, 200, 211, 224
Main family, xii
Malham Cove, 219, 291
Manchester Hole, 67
Mansfield, Ray, xii, 17, 329
Marriott, Gordon Ingram, see Ingram-Marriott
Marsh, Steve, 326
Marshall, Peter, 201
Martel, Edouard-Alfred, viii, 3, 8
Martin, G.M., 283
Mason, Dorrien, 135
Mason, Edmund 'Ted', xi, 75, 91, 93-95, 135, 221
McBratney, Frank, xii
McDonald, Mike 'Trebor', 294, 329
McManus, Stuart 'Mac', xii, 265, 266, 270, 274
Mead, Dennis, 153

Mendip Karst Hydrology Research Project (MKHP), 175
Mendip Nature Research Committee (MNRC), vi, 7, 8, 27, 176, 220, 221, 313, 315
Mendip Rescue Organization (MRO) now Mendip Cave Rescue (MCR), 149, 163, 166-169, 171, 230, 261
Meyrick, Gay, 269
Michelmore, Cliff, 100
Miles, Stanley, 125
Mills, Alan, 196
Mills, Martin 'Milche', 267, 268
Milne, Chris, 193, 207, 210, 229, 250, 252
Miners, 215, 217, 274, 275
Mitchell, Nick, 294
Montespan, France, 3
Moody, Alison, 316, 317
Moody, Pete, 316
Morgan, Tim, 334
Morley, Russell, 169
Morris, Dave, 128, 209, 231, 240, 241, 244, 286
Morrison, Dave 'Tuska', 320
Moss, Neil, 149
Mulholland, Pete, 312, 323, 332
Murrell, Hywel, 51
Murrell, Ruth, 25
Musselwhite, Nigel, 272
Myers, Jack, 63, 293

National Coal Board Rescue Team, 185, 201
National Diving Centre, 307
National Speleological Society (NSS), 257, 262
National Trust, 177
Navrady, Sandy, 200, 201
New Inn, 157, 177
Newman, Gavin, 306, 321, 329, 332, 334-336, 338
Newport, Aubrey, 196
Nolan, Gordon, 144
North Hill, 8, 320
Northern Cavern and Fell Club, 9, 10, 12, 66
Notts Pot, 179
    Oliver Lloyd Aven, 179
Nunwick, Ray, 58, 61, 63, 64, 66, 68, 84

Offer, Bill, 55, 56
Ogof Agen Allwedd, 160, 201, 330
    Sump 4, 201
Ogof Capel, 330
Ogof Daren Cilau, 330
Ogof Draenen, 304
Ogof Ffynnon Ddu, 69, 91, 191, 228
Oldham, Tony, 170
O'Neill, Mary, 65
O'Neill, Ray, 65
Operations
    Avanti, 79, 83
    Avanti Two, 83
    Bung, 57, 72, 74
    Flippers Magnum, 76
    Innominate, 57, 72, 84
    ITTL (In To The Limestone), 84, 89
    Janus, 76, 78, 79
    Linlay, 82
    Muckment, 75
    Nitrate, 81
    Prehistory, 75
    Rearguard, 81
    Sandblast, 75, 99
    Scratch, 74
    Stockpile, 82
    Swansong, 90
Otter Hole, 207
Oxford University Caving Club (OUCC), 255

Paddock, Norman, 315
Paganuzzi, Marco, 179, 258, 260
Palmer, Rob, 245, 246, 264, 280, 330
Parker, James, 219
Parker, John, 83, 187-189, 191-197, 199-203, 206, 207, 212, 214, 221, 222, 224, 235, 247, 286
Parker, Rob, xii, 230, 234, 245, 246, 253, 255, 256, 258, 260, 262, 263, 286, 289, 303, 304, 321, 327, 329-331, 334
Partridge, D.W., 280
Peak Cavern, 84, 122, 133, 149, 251, 291
    Swine Hole, 122
Pearce, Ken, 144, 147, 161, 178, 181, 183
Pearson, Ken, xii
Pena Colorada Cave, Mexico, 253
Penhale, Ron, 134
Perou, Sid, 121, 323, 330
Perry, Bruce, 163
Phillips, Jeff, 206

Picos de Europa, Northern Spain, 255
Pinkstone, Ian, 331, 332
Platten, Gerard, 47, 51
Pointing, George, 178
Pope, Russell, 191
Porth-yr-Ogof, 187, 191, 196, 303
Potter, Keith, i, 229-231, 245, 255
Powell, Penelope 'Mossy', vi, 19-22, 29, 31, 39-41, 47-51, 54-56, 191, 323
Poza la Pilita, 281, 284
Poza Verde, 284
Pozu del Xitu, 230
Prewer, Brian, xi, xii, 201, 211, 214, 225
Price, Duncan, v, 299, 309, 329, 333, 337
Price, Jeff, 178
Priddle, Colin, 83, 192, 193, 196, 200, 201
Priddy, ix, x, xi, xii, 4, 7, 8, 10, 17-20, 31, 54, 55, 71, 92, 134-137, 143, 144, 147, 157, 159, 166, 169, 171, 177, 187, 198, 204, 219, 223, 224, 231, 239, 265, 296, 313, 317, 320
Priddy Green Sink, 296
Priddy Minery(ies), 17, 19, 20, 22, 47, 50, 54, 92, 134, 137, 147, 156, 171, 270, 273
Pridhamsleigh Cavern, 228
Pwll-y-Cwm, 330
Pyatt, Edward, 10
Pyke, Bob, 147, 157, 177

Queen Victoria Inn, 71

Rawlins, J., 278
Read's Grotto, 71
Reid-Henry, Charlie, 334, 335
Reinold, G.A., 280
Reynolds, Tim, 190-196, 198, 199, 286
Rhinoceros Hole, 19
Richards, Simon 'Nik Nak', 317
Rider, Helen, 243, 262, 286, 336
Roberts, Geoff, 135
Roberts, Sherry, 135
Robertson, Struan, 166
Rodney Stoke, 17, 144, 152
Rogers, Allan, 149
Rolland, Ian, 255, 260-262, 286, 330
Romford, Phil, 144
Rose, David, 255
Rouquayrol, Benoit, 58
Rowsell, Phil 'MadPhil', 299, 313, 317-319
Royal Marines, 76
Royal Navy, 141, 245, 247
Russet Well, 251
Rutter, Fenton, 230

Salt, Frank, 241
Sandpit Hole, 223
Savage, Dave, xi, 169, 172-175, 178, 181, 182, 187, 188, 221, 286
Savory, J. Harry, 4, 220
Savory, Keith, 179, 290, 296, 301
Scoones, Pete, 255, 325
Setterington, Tony 'Sett', 82, 83, 93
Severn Tunnel, ix, 59
Sheffield University Mountaineering Club, 103
Shepherd, Bob, 228
Sheppard, John 'Jack', xi, xii, 2, 9-12, 14-17, 20-22, 25-27, 55-57, 72, 77, 91, 119, 122, 124, 131, 149, 155, 286
Shepton Mallet Caving Club (SMCC), 268
Short, Phill, 302, 304, 317
Siebe, Gorman & Co. Ltd., 16, 30, 32, 47, 59-61, 72, 125
Simes, William, 219
Simmonds, Vince, 318
Sims, Robin, 169, 170, 175
Sinoia Cave, Zimbabwe, 241
Sisman, David, 278, 280, 282
Smale, Chris, 170
Smart, Chris, 201
Smith, Court, 257
Smith, David Ingle, 175, 193
Smith, Ken, 11, 15
Smokham Wood, 191, 216
Solari, Roger, 201, 202
Somerset Fire Brigade, 272
SONAR, 279
South Bristol Speleological Society, 294
South Wales Caving Club (SWCC), 69, 169, 321
Speedwell Cavern, 128, 333
  Main Rising, 128
St. Cuthbert's Swallet, xii, 19, 27, 69, 120, 134, 219, 223, 265, 266, 320
  Beehive Chamber, 270
  Cerberus Hall, 69
  Dining Room, 267
  Gour Hall, 269, 270
  Kariba Dam, 272
  St. Cuthbert's Two, 257, 268, 270, 272, 275
  Sand Bag Dam, 270
  Sump 1, 266, 267, 270
  Sump 2, 266-268, 270-272, 275
  The Duck, 266
St. Dunstan's Well, 175
St. Dunstan's Well Cave, 277
Stanbury, Harry, 69, 76, 79, 92
Stanton, Rick, 327, 329, 330-332, 335-339
Stanton, Willie, xi, xii, 119, 172, 199, 201, 213, 214, 219, 224
Stapleton, Alf, xii
Stark, Jim, 120
Statham, Oliver 'Bear', 125, 187, 199, 202, 205, 207, 208, 239, 280, 330
Stead, Ray, 230, 233, 235, 236, 245
Stell, Clive, 296, 298, 301, 304, 310
Stenner, Roger, 298
Stevenson, Richard, 201-204, 207, 208, 241, 280
Stewart, Andy, 312, 325
Stimpson, Vincent, 315
Stock Hill, 8
Stock, Stanley, 159
Stoke Lane Slocker, 119, 134, 162, 168, 171, 181, 331
  Stoke Lane Three, 176
  Stoke Lane Five, 172
  Stoke Lane Six, 172
  Stoke Lane Seven, 172
  Stoke Lane Eight, 172
  Sump 2, 147, 158, 159, 176
  Sump 4, 159
  Sump 6, 331
Stone, Bill, xii, 253, 255-257, 260-264, 278, 330
Stone, Pat, 258, 260-262
Surrall, Alan, 315

Sweet, Dale, 257
Swildon's Hole
    Balch's Forbidden Grotto, 179
    Birthday Squeeze, 179
    Blue Pencil Passage, 130, 155, 171
    Bob's Bell, 88
    Buxton's Horror, 152
    Cowsh Avens, 157, 296
    Great Bell, 26, 27, 170
    Little Bell, 170
    Paradise Regained, 130, 149, 151, 154, 155, 171
    Shatter Passage, 175
    St. John's Bell, 170
    Sump 1, 11, 15, 16, 24-27, 120, 121, 150, 165, 166, 171, 174, 176, 177, 220, 279, 293, 296, 301, 323
    Sump 2, 24-26, 120, 128, 147, 151, 161, 164, 169-171, 175, 294, 301
    Sump 3, 26, 27, 122, 155, 156, 169-171, 175
    Sump 4, 130, 154, 157, 169, 171, 173
    Sump 5, 152, 154, 155, 162, 171
    Sump 6, 125, 155-157, 162
    Sump 7, 156-158, 162, 171-174
    Sump 8, 172-174
    Sump 9, 173, 174, 176, 296
    Sump 11, 174
    Sump 12, 174, 177, 296, 301, 323
    Sump 12a, 301, 302
    Swildon's One, 25,130
    Swildon's Two, 25-27, 66, 156, 170, 171
    Swildon's Three, 175
    Swildon's Four, 88, 130, 149, 152, 155-157, 161, 170, 175
    Swildon's Five, 129, 152, 154
    Swildon's Six, 126, 129, 130, 155, 171, 175
    Swildon's Seven, 131, 155-157, 161, 172
    Swildon's Eight, 157, 158, 172, 173
    Swildon's Nine, 157, 173
    Swildon's Ten, 174
    Swildon's Eleven, 174
    Swildon's Twelve, 174, 228, 320
    Tate Gallery, 174
    Tratman's Temple, 130
    Upper Series, 4, 169

Tankard Hole, 177
Tapley, Carol, 179, 296
Taviner, Rob 'Tav', 317
Taylor, Alan, 290
Taylor, Roy, 136
Templeton Pot, 320
Terrell, Mark, 125
Thomas, Mike, 301, 304, 310-312, 322
Thompson, Jack, 120-122, 125, 128
Thompson, Mike, xi, 131, 133-139, 141, 142, 144, 146-148, 152, 155, 156, 159, 161, 168, 172, 177, 249, 266
Thomson, Donald, 163
Thornycroft, L.B., 75
Threaplands Cave, 158
Tooth, Terry, 214
Tratman, Edgar K. 'Trat', 75, 98, 99, 130, 135, 167
Tregonning, K., 283
Tricouni Club, 10
Tringham, Derek, 204
Troup, Reginald, 219-221

Trowbridge, Laura, 317, 331, 334
Trower, Martin, 9
Tucker, J.B., 280
Tucknott, Bill, 20, 31, 55, 56
Turner, Dave, iv, xii, 157, 159
Twin T's Swallet, 223

University of Bristol Spelaeological Society (UBSS), 8, 147, 168, 249
Urwin, Maire, 191, 192, 195, 196, 286
U.S. Navy Experimental Diving Unit, 257, 264

Vaucluse, France, 257, 280, 281
Vaughan-Thomas, Wynford, 200
Verne, Jules, 59
Vobster Quarry, 307
Volanthen, John, 300, 329-331, 333, 336-338

Waddon, Dorothy, 135
Waddon, Jack, 134, 136, 137, 140, 141, 147, 158-160
Waite, Vincent, 75
Wakulla Springs, xii, 277, 279, 280, 283, 285, 330
Waldegrave Swallet, 8, 17, 18, 20, 47
Waldegrave Pond, 17
Walford, Julian, 199
Walker, Julian, 260-263, 286
Walker, William, ix
Wallington, John Frank, 130, 152
Warburton, Denis, 112, 314, 315
Watford Underwater Club, 152, 153
Watts, Pete, 316
Weaver, Bill, 79, 91
Websell, Rich, 301
Webster, Martin, 195, 196, 198, 201
Welch, Francis, 17
Wells Journal, vi
Wells Museum (later Wells and Mendip Museum), viii, 10, 48, 73, 97, 121
Wells Natural History and Archaeological Society (WNHAS), 7
Wells, Oliver, xi, xii, 21, 109, 119, 121, 124, 126, 127, 134-139, 141-143, 152, 154, 155, 160-162, 167, 168, 174, 181, 277
Wells, Pamela, 135
Wessex Cave Club (WCC), iv, 8, 15, 20, 63, 69, 72, 134, 147, 157, 168, 169, 172, 176, 179, 221, 222, 226, 316
West, Rich, xii, 182
Western Electric Company, 39
Westlake, Clive, 175, 176, 201, 236, 237, 251, 336
Westwood, J.D., 280
Wharton, Barry, 316
Wheel Pit, 17
White Lady Cave, 187, 228
White Pit, 320
White, Ross, 294
White Scar Cave, 202
Whitehead, Roger, 322, 323
Whybro, Paul, 238, 250, 251
Wigmore Swallet, 294
Wilkins, Adrian, 199, 200
Willcox, Thomas, 7
Willett, Mike, 296
Willis, Brenda, 136
Willis, Dave, 112
Wilson, Linda, xii

Winchester Cathedral, ix
Wisely, Jo, 306, 323, 332
Witcombe, Richard, v, 313
Wooding, Mike, xi, 157, 169, 172-178, 181-184, 197, 286
Woodward, Brian, xi, 187, 188, 192-195, 198, 199, 238, 241, 244, 250, 286
Woodward, Janet, 191
Wookey Hole Cave(s)
  22 Inlet Passage, 312
  2W's Extension, 238
  Attila the Hun's Sardine Cannery, 236, 250
  Bear Pit, 83, 105, 106, 107
  Beyond the Thunderdome, 238
  B-Reach, 74, 75, 90
  Cam Valley Crawl, 199, 205, 236, 248, 295
  Chamber 1 (One, First or Witch's Kitchen), 54, 73-75, 91, 95-97, 99, 100, 301, 302
  Chamber 2 (Two, Second or Witch's Hall), 75, 97
  Chamber 3 (Three, Third or Witch's Parlour), iv, vi, x, 20, 21, 35, 39, 54, 72, 74, 76, 83, 87, 97-99, 101, 103, 107, 110, 134, 162, 177, 187, 188, 193, 212, 228, 286, 294, 323, 333
  Chamber 4 (Four or Fourth), vi, 20, 34, 51, 75, 97, 98, 99, 178
  Chamber 5 (Five or Fifth), 18, 21, 35, 36, 77, 134, 200, 201, 214
  Chamber 6 (Six or Sixth), vi, 15, 21, 35, 36, 41, 78, 79, 81, 85, 200
  Chamber 7 (Seven or Seventh), viii, 22, 27, 76-78, 87, 103-106, 122, 182, 199, 200
  Chamber 8 (Eight or Eighth), 78-80, 87, 105, 106, 199, 200, 221
  Chamber 9 (Nine or Ninth), x, xi, 69, 72, 76, 78, 79, 81, 83, 86, 87, 90, 103, 104, 106-108, 110, 115, 119, 120, 138, 141, 177, 182, 183, 185, 188, 190, 195, 196, 201, 211-215, 217, 228, 230, 236, 238, 244-246, 261, 263, 264, 294, 297, 303, 308, 311, 312, 320, 323, 325, 332, 333, 338
  Chamber 10 (Ten or Tenth), 85-87, 111, 131, 249
  Chamber 11 (Eleven or Eleventh), 72, 83, 84, 109, 110, 116, 119, 122, 126, 144, 265
  Chamber 12 (Twelve or Twelfth), 113, 115, 131, 236, 249
  Chamber 13 (Thirteen or Thirteenth), 109, 113, 116, 136, 152, 161, 177, 181, 188, 221, 236, 237
  Chamber 14 (Fourteen or Fourteenth), 128, 129, 131, 146, 181-184, 300
  Chamber 15 (Fifteen or Fifteenth), xi, 131, 133, 139, 162, 177, 178, 181-185, 301
  Chamber 16 (Sixteen or Sixteenth), 182
  Chamber 17 (Seventeen or Seventeenth), 99, 182
  Chamber 18 (Eighteen or Eighteenth), 183-185, 187, 221
  Chamber 19 (Nineteen or Nineteenth), 188, 259, 302
  Chamber 20 (Twenty or Twentieth), 187-190, 193-197, 199, 215, 216, 228-231, 236, 238, 245, 250, 255
  Chamber 21 (Twenty-one or Twenty-first), 301
  Chamber 22 (Twenty-two or Twenty-second), 197, 198, 203, 215-217, 229, 231, 236, 239, 240, 245, 261, 294, 296, 298, 299, 306, 309, 310, 312

  Chamber 23 (Twenty-three or Twenty-third), 204, 229, 231, 248, 261, 310
  Chamber 24 (Twenty-four or Twenty-fourth), 205, 210, 215, 216, 230, 231, 233, 234, 237, 239, 240, 244-246, 256, 258, 260-264, 283, 290, 303, 305, 306, 309, 311, 312, 325, 326, 332-334, 336
  Chamber 25 (Twenty-five or Twenty-fifth), 208, 233, 239, 240, 244-246, 253, 258, 262-264, 304-307, 326, 327, 329, 331, 333, 334
  Charon's Chamber, 73, 221
  Charybdis, 44, 78
  Coase's Loop, 83, 119, 128, 192, 294
  Coase's Loop Extension, 200, 249
  Crocodile, 177
  Deep Route, 195, 230
  Edmunds Chamber, 237, 252, 311
  First Deep. 78, 137
  Genghis Khan's Executive Thunder Box, 296
  Genghis' Revenge, 294
  Giant's Staircase, 78
  J-Reach, 76
  Kilmersdon Tunnel, 201, 203, 215
  Mongol Hordes Information Office, 248, 251, 294, 295
  M-Reach, 75, 90
  Mudball Alley, 104, 107
  Nine-One (9.1), 83, 85, 90, 107, 110, 113, 114, 119, 122
  Nine-Two (9.2), 83, 85, 90, 110, 111-115, 119, 122, 126-129, 144
  Pleasant Valley Monday, 251
  Pleasant Valley Sunday, 251, 252, 309
  Scylla, 44, 78
  Second Deep, 137
  Shallow Route, 195, 196, 230
  Static Sump, 299, 307, 308
  Sting Corner, 206, 215, 226, 232, 234, 238, 251, 308, 309, 311
  Sting Corner Sump, 309
  The Lake of Gloom, 208, 292, 240, 304, 328
  The Well, 240, 244, 290, 291, 301, 304
  The Witch, 91, 95, 96, 99-101, 116
  Wall Bypass, 306
  Witch's Scullery, 75, 96
Wookey Hole Paper Mill, 18, 200, 215, 220, 224-226
Wookey Hole Ravine, xiv, 19
Worcester, William of, 219
Workman, Brian, 274
Worsley, Francis, vi
Wright, Jerry, 152
Wynne-Roberts, Steve, xi, 141, 144-147, 150, 155-157, 159-161, 172, 177, 183, 227, 249, 266, 286, 331

Yeadon, Geoff, 125, 187, 199, 202, 205, 207, 225, 239, 280, 330
Yeandle, Dave, 225
Yeisley, P.A., 280
Yorkshire Dales, 9, 61, 147, 152, 219, 289

Zacatón, 257, 284
Zumrick, John, 257, 258, 260

Photo by Gavin Newman